Susan —
for your
country dreams
turned reality —
love,
Amy

COUNTRY
WOMEN

ANCHOR·PRESS·
·DOUBLEDAY·

COUNTRY WOMEN

A HANDBOOK FOR THE NEW FARMER

Written by Sherry Thomas
Jeanne Tetrault

Illustrated by Leona Walden

Photographs by Sally Bailey

ANCHOR BOOKS
Anchor Press / Doubleday
Garden City, New York
1976

This book is dedicated to every woman
who has shared or will share
this dream...

Lea Bartneck
Bobbie Bishop
Virginia Butler
Harriet Bye
Ellen Chanterelle
Carmen Goodyear
Kathryn Kilgore
Julia Kooken
Lynda Koolish
Mary Mackey
Jean Malley
Jean Mountaingrove
Ruth Mountaingrove
Marnie Purple
Jenny Thiermann
Sam & Thomas

JEANNE TETRAULT grew up on the East Coast. After attending college, she spent seven years homestead farming with another woman, raising dairy goats, sheep, and poultry. During that time, Jeanne Tetrault belonged to the feminist collective which began and nurtured Country Women Magazine. She is presently working in the San Francisco area as a professional carpenter and continues to contribute to Country Women Magazine.

SHERRY THOMAS was raised in the suburbs of Washington, D.C. After leaving college, she worked a variety of jobs – ranging from waitress to state park employee. In the last six years, she has built and run a California sheep ranch, first with her husband, then with her sister, and now alone. Currently, she is also involved in the writing and editing of Country Women Magazine.

OTHER CONTRIBUTORS

Lea Bartneck - 'Raising a Weaner Pig'

Bobbie Bishop - poem: 'Forest Flatlands'

Virginia Butler - portions of 'Butchering'

Harriet Bye - poem: 'Bare Feet'

Barbara Champion - 'Drying Foods'

Ellen Chanterelle - 'Butchering Chickens'

Carmen Goodyear - graphics and text for 'Splitting Wood,'
    'Planting and Transplanting Hotbeds,' 'The Basic Chicken House,'
    'Choosing and Using a Cream Separator,' graphics for 'Wood Cookstoves'
    and text for 'Milk Products'

Kathryn Kilgore - poem: 'The First Growing Season'

Julia Kooken - poem: 'Outrageous Old Woman'

Lynda Koolish - photos pages 23, 91 (top), 96, 120, and 306

Mary Mackey - poems: 'In Indiana' and 'Grandmother Poem'

Jean Malley - main portions of 'Carpentry Tools'

Jean Mountaingrove - poem: 'Elegy for Bees'

Ruth Mountaingrove - poem: 'Naked Summer'

Marnie Purple - 'Recycling Wood'

Jenny Ross - 'Chainsaws' and 'Hammers and Nails'

Nancy TeSelle - Calligraphy

Jenny Thiermann - 'Diane's Poem'

Sam Thomas - 'Fighting Fires' and portions of 'Outhouses'

Diney Woodsorrel - 'Choosing Your Land'

Cat Yronwode - 'Raising Rabbits' and 'Rabbit House'

TABLE OF

CONTENTS

SECTION III ANIMALS AND POULTRY 183

Special thanks

to Sam ♀ who kept the farm running
        and without whom ...

to Carmen ...

and to the women of Albion, for the book
        truly belongs to all of us in our collective
        strength and visions and discoveries.

We would also like to thank Pat Cody,
Ruth Gottstein and David Charleson
for help with the technical production.

INTRODUCTION

This book has its roots in the earth, in a small piece of the country bounded by ocean and redwoods. A few years ago, a small magazine grew from the same area. This was the vision and the voice of a group of country women who had come together to explore "consciousness raising." We were shy of one another, of ourselves, and of the whole concept of woman identity. Most of us were preoccupied with the demands of living in a completely new environment. We were somewhat surprised to find ourselves drawn together—surprised, curious, attracted, unsure. We met together week after week for almost two years, slowly and often painfully searching out who we were and what we wanted for our lives. We talked, laughed, and cried together; we taught each other to believe our dreams and helped each other to live them. That small group was the nucleus of a change that spread woman to woman, acre to acre, gradually touching the whole area and then reaching tentatively beyond. We heard of other small groups in other isolated areas and began to realize how much women needed to be in touch with one another in the "new communities" of the back-to-the-land movement. We talked of a country women's newspaper to share our new understandings, our country skills, our questions.

As we talked, our visions and excitement grew. Almost before we knew it we had talked our newspaper into a national magazine for women living in the country—a network to stretch across our chosen isolation. We were naïve then and in a funny way courageously humble: we had no experience of writing or publishing and no great external standards to live up to. We had only our eagerness and enthusiasm. And those first issues of *Country Women* magazine clearly showed it; they were simple and funky and plain crude, but there was also something real and touching about them because we weren't writers, just people. The magazine was a blend of the practicality of the work that we do with the personal experiences of consciousness raising. The first issues tackled chain saws, living with children, raising goats, and women's groups, in an unlikely but also logical combination. The magazine grew and spread bit by bit. We sent it to friends who sent it to friends. A bootlace budget, loans and a lot of work kept it going at first. We were amazed as each issue sold out, as subscriptions began coming in, and as bookstores sent increasing orders. *Country Women* reached beyond our area, then beyond our state. Letters began pouring in, opening up the communication we had dreamed of. Women on farms in the Midwest, communes in Vermont, homesteads in Missouri began writing letters of support, surprise, encouragement.

With more and more requests coming in for sold-out back issues, Jeanne and Sherry began talking about an anthology of *Country Women*. Idle talk at first for both of us were overworked already with our respective small farms, our women's group, the magazine, and more. Gradually, though, we began to take each other more seriously and to really talk about "the book." Others in the magazine collective declined to take it on: too much work already. The seed was planted, though, and inevitably it began to sprout.

A book for the new farmer, for the woman who would come to the land, is what we wanted. Years of struggling to learn how to do a million unfamiliar new things taught us the value of sharing. The process as well as the information seemed important; many times we discovered ourselves and other women mystified by the first steps of learning. We were determined to put together a book that would be an encouragement as well as a tool. We began to write for the book, expanding old articles from the magazine, writing new chapters. Beyond a book for the newcomers to the country, we wanted something for women already on the land. Our book idea was growing; we wanted not just an anthology but a compilation, a reference, a definitive work. Our expectations grew and grew; it would contain all the things we'd needed to know or wanted to read and couldn't find; it would be *the* handbook for the new farmer.

With our grand ideas blazing, we came up against many problems. Our idea sometimes made for ponderous paragraphs and lengthy inch by inch explanations as we began to write out what we were doing and learning. We were stumped by orientation: what was too much detail, what was obvious, what would be over-simplified or too obtuse? We prepared endless outlines, worried over them, revised them. We came up against all the areas in which we had no personal experience—"But what about *cows?*"; "We need something about growing hay." We wrote endless letters, made pleas, ferreted out possible writers. We were disappointed when other women didn't have our enthusiasm to share, hadn't the experience, or, more often, didn't have the time. Sometimes we were wonderfully surprised by a positive response, and a new voice blended into our book, adding what was missing and instructing us. The voice of the writer became a problem, as each of us who wrote did so in a direct and personal manner. The "I" who explains how to repair a pump lives in coastal California; the "I" of raising rabbits homesteads in Missouri; some of us live on communes, some with partners, some alone. We decided to let each narrative stand as an individual as well as a collective voice. Thus, the "I" becomes a single country woman, and all women. Our experiences in consciousness raising validated this, for much of what we learned was personal/collective, unique/universal.

Many women have a part in this book. Four of us form its core. Jeanne and Sherry dreamed it, planned it, organized it, did most of the writing and all of the editing. Then they were joined by Leona and Sally who added images to our words and ideas, giving the book shape and form. Many months and then seasons wove through our work. We went through high and low work periods. The book changed a hundred times, seeming to take on its own energy and direction. Its structure became more clear, a weaving together of three separate strands that we hoped would reflect more wholly the reality of country life: The first strand is the journal Sherry wrote of one woman's growth through feminism and country consciousness, a reflection of our personal experiences and changes. The second strand is the poems of many women living scattered across the country. The third makes up the bulk of the book, the practical how-to information needed for country survival, written by Jeanne and Sherry with help from others. The manuscript grew to over five hundred pages and had to be patiently and with much difficulty cut back. We were plagued by doubts and afterthoughts, still adding new chapters as we cut out favorite old ones. Half-jokingly, half-seriously we consoled ourselves with an imagined "Volume Two." We tucked away all of the extra illustrations, the sections that just wouldn't fit, the photographs for which there was no space.

The book did not turn out to be the perfect reference book for new farmers, but it is a good guide to what you need to know to begin and to run a small homestead-type farm. We use the term "homestead" to mean a small, primarily not commercial, diversified, and, hopefully, self-sustaining place. There are many underlying and probably some unconscious assumptions in this unperfect farmer's handbook. We assume most readers can't afford large pieces of land or already developed farms, so we have written our animal feeding and housing sections for the homesteader with a small piece of land and little money. But we haven't written for those who want to live with almost no money; we haven't included how to split boards with a hand adz or get rennet from a calf's stomach. That level of "self-sufficiency" belongs in another book. We haven't written about the new technologies of alternative energy sources, either; though we see their development as essential to the earth's future, we have no personal experience in their use. So in some ways, this is a very conventional book: We assume that you will probably buy most of your lumber and that you may have electricity, or at least gas and kerosene. We use antibiotics to treat diseases when they're needed and herbal remedies when they work. We have a basic belief in organic methods of enriching the soil, which we haven't bothered to justify since so much other writing has been done on the subject. But our experience bears out our faith, and it is no small satisfaction to see the soil getting richer and richer year after year despite abundant crops. We have a strong prejudice in favor of goats as dairy animals for a small farm because of their feed conversion (more milk per pound of feed than a cow), the quality of their milk, and their wonderful personalities. But we would gladly have included a section on cows if we could have gotten one in as much depth as the one on goats. Sections on some other farm animals were cut in the final editing when we had to eliminate a lot of material. The choices were hard and we tried to make them on the basis of what seemed most essential. We decided to include a chapter only if it tells you what you really need to know about that subject. Far too many back-to-the-land publications haven't followed that principle and while they look good to the city neophyte, they aren't much use out on the farm. We have also had to abandon dreams of including sections on more conventional and large-scale farming techniques such as mechanized cultivation and grain and feed crop production. We simply don't have enough experience to write about them yet. So, this is not a book for the commercial agriculturist. It is for the new farmer whose small-scale productivity is as old as America itself.

One of the most fundamental facts of country life is that one is always learning. This fact makes the act of publishing a book presumptuous and almost embarrassing. Every day we learn something new that this book would benefit from. It is a better book than it would have been if we had finished it a year and a half ago as planned for we have grown a lot in that time. It would be a better book yet if we waited twenty more years to write it. It is the best we can give now—its information has been tested by our years of learning. We don't know everything and we have not yet forgotten what it is to know nothing.

Coming close to the finish of this book, we made a critical discovery. In the country, there is room for a women's renaissance. The space and time is there for a total redefinition of ourselves, our relation to the earth, our relation to each other. This has become very clear in the making of this book. We have the skills, the enthusiasm, the resources to build our dwellings, raise our food, take care of our most basic and real needs. This self-reliance fosters an incredible strength. It is a hard and demanding way to live, but one rich and sustaining.

We are still hesitant to let go. This book, with its solitary hours of writing, its meetings crammed in between kiddings and crops, its endless problems and joys, has become a symbol of all our strengths and shortcomings. It is alternately finished and incomplete in our minds. We offer it as a beginning. We hope it will encourage every woman who picks it up, will help a whole wave of new farmers rediscover the land. We dream that sister volumes will spring from all over the country so that we can all share and learn.

This book is dedicated to every woman who has shared or will share this dream . . .

Jeanne Tetrault
Sherry Thomas

SECTION I

BEGINNINGS

choosing your Land

Sitting at the writing desk, looking out over miles of Mendocino hills and woodlands, I think over the numerous aspects of looking for a home in the country—the things we carefully checked out and those we overlooked. "Babes in the woods" we often called ourselves during the legal transactions. Our position could easily have been taken advantage of unless we were armed with a lot of information. I'm going to cover here what I learned in our own dealings and what has come to our attention during the more than two years since. It is by no means all-inclusive, but perhaps can be of help to those now looking for land to buy or dreaming about it for some later day.

The dream is, of course, what comes first—and how sweet it is. No matter how much harder the reality hits later, the dream is the strong starting energy—and often what keeps you going. I see no harm in formulating the ideal country scene, to be used for enthusiastic searching and bearing in mind that you *probably* won't find it but *will* find a more possible dream for your area. I was looking for a gentle, lush Vermont farm in the California ranch lands—which are hardy and summer dry—and found one about as close as you can get. The images of rolling hills, big open meadows, quiet woods, mountaintop air, and panorama were realized. Undeveloped bountiful water including streams were not—except in the valley 2,000 feet below.

If you know the general geographical area you are interested in, you've got a good start. Our choice was determined by weather, terrain, accessibility when desired to city activity and market (about 160 miles), land prices, general direction of the country's changes (i.e., lots of people like us). Second, have an idea of what you can afford compared to general prices of the area. This means cash down payments plus yearly installments. Do you have enough savings to pay the installments, or will you have to earn it in order to keep the land? The latter means: (1) jobs to acquire a

5

savings before moving to the land, (2) jobs outside the land after moving there (can be a big energy drain), (3) agricultural earnings from your land (usually takes a long time to earn a lot of money). It's worthwhile to check many sources about average land prices so as to avoid innocent overpayment. If you're buying with others, get it straight now how communal land payments relate to individuals.

Usually corresponding to what you can afford is how much acreage you want—the larger the piece of land, the lower the price per acre. We had the pleasant surprise of getting three times the acreage for one third the average price per acre. The extra two thirds is largely sloping unfarmable land that can remain for exploration and appreciation of its mountained natural state (or hopefully returned natural state after last generation's lumbering and sheep grazing destruction). Remember also that acreage is recorded from the air as if it were flat, so hill country actually gives you more acres than the official computation states. What topographical features are you looking for? Do you prefer valley land with river-bottom soil and upward-looking view, or woodland with acid soil and little direct sun and enclosed existence, or hillside land with a mixture of these and lots of slopes to live on, or hilltop land with full sun and space and out and downward view and the likelihood of being *above* your water source (i.e., no easy water system)?

Very importantly—do you want developed or undeveloped land? A place that already has a house and barn and fenced pasture is more expensive and gives you a good start. The land we have had nothing but an old logging/hunting cabin whose insides needed complete redoing and an outhouse. We worked frantically that first summer to shelter and provide for ourselves for an unknown winter.

In this county, virgin redwood acres cost about sixteen times as much as logged acres and are getting impossible to find. Also, at this writing, there is an annual tax per tree if they are of timber market value. I often lament the many enormous redwood stumps, but know, ironically, that we could never have afforded the land otherwise.

Next, why do you want to live in the country? To farm grain, vegetables, legumes, Christmas trees? To raise goats, horses, chickens, cows, bees, llamas? To have a rural base and work in town? To log or mill? To just hole up in the woods and live as simply as possible? To do you-don't-care-what just to get Away From The City? Each of these makes a large difference in what you're looking for: soil, water, sun, vegetation, pasture land, access to a market for your product or to your job, seasonal changes and weather severity, personal survival needs. If you're going to live collectively, get these thoughts started first. Is there a general direction you all share for the land? What are each person's commitments at this point? The workable commune question is, quite obviously, an entire matter in itself.

Do you have dreams of self-sufficiency, or is this not a concern? If so, you will probably find it a big shock to learn just what that means. Shelter and heat are not hard to come by. As a novice, I was at first awed and frightened by the prospect of construction and a constant wood supply. You'll be happy to learn it's quite within your range once it's your active direction. Self-sufficient food is harder and will take many years to approach realizing. Even without gas or electricity, you still have to run your lamps on something. And even most candle wax is petroleum based. The only thing I can think of is many years of many hives for beeswax candlelight! Besides the basics, there are needs for tools, animal food and care, taxes, fencing, water systems, medical care, agricultural investments. Realistically, it would take a huge amount of time and work for a majority of your needs to be self-met. I think the more immediate concern is how to enter this new breed of consumerism the most ecologically. In any case, you'll need capital beyond your down payment, unless you can get, and are satisfied with, medical and food provisions and truly have no desires for land or animal improvements or more than the simplest of personal needs.

If it sounds like much more than you can possibly get together in your head at this point, do take comfort in the fact that most of what did come together for us before buying happened *during* the search period. What we saw gave us better ideas of what we were looking for. So—with your basics in mind—hit the road.

THE HUNT

Do hold on to your dreams until you're *sure* they don't exist. It takes a lot of time to be careful and thorough—and well worth it. There are different ways of finding out what's available. I had the idea that I could wander about and ask local farmers and storekeepers what was for sale and they would turn me on to old Farmer Molly who was selling her bean farm down the road. Not bloody likely in Mendocino or most popular California counties! Perhaps still in parts of Kansas or Maine, but here almost everything is under the control of the real estate business. And it's quite a business, with money spurring it on—not to be taken lightly. Now there is the possibility, if you carefully check newspapers and bulletin boards, that you'll find some real person advertising to sell outright or to buy shares or to join a group who is buying. I'd check those out first. Unfortunately, I must advise *caution* even here. That's how we got started—we followed a notice that sounded very down to earth and ended up dealing with a hard-nosed, scheming real estate dealer whose only difference was his long hair and funky appearance. We liked one place he told us he had for sale, but it turned out he was hoping to buy it by getting our money. If you deal with individuals, make sure they at least own the land they're trying to sell! It did start the process of lead leading to lead leading lead for us, however.

Spend as much time as possible in the area in which you're interested. Make sure it's what you want. Drive and camp around. Look for "For Sale" signs and follow up every interesting one. Go to real estate offices and describe what you want. Occasionally you may hear about land auctions or pieces being sold for the price of back taxes. We started looking on February 12, 1971, after a year of dreaming, and first saw our home on May 3. Dealings were not over until July 15. I'd recommend looking during the most unattractive season possible. If you love it then, you know you're right, for every place is beautiful in the spring. Take time to really explore any place in which you're interested. Walk all over. Camp there. Don't let the agent rush you. You're bound to get the "so-and-so has offered me such-and-such already" line. But stay cool. If you're really interested and get that pressure, work quickly but carefully. We dealt directly with the owner but heard about the place through an agency. He was a fast talker, but we ended up with a good deal. Except we probably should have been stronger in settling on a longer-term payment.

If you fall in love with a place, recheck your limits. Can you really afford it? We almost got ourselves into a situation of *having* to sell off parts as soon as buying. Keep looking. When you find the place that's meant for you, you'll know.

So—you've found a place you love and can afford. What are its connections with town? Are you nine miles up a dirt road you can't drive in winter? Is that what you want? Careful thinking about your

Living in the country has meant beginning all over again—like a child learning how to survive. Nothing that I know from twenty-four years in cities and three years in college feels useful; everything that I don't know feels necessary. My self-confidence is shaken, and I look to Peter to lead me and direct our life.

Somewhere back in Pennsylvania among five men burning out power saws, pounding nails, and ego tripping, each with his own Grand Design, I began to retreat from my eagerness to learn and to try. I found myself and Claudia, the only other woman there, unconsciously gliding out into the garden, attracted to its peace and the simplicity of learning to pull weeds.

Now, alone with Peter on our own land, without even thinking about it, I've been consolidating my own country skills. I gather confidence from what I have been able to master safely: the garden, food gathering and preparation, the intricacies of wood-stove operation. I am wrapped up in more domesticity than I have ever before experienced in my life. Jealously, I guard the secrets of plant nourishment, my one area of expertise. Kathy, Marian, and I travel each day to watch the progress of each other's gardens, as women in suburbia watch each other's children.

transportation situation will save you from grave frustration later on. Some roads absolutely *cannot* be driven in the winter. In California, they're flooded, develop impassable ditches, become muddy and slippery and dangerous to drive. In other places, you will be snowbound. Much of our winter access is hiking in and out several miles. Even in dry California summer, the roads can be full of ruts and holes. Road maintenance is costly and/or hard work. Ours needs nine miles of cross- and side-ditches every fall by the people who use it. These ditches are graded away every spring by an expensive hired machine and driver, making the road smooth for a while, but extremely dusty and causing worse damage each winter. Each grading also widens the road. To rock it would cost thousands of dollars. Country roads are extremely hard on your vehicles. They too will need constant maintenance. How do you feel about your closeness to other people, to town itself? Are your neighbors friendly? Do you want a phone? Electricity? Is that easily and cheaply installed or will it cost $40,000 (literally)? Do you want metered gas?

What is the water situation? Good water is ES-SENTIAL. Where is it located and in what form? Is there plenty *all year*? See for yourself what the source is. Talk to area people about their water and find out if they know anything about your land. Ask the previous owners about your water. Look ahead to expansion of your water needs. The spring by our house was sufficient for the first summer. Now that we have a big garden, two horses, a mule, ten goats, many chickens, and a dozen fruit trees, the spring is not nearly enough. We must find well water or develop the springs down the slopes to survive as we are now, let alone do any irrigating.

What kind of soil does the land have and how deep is the topsoil? If you're thinking of agriculture, this is of utmost importance. The agricultural adviser of this county tells me a major problem is locating soils and adequate water to allow desired crop production. Get a vegetation and soils map of the area. They are available through your agricultural extension service. There is a rating of the different soils on the map, for timber or grass, all available in this county for fifty-five cents. Here, you can only get a soil test done after you've bought the land. It's possible you could get one done before in other areas. Also available are topographical maps, acreage computer grids, maps from the tax assessor's office telling who owns the land, and detailed road maps. The vegetation-soils maps and

legends show dominant plant species and miscellaneous elements with a tabulation of sprouting nature and browse values, the type and depth of soils with identifying characteristics, and the capacity of the land for growing timber crops and the size averages of the trees, and ratings of the relative suitability of the soils for grass production as well. The illustrated glossary will also help you recognize various plants.

Finding out about the weather is very helpful. You know generally what to expect for each season and how to prepare for it, how to judge your garden by it, what is meant by "good road access," etc.

You'll always, under this government, have to pay taxes. Find out how much they'll be. Can you afford it? We have put our land into an agricultural preserves act, called the Williamson Act, which lowered the taxes considerably. To qualify, you must have one hundred acres or adjoin land already in the act. It requires, essentially, devoting your property to agricultural use and preservation. Go to the tax assessor's office and find out about the likelihood of your taxes changing and about possible tax-exemption acts. Another possibility is the homeowner's exemption. If you live on the land you own *in a code house*, you automatically qualify. Most likely if you're building to code you will have gotten a building permit—which presupposes a visit from the building inspector. He (most likely he) comes three times to check out your building. If approved, the assessor comes to evaluate your taxes under homeowner's exemption. If, however, your house is not to code and you've invited the inspector, you're actually in violation of the law unless you can convince him it's not your permanent living quarters. *Don't* invite the assessor or building inspector unless you want this exemption and are sure you qualify. The assessor has the right to come every three years, mostly to look at the *land*. The building inspector must be invited unless he has a warrant.

It is wise to check out what's happening to the surrounding areas. Are they going to lot your favorite view? Do all your neighbors run generators and chain saws when you want to live quietly? Are developments going up around you so that in three years you'll be looking at roads and houses and lights? Who owns your neighboring land and what are their plans? What are your boundaries? Have the agent, or if she/he doesn't know, the previous owner, go around and point out the boundary markers. If it's unclear and makes a difference to you now and you're definitely buying, you can get it surveyed at very high expense (seller might pay). Sometimes it doesn't matter about exactness yet—but very well may later. In any case, have an approximate idea of every boundary. Are your borders open or wooded? Recognize what this means in terms of your neighbor's developments affecting you.

One other interesting thing to find out is what, if any, natural dangers exist on your land. We have rattlesnakes and black widow spiders and tarantulas on a very small scale—but it's good to know about. And don't forget that trees fall down!

MAKING THE DEAL

So now you've spent a lot of time checking out your dreamland and are madly in love with it and have magnificent energy to begin your new life! Now it's time to make your deals. Above all—use caution along with your exuberance. Something to find out is if you can rent for a year before buying (rent to go toward purchase) and *really* get to know the land first.

You are entering a relationship of ownership. You've probably done it before with a car or a bicycle or a fountain pen or a dog. This feels different. Buying land legally means buying power over everything that goes on there. I feel one of the biggest reasons why we land-loving people—especially women—take this on is to combat the patriarchal trend of heavy developing and misusing and raping of land. Instead, we turn toward care and further-

ing life and happiness. I have heard many of us say, "We're buying to save and protect this land." It is an incredible power to *think* that because you own land, you can change in any way you desire all other existing life. Just walking around alters the natural balances. We must remain highly conscious and loving and comprehending to use this human tool.

A woman, especially a young one, is considered more of a pushover than a man. Be firm and convinced. Use all the help you can get. It is hard to go through hassling and sometimes dirty dealings for something beautiful. This communal land was bought with my name on the deed (the money came through me) for reasons of expediency, stability, and an inability to find at that time a suitable group plan. Now that we're more of an established family, we know better what we need for legal collective ownership. If you are buying collectively (as one purchase, not shares), I would strongly advise finding a plan that gives power and stability to the entity, the total group, and allows the individuals to change and move within it. Check out property laws in your area. In California the property of a married person is half the spouse's, whether or not they are together.

For this and for everything else, you'll need a real estate lawyer. This is very important and absolutely worth the expense if you can't find free counsel. There are many, many details to carefully explore that you probably won't know about unless schooled in real estate law. And a lot of sellers will,

I went walking today out across the meadows toward the creek, going nowhere. The hardest thing about living here has been adjusting to the absence of almost everything I've known—cars, phones, people, noise, books, newspapers, television: contact and distractions. Here there is only the land and the inside of my own head. I invent schedules, programs, activities for myself to fill my time and give it meaning, feel panicked when I have nothing "to do."

Today was like a revelation, walking nowhere, just being there. I am learning to see and smell and feel all over again. I am moved almost to tears by the tiniest details: the hint of purple in the spring green grasses, the curve of a foxglove blossom against an ancient redwood stump, the taste of sun-warmed blackberries. Each tiny impression of beauty is etched into my cells. The longer I live here, the more my senses keep expanding.

unfortunately, take advantage of this ignorance.

It would be wonderful if you could trust the people you're dealing with. You'll probably know. The lawyer helps here. We worked with an agent, the owner, and a lawyer (young and more into drug cases than real estate and not much good for us) who constantly bad-mouthed each other. Get outside references if possible about the people involved. If your agent is a wheeler-dealer, at least you're armed with that information.

So you sit down with the seller and figure price. Make low offers! You already know the approximate range, but this is for final numbers. It is *hard*. Be firm! Find out beforehand what the normal interest rate is on installments, what the average down payment percentage is (this varies widely, depending on the mortgage period). Interest is paid for the periods following the last payment and only on the remaining amount due. We had to pay the seemingly huge amount of 37 per cent down. The highest amount before the seller has to pay taxes on it is 29 per cent. Talk price and determine the length of time to pay off the mortgage. The longer the mortgage period, the better for you since it lets you use your capital for starting-out needs or lets you generate an income with it. Whereas the amount of interest gets higher as the term gets longer, inflation causes the value of the dollar to continually decrease.

Find out what is included in the purchase (equipment, other rights, all land improvements, etc.). Make requests. Especially, learn what liens if any are attached. Liens are holds on the land that are transferred to you in the purchase unless specifically stated otherwise. These include such things as unpaid back taxes (these are *not* your responsibility), Joe Green's hunting rights on your land, Sally Shoe's right to tear down your buildings and take the lumber, Wilbur Mush's mining and mineral rights, R. T. Hit's water rights, P. G. & E.'s power-line rights, and other possible goodies. Just ask, "Are there any liens attached?" That's all you need for now, and you can get rid of any you don't want by writing it into the contract. If the seller insists, for instance, on keeping the mining rights, then you've got another *written* deal to make to satisfy both of you by setting up limits.

The next matter of absolute importance to discuss is access. This was our point of contention. Do you have full right of way over the roads to your property? Get it guaranteed in writing. Get copies of the right-of-way papers. Find out the nature of the access—is it deeded or prescriptive (by virtue

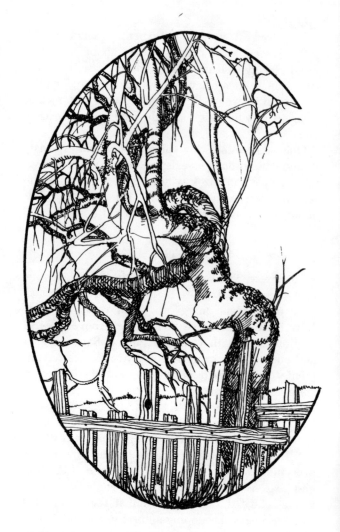

of long use)? If deeded, you're fine. Get copies. If prescriptive, get affidavits proclaiming that use and its transference to you. Does anyone have right of way *through* you, and how do you feel about this? Discuss each of these things with your lawyer and find out their meanings. If you really can't get a lawyer, go to all the claims offices yourself and check everything out. Don't be thrown by the seller's "Don't you trust me?" Just say you want to make sure of everything for yourself.

There is a legal procedure provided to make sure all these things are taken care of. It's called a title search, your title insurance. A third party, a title company, holds the seller's deed and the buyer's money in escrow while checking out any liens and making sure the seller has the right to sell. Usually the buyer has to pay for this—but the seller might. At the close of escrow, if the land is free to change hands and the buyer has met her/his requirements, the title company gives the buyer the deed and the seller the money. Do it! It's a necessary safeguard.

O.K. You're ready to sign a preliminary contract, which is also a receipt of your deposit (fraction of the down payment). It is legally binding, setting up the terms of the agreement (seller might have requirements to meet, too), the property description, any liens, the terms of escrow, other conditions. Make it *subject to the approval of your lawyer.* If not, then make another preliminary contract with new terms. Money and deed go into escrow, you sign the last real paper, and wait for the close of escrow, at which time you'll be presented with the deed. The land is yours except for the deed of trust, which belongs to the previous owners until the mortgage is paid off in full. This is their claim to the land if you don't pay up.

At this point, you should have copies of preliminary contracts, mortgage contracts, any liens, rights of way, seller's deed of trust, preliminary title report when escrow begins and the policy of title insurance, and the deed in your name when escrow closes.

And there you are—a landowner. Full of love and enthusiastic energy and wonderful dreams, you begin the process of country living. You start your garden, find old cabins in the area to tear down for lumber, look for farm animals and build your shelters, and wander in the woods and breathe fresh air, and work really hard. One last piece of advice is to post your land against hunters if so desired. They are legally trespassing if the land is properly posted.

You begin that magnificent rediscovery of natural cycles and rhythms and find yourself slowing and calming down. You sit under your favorite tree and watch the scampering lizards, listen to the singing wind, and find your bond to the land already growing strong.

the ECONOMICS OF BUYING LAND

As magical and beckoning as the land might be, the actual buying and financing of even the smallest acreage may seem overwhelming. In the beginning, it is enough to just *be* in the country. As you consider a more serious relationship with the land, you may find yourself wanting the stability and future of owning or sharing ownership of your farm or home. This "owning" is a reciprocal arrangement between yourself and the land, involving responsibility, commitment, and exchange. In the preceding pages, the process of finding and choosing land is discussed. Here, I would like to explore some of the purchasing problems and the feasibility of owning your land.

Wander into any country real estate office with your homesteading dreams in your pocket, and you will find the economics of buying land as variable as the land herself. Land is not cheap, and land that your farming dreams or wilderness yearnings can flourish on is not easily found or easily purchased. Years of planning, saving, and thinking about the venture may precede your actual search for your land. Or you may find yourself hastily, impulsively turning all of your plans upside down to buy that perfect, coincidental find! Whatever you plan for or actually do, you need *not* be incredibly wealthy to buy land. You *do* have to be willing to struggle, invent, endure, and improvise. I'd like to share some of my own experiences and those of friends in land buying on a low budget. It is not an easy thing to undertake—but it is possible. I hope to encourage other small farmers to grow, too. The land is rich in rewards and sustenance; your economic groundwork can be simple and solid.

The first year after we began to buy our land (usually I say "bought"—yet here I am nearly five years later still making land payments) my advice to city friends was, "don't try to live in the country and pay for land." Now years later, I have no clear advice to give. I wouldn't have traded these last few years of country living for equivalent time earning more money easily in the city. But I would like to talk about some of the real economic problems for anyone wanting to buy land.

When we came here, my husband and I had both been working full-time jobs for about a year and a half trying to save money for land. We had over $4,000 saved and wanted a place to settle into and to grow with. We found that small pieces of land were hard to find in this area (where we wanted to live) and exorbitantly high in cost (two to three thousand an acre). We began to talk to friends still in the city about buying a larger piece co-operatively or communally. No one we knew well was ready. At this point, my parents offered to help us by buying a forty-acre piece, selling ten acres back to us at the same interest rate they were paying and keeping the other thirty as an investment. This meant no down payment for us (usually 10–25 per cent of cost), leaving us our capital to get a home together before winter. That first six months was hard—$3,000 disappeared into the land in the form of a septic tank, well rings and pump, electricity, road grading, and gravel. Almost all our savings were gone and nothing showed above ground except the pressure tank on the pump. We were not extravagant nor were we urban neophytes. We did what was minimally necessary to free ourselves from investigation by the building inspector (a problem in this area). We dug the well and ditches for the electric lines ourselves, we did all the wiring and plumbing ourselves, we worked with a backhoe driver spreading gravel on the road and in the drainage lines for the septic tank. That $3,000 was for basic materials. Meanwhile we were both still working full time to have enough money for land payments (higher than our previous rent) and lumber for a cabin. In the evenings we multiplied payments by months by 7½ per cent interest and discovered we would spend an entire year paying not for land but for borrowing.

What you want and what you need will determine how similar your experiences will be to ours. We chose to live on the California coast where land values have been inflated by real estate developers. We chose land that was suitable for farming (rolling grassy meadows) but too exposed to be free from the restraints of the building code. Also, we wanted electricity, which meant improving the place to the building inspector's standards (or running a gas-driven home generator). For all of this we paid $900 per acre. *Prime* farm land in other areas is similarly expensive. I have seen good farmable land (good soil, level ground, pastures, water, maybe fences, maybe barns) in parts of Pennsylvania, Massachusetts, Utah, and the Midwest selling for $600–$1,000 an acre. On the other hand, there are still whole areas of the country (Appalachia, the Ozarks, the Northwest) and parts of every state, where farmable land is *much* cheaper—$150–$300 an acre. These may be areas with more severe winters or where the land has little capitalist commercial appeal. The steady decline of small farms in America has lowered the price of acreage in areas where large corporate farming is not profitable. Rundown or abandoned farms are also available at relatively low prices. They take energy and devotion to reclaim, but they can become richly productive again. Cheaper land sometimes means it is suitable for only partial subsistence (it will support a garden and partial feed for animals but won't produce cash crops to help buy the seeds, tools, lumber, and feed you'll need). Cheaper land may also mean greater isolation, which is a mixed blessing. Privacy, peace, no contact with restrictive society are its assets. *Real* isolation (no friends, no help, no contact with urban resources), not to mention bad roads and no electricity, are its drawbacks. If you are a community of people, you may be able to meet many of your own support needs and will welcome the isolation. Where you choose to look for land and which piece you buy will reflect a delicate balance between your needs and your resources: what you want in terms of climate and community, what you plan to do with the land, and how much money you can afford to spend.

Unless you have inherited money, buying land means saving money—either one or two people saving a lot of money (several thousand dollars anyway) or a group of people saving smaller amounts. Banks don't lend money on land; they lend money on houses. In most instances you must either pay cash for your land or arrange a mortgage

This job as a waitress is eating up my life. Smiling and ingratiating myself, smiling and hustling, I struggle for a little bit bigger tip to supplement the below-minimum wages in this ritzy hotel. I serve bright young lawyers, my own age, who speak arrogantly of hippies and casually rub their hands against my body at breakfast. Their wives, subtler and more sophisticated versions of myself, sit smiling.

When I get home after eight hours, I take several hours more to wipe the mask off my face: to transform a piece of female flesh into this skinny, plain country woman. And I take even longer to rediscover my own voice and personality, to remember I have a life that does not wait on others. Living on land while working to pay for it is a contradiction in terms—I don't live here, I only sleep here.

with the seller. This means a higher cash down payment than you would have to make for a city or suburban home. So, before you try to buy a piece of land, you need enough cash for the down payment and the assurance that you will be able to earn enough for the monthly payments. I strongly recommend whenever you can afford it (emotionally as well as financially) that you have the down payment, one year's worth of land payments, and *at least* $2,000 for initial improvements saved *before* you move onto the land. I can't say strongly enough that the simple life will cost a lot more than you expect.

But I also want to say that people are buying land and you can do it too. What I want to do here is describe the cost and method of payment of three pieces of land—I am ignoring cases where people have bought land with inherited wealth, since for them the only problem was finding the right place.

My own situation first. We were lucky in not having to make a down payment. If we had, we would have had to spend at least another year earning and saving money, or we would have had to live very primitively while still trying to work. My husband and I (and later I alone—we separated two years ago) have worked full- or part-time jobs for the last five years. Our land payments and taxes are $1,500 a year—more than two thirds of my average annual salary (working full time seven months a year—there being almost no work here during the rainy season). I have four more years of payments to make at that rate before I have clear title to the

land. I am beginning to raise sheep for my livelihood and hope by next year to make enough from them to meet the land payments. This will not free me from having an outside job, but will let me use my salary for other improvements on the land which may eventually provide my support. This is a relatively hard way to buy land—it means a lot of work and struggle and perseverance. It has also meant that the land is my own to create visions on, to let myself belong to utterly.

This reminds me of something else which is important: owning land is a long-term relationship—for the rest of your life if you want. Human relationships are not always so enduring. Make a very clear agreement with your buying partner(s) about what to do if either (any) of you leave. Even (especially!) if you feel you will stay together forever, make the agreement before you even move onto the land, while you are feeling clear and trusting and loving together. Does someone leaving have the right to get her or his investment back? If so, when they leave or when the land is sold? Is it the responsibility of those remaining or of newcomers to pay that debt? These questions are very important and need to be agreed upon. I am paying my husband in monthly payments for what he put in (we had a hard time agreeing on exactly what that was, too!). Another friend has agreed to pay her former partner when (if) the land is sold, so that both will benefit from the hypothetical future profit. Some friends who own land communally have agreed that the land is not an investment—no one may withdraw his or her money unless everyone simultaneously agrees to sell.

There are ways of buying land that are much less difficult than the individualistic pattern I have followed. The friends that I mentioned above have begun buying forty acres of beautiful woodland with a creek, several springs, and a large ten-acre meadow. This land has poor road access, no electricity, and is in a much less commercially desirable area than mine (away from tourist and vacation home development). There is a community of ex-urbanite neighbors in the surrounding ten miles. Their land is not suitable for animal raising beyond their own needs, but they have begun a large market garden. They paid $11,000 for their forty acres ($2,200 down). Split six ways, this was $370 per person down payment. Monthly land payments and taxes are $20 per person per month for ten years (this is at 7 per cent interest). They also made an agreement before they began to buy to have one year's land payments put away. My friend Marnie writes: "This money is inviolate and somehow makes life much easier, knowing if times get really hard we've got a year's grace to figure something out." Three new people have joined the original six and they also make $20 a month "land payments." Currently their money is going into a building fund to get a main house together before winter. After that their money will be used to pay off the land more rapidly (and therefore reduce the interest owed). There are fewer jobs available in their area than in my area, and all of the original six members have gone back to the city periodically to take high-paying jobs (from three weeks to five months long). Marnie says, "We *prefer* not to go to the city; it's hard on our life together here and the transition is hard on each of us personally. We do it because of desperation, discomfort with indebtedness to each other (though we do loan and borrow easily and often from each other), or because opportunities arise that we can't afford to miss. David doesn't seem to mind as much as some of us—the criteria seems to be whether or not it's a high-pay, short-term job. We are very aware of (but not yet resolved) the fact that some of us can earn five to eight dollars an hour and others of us $2.50—mostly a difference between men and women." They hope eventually through garden crops and some craft work to meet their cash needs at home, but they haven't yet. As a group, they have a clear but unwritten agreement about land use: which is community space, which must be left wild, which is available for private cabins. They also share an ecological and consumer consciousness: no power

tools (chain saw and Rototiller for example); no electricity (they hope to get a windmill); no hunting or inorganic fertilizers. Their land is beautiful, quiet, and private; nine people, who have no money other than what they earn, are making their home there.

The best co-operative land arrangement I know of is also relatively rare: the land trust. I can only hope we will see more and more. My friend Judy was forty-three years old, had worked as a secretary and a waitress for twenty-one years dreaming hopelessly of owning a farm, when she heard of the Greenland Trust in northern California. This was a five-thousand-acre ranch that was bought by an ecological corporation and sold in fifty-acre pieces (at cost—$125 an acre) to individuals. Really large pieces of land often sell for such a small cost per acre because of the large sums involved. Judy's piece of land is as beautiful as mine—rolling, grassy hills with oak and pine woods, an old (crumbling) hunter's cabin and a year-round spring. She got clear title to her land, but all members of the land trust community also signed a binding legal agreement. This agreement prevents further division of the land, limits the number of persons living on any one piece, and prohibits pollution of water sources, hunting, use of pesticides or inorganic fertilizers, etc. In general, the agreement provides for good, loving use of the land. (Though it's interesting that the hunting prohibition is now a source of controversy as the deer are proliferating and feasting on gardens.) In this land trust, members had the option of paying cash or making a down payment and payments. Judy did the latter, but found she was unable to manage alone and now shares the land cost and ownership with a friend. Unprepared with skills, fuel, or money, she found the first winter in her still leaky cabin really rough and nearly gave up in despair. She still finds it almost unbelievable that she should be building her own house on her own land. For information about forming land trusts write: International Independence Institute, West Road, Box 183, Ashby, Maine 01431.

Buying land is not easy, but it is possible even for those with very little money. And it is joyous. Owning land is letting your heart and soul love it unreservedly, is sending down roots that seem to touch the earth's core, is knowing that only what you believe in and are committed to will happen there. My relationship to my land is as profound as any other I have ever experienced. Viewed that way, years of working a job for someone else is *not* too high a price to pay.

OUR CLOTHING, OUR BODIES

The country has changed my whole relationship to my body—it is no longer a vehicle to get me around but has become an extension, an expression of myself. Its limits have expanded as my own have, muscles able to lift or reach, as my mind also lifts away fears, reaches out to try. But this has not been an obvious or self-conscious process; the changes have been subtle. Along the way I have at times badly misused my body or ignored its pains, following an old pattern of seeing it merely as a necessary but unimportant appendage to my "real" self. Now, body consciousness is with me all the time from aching feet to muscles stretching loose and free, naked to the sun. My body is my most precious tool.

A friend wrote to me recently from her farm in Arkansas and said, "When we first moved here, I tried to lift bales of hay by myself and couldn't. So

I've been letting Paul do it ever since. Two months ago I tried again . . . somehow I found it well within my capacity to pick the bale up. I had grown stronger without ever retesting my assumption of incurable weakness against reality. It is still not an *easy* thing to do, but I do it often enough now to know exactly how hard it is . . ."

What my friend discovered has come to me time and time again too. Strength does not happen overnight—it's a gradual process of using your muscles. There's a fine balance between accepting the current limits of your body (not overstraining) and pushing yourself enough to grow stronger. The most important part of this balance I've discovered is listening to my own body rhythms, instead of following externally imposed standards. When I'm doing something I've never done before, I work until I start to feel tired (sometimes only five or ten

16

minutes), then I rest until my breathing is normal and my muscles relaxed before starting again. Working when I am too tired is always a mistake—my arm will shake while hammering and I'll bend nails, or I will hit myself with the mattock through inattention. With my stop-and-go process, I can work long hours and enjoy myself without ever feeling pushed to my physical limits. Over the years as I've grown stronger, I've learned a new kind of tiredness, that which comes from a long, hard day but leaves the body feeling slow and relaxed all over. This is really different from the kind of strain I feel when I push a muscle to do what it's never done before.

Using the right tool is almost as important as the pace at which you work. These tools will change as your body does. I learned to drive nails with a sixteen-ounce hammer, and now I use a twenty-ounce one for heavy work—its extra weight drives the nail without my having to use so much force. But when

I was learning (and getting accustomed to using my arm) that heavy hammer placed such a strain on my wrist that it took more energy to hold my hand steady than it did to drive the nail. Tools should be *tools;* they should help you do things. This has been especially important to me as a woman just learning to use my body—I don't have brute strength to fall back on and so must use tools as tools. Leverage is really important here, so that the motion of the tool does the pulling or lifting, not your arm. Use long wrenches (you can extend them with a piece of pipe) or use socket wrenches with a long-armed ratchet wherever you can. My extra-large screw driver is one of my most prized possessions and it fits many relatively small screws. So are my large pipe wrenches which have broken open joints I could never have gotten with force alone. Heavier is not always better for a tool. I learned to split wood with a very heavy ax and have only recently discovered that a lighter weight one is sometimes

Working on the drag boat has transformed Peter. He is filled with confidence and joy. Something about the primitive struggle with sea and wind and fish makes him radiant. He has grown a beard and now smokes a pipe and tells tall sea tales of dolphins and albatross in a gentle, believable voice. Whatever the rituals are in that tribe of fishermen, he has joined them. He has found his place.

Today was our first day home together in several weeks. We spent it playing and working in the woods—searching for redwood orchids and late mushrooms we couldn't have identified even if we'd found them. Peter taught me to split wood. My muscles found his instructions unintelligible, not knowing how to translate them into actions. So I spent the morning mostly lying on the ground letting frustration drain through me into the earth, watching Peter split the wood and seeing sunlight filter through the tree trunks. Then this afternoon, faced with a knotty round, I felt my wrists give the ax a snap that sent it falling swiftly, easily, through the wood, heard the soft crack of wood splitting clean. And I laughed and we hugged and danced.

Now we have a new division of labor: Peter cuts the wood with the chain saw, his new toy, and I split it with the ax. I am so excited to feel that I can do it. There is such pleasure in the work and constant motion—tensions become grounded. And the woodpile grows much faster than the garden!

I take this insane delight in just being in the country sometimes. As if, by some wonderful accident, I will never grow up to a city job and suburban home. I laugh with sheer pleasure at splashing through mud puddles in my rubber boots, hiding in a secret house beneath low-hanging cypresses. I will never have to lose my childlike wonder.

more efficient. The point is to find the tool that will give you the most advantage while placing the least strain on your body.

When I work together with a friend here, I try to regularly trade off positions and tasks. The person who was holding the board trades with one who was nailing; the one who was shoveling trades with the one who was tamping dirt. It's not that one job is necessarily easier, but each uses different muscles. By trading, both of us get stronger without overtiring one particular part of our body.

The country has also been teaching me a lot about relaxing—though this is not so obvious as it may sound. Coming here—for the first time in my life to a whole vision I believed in—it has been easy to throw myself into working all day long. There is so much to do, and it is never done. But I have learned that doing some new activity—when I finish working—that relaxes the muscles I've been straining and stretches the ones I haven't been using helps my body feel balanced and gives me new energy. My favorite thing is swimming—diving in the pond after hours of clearing brush or setting fence posts leaves me feeling almost rested. I also do yoga, dance, running—anything that shakes me loose.

The back is probably the most important and most vulnerable part of a working body. For the first two years of living here, I listened to friends' pains and their warnings about lifting things properly. But with my usual casualness about my healthy, uncomplaining body, I didn't really hear them. The first time I heard what they were saying was one winter day last year when I bent and lifted an eighty-pound grain sack—as I had dozens of times before—and felt a sharp, stabbing pain in my lower back. My back has never been the same since. It doesn't always hurt, but it does every time I have to do a job that takes bending or every time I stoop to lift something. That carelessness is something I will always regret.

I feel so isolated here, even surrounded by friends. I wonder if I am crazy or a misfit. Coming to this area, where I can be part of a community while still living in the country, felt so right. Most of all, I was attracted by the women here, by the chance to have women friends again. These last two years, Peter has been my only nearby friend, but there is still so much he doesn't understand, so many spaces between us.

Yet even here in this circle of women, I feel separate, and I wonder what is wrong with me that I can't accept things, that I am not contented. Today, I was talking to Sarah about the *Politics of Housework* which I had just read and about hassles with Peter over the shopping. She gave me the blankest look, as though I were speaking Turkish. There are no politics to housework here, the roles are so completely set! Women talk of gardens, goats, crochet, and canning. Carol has even wallpapered her outhouse! I am ambivalent; I long to be a part of that cozy circle yet feel myself an outsider.

Alice just sent *Sisterhood Is Powerful* and some magazines about the women's movement to me from the city. It helps a lot to know there *are* other women out there somewhere who want to share their lives equally with the men they live with, who are exploring the subtleties of equality: such as who cleans the toilet or makes the shopping list, not how to divide the dishwashing. It's so hard, trying to learn and grow in a vacuum—especially when every man and most of the women Peter meets tell him that he's crazy "to put up with" my demands at all! I worry that someday he will leave me for a woman who will get up at 6 A.M. to pack his lunch; that this struggle is not really ours, but mine—one he's committed to only as long as he needs me.

Thru many moons the mountain was waiting:
　　For a dancer to float above its fog.
　　For a spirit to soar high as its hawks.
　　For a body to roll thru its flowers.
　　For a laughter to echo its joy.
　　For a soul to drink of its secrets.
　　And for a smile to soothe its storms.
Now, in this thirsty year of the big snow
　　　　the mountain found
　　　　a woman Diane
　　　　who calls it home.

The most important thing to know about backs is not to stoop to lift—bend down from the knees and hips to a crouching position and then lift up with your back relatively straight. This takes no more time or energy than lifting the other way—it's a matter of breaking habits. If you take the energy to retrain yourself to lift the other way, it too will become a habit and you will never know what "bad back" means. With heavy objects, I begin by lifting very gradually—testing whether I can really handle it—and by being careful to keep my balance. If I am carrying the load in my arms, I hold it close to my body so there is no pull on my back. Whenever I can, though, I carry heavy loads on my shoulders or my head so the weight is carried straight down my spine and through my legs. Plywood, for instance, is really easy to manage on my head and I can carry ninety pounds roofing paper on my shoulders. (I only weigh 115 pounds myself.) What is important is raising or positioning the object so I can get my head or shoulder under it without bending my back (stooping over)—the bed of the pickup is especially good for this since I can crouch down and come up under what I want to lift.

Having my body feel balanced and on center is really important. I like to carry equal loads on each shoulder or in each arm when I can get someone to help me load up. Several years of hauling water out to animals has taught me that I can carry *more* weight more easily by using two containers, one in each hand. I also like to lift with a partner for really heavy objects—one of us on each side. I use this method for stacking heavy three-wire bales of hay or moving logs for sawing. Formerly, I might

have pushed myself to do such things alone, now I prefer the help of a friend.

Needing my back for work has also made me change my attitudes toward posture. A childhood of being told to "stand up straight" and a college that required a successful "posture picture" for women graduates had made me a rebellious and confirmed swayback. Now it is my back that tells me how to stand—that feels comfortable only when it's straight, supported by the muscles, not supporting them. From my friends who have had bad backs longer than I, I've learned to sleep on a firm mattress with a low pillow under my head and a pillow under my knees. This straightens out your back if you sleep on it—the alternative is sleeping on your side with knees bent. I have found this lying on my back with a pillow under my knees is a wonderful way to relax both back and legs—even five minutes that way when I am really sore and aching helps a lot.

Being in the country has also changed my feelings about dirt. It's hard even to write this down—for though I feel comfortable with my new self, I am still enough the daughter of a good middle-class mother to feel embarrassed about publicly confessing my unbathed state.

My first discovery about dirt came when I first crumbled soil through my fingers and looked at it closely. There were tiny stones, sand, bits of grass and leaves, but where was the *dirt?* Later came my fascination with manure, another embarrassing confession. Right from the start I liked it—liked the smell, the heavy sloggy work, the richness of its potential—a fantasy taste of life as a farm hand. But after the first garden grew like a tropical jungle

on this cold, foggy coast, then I *really* got into manure. I mashed it in my hands, looked closely and saw—what! well-digested grass. There I was with a handful of the most taboo of all dirts, and it was nothing but a bit of grass.

Simultaneous with these discoveries about the dirt itself came a year of living with no running water. I learned then that I could keep my body and hair clean on four gallons of water heated weekly on top of the wood stove. And later with a well which got perilously low in summer, I discovered biodegradable soaps and the nearby pond. So that even now with abundant winter water, the habit of less frequent baths persists. There's another consideration too, besides the lack of water—the whole question of why I bathe at all. If it's to remove body smells, I hardly ever have any that I don't like. There are few smells I like as well as sweat on sunwarmed body; it's a sweet, rich smell. Bathing to remove "dirt" often seems a futile effort too. I live and work with dirt all the time; my life is never "clean." About the most I can hope for is that a bath at night will leave me in the same

state the next morning, *before* chores. For the same reason (plus a shortage of clothes) I tend to wear the same clothes for days at a time—one set for daytime work, one for evenings. What's the point of putting on clean work jeans when in two hours they will be like the rejected pair?

I must also tell you, though, that few would guess my perpetually dirty state. I do not "smell" in the traditional American sense (but then I know very few people who do), and I do not look like a slob. There are, however, many things I do find worthwhile washing off: grease from the truck engine, dust and grit and sawdust on my hair and skin, the essential oils of poison oak. And I do still find peace and therapeutic relaxation in a long hot soak, but I'm needing to relax less and less. What I've really learned from living here is to see and smell and feel everything around me, from my own body to the grains of soil. And to respond to what I experience, to no longer recognize the concept of "dirt." The American habit of daily scrubbings (not to mention deodorizing and perfuming) strikes me now as wasteful, unnecessary, and absurd—the sad reflection of a culture so divorced from its bodies and from the very soil it rests on that it couldn't recognize the true smell of either.

I don't know if my body will ever unlearn the lessons the country has taught it; I hope it won't. When I go to cities now, I walk for hours trying to use my pent-up energy or I find food filling up my stomach pointlessly. Growing stronger has meant having more energy. Having discovered the power in this body, I want it to carry me along. I won't return to a life where I have to carry it.

WORKING CLOTHES

Wearing clothes to work in has been a new experience for me too. Before I came to the country, the question of what to wear to work required only a personal choice between my taste and what was fashionable. Even when I wore denim or bandanas, it was more a matter of personal style than of need or purpose. Some of the first pieces of clothing I made after I moved here were long skirts—fulfilling childhood fantasies of pioneer days and current ones of the homestead life. My very favorite was made of very heavy pink cotton with seven gores and a three-yard ruffle around the bottom, perfect for dancing or playing the pioneer lady. I put that skirt on one morning when I was feeling in a very ladylike mood and went out to start morning chores. The day began by my fa-

vorite goat leaping to greet me—leaving manure prints on the pink skirt. At ten o'clock that morning I was off to town to deliver milk and get the mail when the clutch pedal went out on my truck as it was wont to do. There I was, underneath the truck on a January day and a dirt road, fixing the clutch with baling wire. That was the last time I wore the pink skirt—the only place there is in my life for one is after evening chores (about 8:30), and I'm usually too tired to bother.

Finding clothes that are good to work in is partly a matter of personal choice, but there are real matters of safety and practicality too. Skirts can be worse than a nuisance—they drag on the ground, catch at your legs when you need to run, get caught in saws and cracks. They can cause you to fall, trip, or get caught in a moving tool.

I have come to prefer the really heavy jeans for most work since they don't snag easily and present some sort of a barrier against berry vines and barbed wire. I try to buy them loose enough that I can move really freely without having them hang off of me (in fact, I keep two sizes for my different

weights). Overalls are my other favorite piece of clothing—probably the most functional and comfortable thing I wear. They hang freely from my shoulders letting my body move without any binding or constrictions. They protect the bulk of my body too—and have a pocket or place for everything I need.

For T-shirts, turtlenecks, and work shirts, I choose men's cotton or flannel rather than the synthetic blends so often sold to women. Cotton absorbs sweat and leaves you cooler; synthetics leave the sweat on you. Wool is also better than synthetics for socks or sweaters since it has the ability to keep you warm even when it's wet. A sweater of unwashed natural wool will actually repel water, too. Long underwear is another find, long known to people who needed to stay warm. I find a cotton thermal undershirt under a wool sweater is about the warmest and most comfortable combination in winter. My levi jacket is no longer an affectation but a tool—good protection when I'm working with barbed wire or clearing brush, as well as the perfect coat on cool days.

Boots have been my other major country clothes investment—heavy work boots with thick leather tops and corrugated soles. My first pair worked their way through four winters before the soles wore flat and the leather began to split—well worth the initial thirty dollars. Work boots need to be well cared for to last. They must be kept clean and oiled or treated with a silicone or paraffin waterproofer. I prefer oiling myself, and it must be done often. Here where the winters are very wet and not too cold, rubber work boots are also a necessity. Rubber conducts cold so they are no good in harsh climates, but elsewhere they will keep your feet warm and dry and save your leather boots for another season. I've never yet had a pair of rubber boots last more than one winter, but they are cheap enough that I'm willing to pay that price to stay dry. I have found, though, that it's important to spend the extra dollar or two to get well-made boots of heavy rubber which really *will* last one winter.

The importance of wearing clothes for protection and safety can't be stressed too much—especially to women and city people who have been brought up to think of clothing as ornamental not functional. I almost always wear my work boots even on the hottest days, and when I don't I'm usually sorry. They protect my feet from cuts and scratches and support my ankles. They are ready for *work*. I can pack down dirt, keep my balance

on steep ground, roll a log out of the way, without hurting my feet. I once watched an experienced country woman chopping wood barefoot. The ax slipped and she gouged her ankle. I have learned to never use an ax, chain saw, or power tool without wearing my boots. In fact, my boots carry the scars of several times when I was not hurt.

From a carpenter friend, I learned not to run a chain saw or other power tool without wearing a shirt. Without one, there is nothing between you and the saw blade—or whatever— that might catch or tangle, saving you from getting cut if it flies back at you. I have also learned from experience not to wear earrings, bracelets or most rings. They get caught on or in things and have pinched or cut me. More importantly, they can be dangerous, catching your hand or face in the way of a moving tool.

It was a car repair book written for hippie men that taught me to be conscious of what I do with my hair. I nearly always wear my long hair tied back. When it's loose, I always have a barrette in my back pocket. Long hair can catch in generator pulleys or fans on car engines, in the chain of a chain saw or the bit of an electric drill, causing you to be cut or even killed. It is in my eyes whenever I bend down in the garden, the field, on a carpentry project. I have more than once smashed my thumb when I couldn't see the nail and once badly burned my eyes with a splash of wood preservative when I couldn't see to lower the post into the barrel. Hair should be worn short enough so that it can't get in your way or long enough so you can keep it out of the way—securely, with a rubber band or barrette. A bandana tied around your forehead is also useful when you're working in the sun, to keep both hair and sweat out of your eyes.

A good pair of work gloves is essential for a lot of jobs. They should be made of leather (preferably) or have leather palms, because leather won't slip and offers better protection against cuts or binding. They should also really fit so you can work in them. Work gloves are indispensable for jobs like stretching barbed wire, handling splintery wood, or unrolling wire fencing. I also use them any time I'm doing lots of hard work with my hands —clearing brush or digging postholes for instance. And when I don't already have callouses built up, they are a god-send.

I've found that having the right clothes for a job changes how I feel about starting and doing it. Wearing a bandana over my hair and my workwoman's coveralls, I'm not at all inhibited about crawling all over my truck engine. I hardly ever work without wearing at least heavy jeans and my work boots, except perhaps in the garden or in the very center of the meadow (and even then I'm bound to step on a thistle). Bare bodies are very beautiful, but they're not very practical for work.

FINDING WATER

More than with any other single element, country life is involved with water. Abundant or insufficient rain means damaged fruit crops, floods, poor harvest, green on what would be golden hills, enough water in the well for baths next summer. It is a source of life and destruction. I am never indifferent to it, never simply annoyed because it interferes with my plans. My life is too deeply dependent on the water that flows out of the earth or comes down from the sky for me not to feel the weather as an extension of myself.

Helping dig my own well has given me another tie to water. The miracle of digging a hole into the ground and having it fill with clear water is reenacted each day. I never turn on a faucet without being aware of where that water comes from. As I drink, I can shut my eyes and still see that last despairing shovel full of clay from under which the water began slowly to bubble and to seep. I am tuned to the levels of the well, to how much and when water can be used, feeling a shock as I watch city visitors run taps on full with no thought for the water flooding down the drain. Water is a finite and precious commodity. It must be conserved even more carefully than the land, for without it there would be no homestead.

Finding and developing an adequate water supply is perhaps your most important action on first moving to a piece of land. Water flows beneath the surface of the ground in small streams. This water can be collected by digging down to the level of the stream (a well) or by finding where the stream comes to the surface (a spring). My favorite source of water is a free-running, bubbling spring because it speaks from the depths of the earth.

SPRINGS

Springs are usually the easiest source of water to develop, but they are as varied in shape and kind as the contours of the land itself. On my land alone there are three distinctly different types of springs. There is a deep canyon one where tiny trickles of water flow out of the steep walls and eventually come together, dripping slowly over a log. This spring looks like only a tiny seep but it kept us supplied with water for a year until we bought more land and dug a well. Then there are hillside springs —flowing out of natural amphitheaters near the hill-

tops and funneling into a narrow gurgling stream as they fall down miniature waterfalls on their way to the canyon bottom creek. The largest spring of all is a hilltop basin of marshy ground covered with horsetail ferns and silvery alder trees. Once while searching for blackberries, I found an old spring box at the edge of this big spring. The box was made of old redwood timbers encasing a hole dug into the spring's surface about five feet deep and filled with clear water. It was an exciting discovery, not just because it meant relief for an overstrained well, but also because I felt a deep bond with whoever built it over a century ago. They shared the same basic needs on this land as I do, and I am still drawing water from the hole they dug.

Springs on your land may not be like any of these—all that really matters about the type of spring is the clarity of the water and the best way to collect it. "Ground water" from springs, boggy areas, or swamps can be brackish, prey to animal and chemical pollution, and full of silt. It can also be an easily accessible source of fine water. Any new water supply (well or spring) probably should be tested for purity. You can send samples in a sterilized jar to your local health department. Don't let them come out and test it unless you want to give them the power to condemn your living situation. Be sure to ask for an explanation of the results, too! On our first farm, we stopped using the well and hauled water because a water test showed 1,000 coliforms per million parts water. Three months later we learned that that was normal surface bacteria for our area.

Usually spring water must be gathered in a spring box or collection pit in order to be pumped or piped to where you will use it. The design of these can vary greatly depending on the type of spring and the orderliness of your soul. If you're lucky enough to have spring water in a narrow stream *above* where you want to use it, you won't need a collection box. Instead, you can funnel the water along a gutter into a pipe and use the pull of gravity to bring the water to you. Such situations are few and far between, but if you have one, rejoice!

By far the funkiest of all our water systems was the deep canyon spring which in its final days resembled a caricature of a back country still. The spring was really a series of rivulets dripping out of

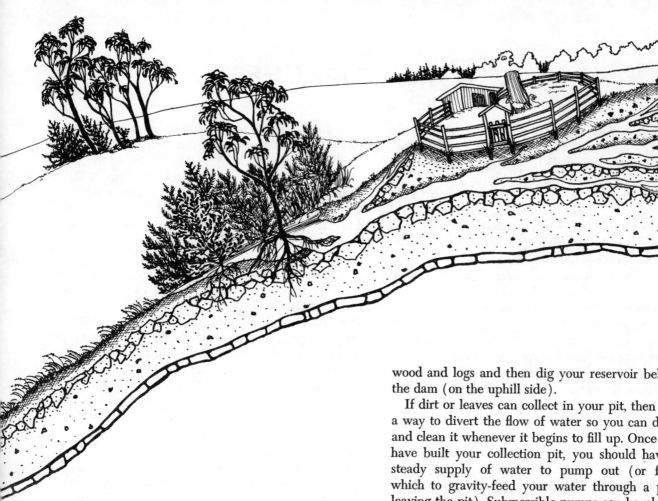

a hillside. Where they came together, we built a dam with a fallen log, sticks, and mud and behind the dam we dug a pit. From the dam, a rain gutter carried the water to an old fifty-five-gallon drum. A plastic pipe with a cheese cloth and wire screen filter on its end (to prevent silt from clogging the pump) carried the water from the barrel to the pump and pressure tank. Periodically I had to dump the barrel to clean the mud and leaves out of the bottom. The one time I forgot to check, sand got into the pump, tearing up its impellers, and I got my first lesson in pump repair.

Essentially, you need some place to collect water. If your soil is fairly hard clay, you can just dig out a hole approximately 5'×5'×5' (or 6') deep. If your soil is loam, sand or some combination of those, you will need to line your pit in some way. This can be with rot-resistant boards (redwood or cedar) as in old spring boxes, a plastic-lined wooden box, a plastic wading-type swimming pool, an old fifty-five-gallon drum or anything else you can think of. Where your water flows together in a narrow channel, you should build a dam of

wood and logs and then dig your reservoir behind the dam (on the uphill side).

If dirt or leaves can collect in your pit, then find a way to divert the flow of water so you can drain and clean it whenever it begins to fill up. Once you have built your collection pit, you should have a steady supply of water to pump out (or from which to gravity-feed your water through a pipe leaving the pit). Submersible pumps can be placed directly in the pit. Centrifugal pumps must be placed above it, covered by a house or box and connected to a pressure tank. Their advantage is their ability to lift water greater distances if your spring is below your house or garden.

If you will be using a pump with your spring, there are two things to be careful of: filtering your water—so dirt can't clog your pump—and regulating your water use—so you never drain the collection pit and run the pump dry. Filter systems can be made from wire screens, cheese cloth, coffee filters, sand traps, or anything else which will catch grit and leaves. Water use can be regulated by adjusting the pressure range on the pump, so that it draws only small amounts of water at a time and by being careful that you use water only at the same rate (or slower) that it flows into the pit. Your pump will come with directions for regulating the pressure range, usually on the inside of the electric box cover.

WELLS

Digging your own well is exciting, frightening, empowering, and mystical. Where to dig, how to

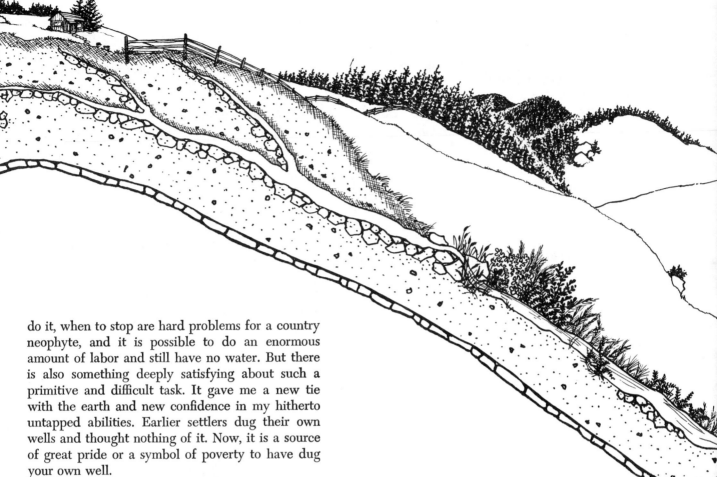

do it, when to stop are hard problems for a country neophyte, and it is possible to do an enormous amount of labor and still have no water. But there is also something deeply satisfying about such a primitive and difficult task. It gave me a new tie with the earth and new confidence in my hitherto untapped abilities. Earlier settlers dug their own wells and thought nothing of it. Now, it is a source of great pride or a symbol of poverty to have dug your own well.

The most difficult and important of all the well decisions is where to dig. Once you have dug a hole like that you will understand why you want to choose the right place the first time. When we were down sixteen feet and about to give up, I realized I could not bear to fill in that hole; it would have to stay a hollow monument to all our work. The surest way to find water is to use a two-inch auger (a long-handled hand drill with six-foot extensions) to drill a sample hole. Augers can be rented or borrowed from neighbors. It is important not just to find water but to know how fast it flows in. What determines a good well is not just how deep the water is but, more importantly, its "recovery rate": how fast the water replenishes itself. With an auger hole, you can use a plastic pipe and a hand pump to test the recovery rate. A good well will recover at two to three gallons a minute; an exceptional well or spring will pump five gallons a minute. Less than a gallon a minute means trouble when your well gets low—overnight waits for water or an altogether dry well.

We did not use an auger when we chose our well site, but used other reliable though less secure signs. We did find water—an adequate but not abundant supply—but I would use an auger if I were to do it again. What we followed were natural signs—the types of plants and trees growing in the area, the shape of the land, where there was nearby water. We asked long-time residents which grasses, flowers, and trees grew in standing water or indicated surface water. Each area has its own. We chose a site in a natural amphitheater filled with swamp grass and wax myrtle, just uphill from a known spring (not on our land). We were trying to tap into the underground feeders for that spring, and we succeeded. Once you get to know your area, you can predict with some accuracy the location of springs and underground water. Now, as I walk, I always notice likely sources of water and over and over again find that certain shapes of land contain springs.

Another way to find water is to have someone witch it—a mysterious, mystic way entirely in keeping with the magic of water bubbling out of the ground. I, personally, have never witched but have watched an old neighbor witch several times, and even to my sceptic's eyes, the branches quivered and pulled down to the ground over veins of water. Where we dug our well on our old waterless place, we did indeed find damp veins where the witched lines indicated, but not enough water to fill a well. A friend has told me of her first witching experience; "When some men came to drain our septic tank, they had to witch for its location. They marked out the six- by six-foot tank exactly and they showed me their mysterious art. Take two pieces of baling wire (three feet each) and bend them so that you have a handle and a long rod. Hold one in each hand, by the handle, loosely with the long part pointing directly ahead. Hold about chest level. Walk slowly across the land. When you are over an underground water line, the rods swing with a strong pull, in your hands. If the water is very close to the surface, as our septic tank was, the rods will swing around in full circles if you hold them above your head. Not so mysterious after all, is it?"

Finding a sure source of water before digging will repay you as long as you live on the land. It is also important to wait until the driest time of year to dig your well. Whenever is the worst possible time for water in your area is the best possible time to dig your well. If it has water in the worst of times, it will have plenty in the best.

Once you have chosen your site and waited for the right season of the year, you are ready to dig. If you have loose soil, a little money, and road access for machines, I would suggest having a backhoe dig the first twelve feet and then going on by hand. The backhoe can do it in an hour or two. My husband and I dug our well in three very hard days— a hole five feet in diameter and eighteen feet deep. The first day we only reached four feet because of the difficulty of clearing ground and penetrating

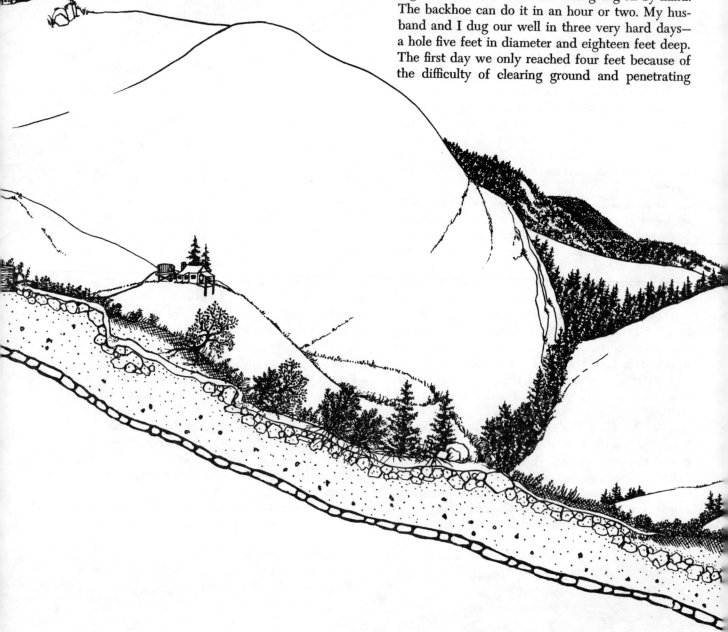

top layers. A backhoe could have done that first day's work in half an hour.

For digging, we used a short-handled, pointed edge shovel, a mattock, and pickax. On the advice of our eighty-year-old neighbor, we kept the tools clean and shiny with steel wool and the points sharpened. Dirt slides right off shiny tools, and you lift only the weight of the shovel—not it plus a load of wet mud. We were lucky enough to be loaned a windlass and rope by a neighbor who had dug several wells; without that luck, we would have had to build one. You need a strong (one inch) hemp rope and a good cranking mechanism to haul up the bucketfuls of dirt out of the well. Once you hit water, the wet dirt almost literally weighs a ton. Our windlass was a smooth log with a crank attached at one end, resting in a crisscross frame and padded with leather and a lot of axle grease—very homemade, but strong and smooth running. A bucket of dirt falling fifteen feet can kill the person standing in the well, so the cranking mechanism is not something to be careless with. The rope must be hemp—not nylon which will slip—and it must be unfrayed so it will not break. We dug our circle five feet in diameter, partly because well rings here are four feet and partly so we would have enough room to swing a pick. I dug two feet of the first four, discovered I was more than twice as slow as my husband, and quit digging in favor of running the crank. Later, I learned that the first four feet are by far the hardest, and I realized that hauling all that dirt up out of the well and away in the wheel barrow is just as strenuous as digging. My next well I will dig *and* crank, but on this one I accepted my limitations as finite and kept in "my place."

Windlass

One of the joys of digging a well is becoming a geologist for your own land, uncovering layer after layer of loam, clay, sand, gravel, and rock and coming to know your land as you never will by walking on its surface. Somewhere around sixteen feet, we hit seashells as well as discouragement, concrete evidence that this rich soil really was once the ocean's floor. At several points we hit cement-like rock which could be broken up with a pick and very hard work. Fortunately, it was never in layers more than two feet thick. Bedrock must be broken with a rented jackhammer, or the well must be abandoned. After sixteen feet, we hit rock again and were about to quit, thinking it was bedrock, when a thoughtless shovel punctured through it into water—not a geyser, just a slow seep, but three feet of it in the well by morning, and we could not bail fast enough to go on digging. Below the water was more rock and old-timers told us that it is important to know when to stop. You can dig too deep, puncture the bed beneath your water and have a dry hole.

A hand-dug well can go as deep as forty-five feet (they say you can see the stars in broad daylight if

Every time we have dinner with John and Marian, Peter and I begin playing incredible games, relating as we never would at home. He becomes "the man" and I "the woman." Peter and John sit talking about fishing and how hard they're working, each one trying to out-tough the other. Marian chops the wood, builds the fire, and cooks. I, helpless—resenting the expectation but not wanting Marian to wait on us all—cook too. John says the most insulting things in a casual conversational tone, and we all nod sociably as though he were speaking of the weather.

While we were there tonight he told us about the terrible pie Marian made yesterday and about how she thinks she's on a diet; he then demanded to know what she had eaten today. Later he asked if she had taken the car to get it fixed, questioning her until she betrayed her ignorance of engines. Sitting down to dinner, he turned to her and said, "Hey, dumb twat, you forgot the chutney." I couldn't believe he'd said that; I couldn't believe he would dare. I was furious and shaking and I wanted to hit him. Instead I told him that I was leaving because I couldn't stand it any more. How can Marian who is so fine live in such perpetual humiliation? I walked home alone. Peter made a joke and stayed to play cards.

He came home just now wanting to know why I couldn't be pleasant and sociable, why I always have to push things. I feel like I've been the resident bitch

you look up from that depth); eighteen feet felt quite frighteningly deep to us. It is safe to dig as deep as you want through clay or rock, and the well only needs to be lined for the first five feet to prevent topsoil and ground water from seeping in. Old wells here were lined with redwood boxes at the top and bare clay below. We intended to rest concrete well rings on old car axles driven into the sides five feet from the top. But about ten feet down we hit sand and had to line the whole well, digging inside rings. Sand can mean a cave-in, so it is terribly dangerous to dig with sand above your head and no well rings to protect you.

We bought precast concrete rings from a dealer in the nearest big town and hauled them on our VW bus to save money. We used perforated rings (with holes in them) for the bottom four and then solid rings above so that we would get deep, clear water, not muddy surface water. The well rings were the hardest part of the whole process. It was sheer bad luck (and expensive) that we had to ring the whole way. Each ring weighing two to three hundred pounds was lowered using our rope

and windlass—the two of us holding the weight with our arms as we centered the ring over the well and gradually let it down. One fell too fast and jammed crooked part way down. My partner had to climb down in and use a pry bar, while I lowered the ring on down, a frightening experience as one slip could have hurt him dreadfully. Digging inside the rings was slower and more difficult but not impossible. As you dig, the rings slip down around you and you add another to the top of the stack.

Once a well has been dug and filled with water, it is important to use it fairly regularly. Water feeds into a well through tiny capillaries and needs the suction of water being drawn out for the capillaries to send more in. If the water in a well stands stagnant for a long time, the tiny streams will gradually find other passages for their water and your well will lose its recovery power. When your well gets low (*if* it does!) it is important to watch its level and never drain it completely dry, which will damage its ability to attract more water. If it is at a critical level, draw out water only as fast as it

everywhere I've lived since college. I can't stand it. Back in Pennsylvania, Claudia and I were always trying to make the men distinguish their fantasies from reality, trying to free ourselves from living out their myths. We tried to make them see that their opening the house to anyone who came wasn't so groovy for us since *we* were the ones who cooked and cleaned and cared for the seventeen-year-olds looking for a new Mom. We tried to make them *see* the gaping hole where six months before they had decided to tear off the roof and build a sundeck. Even though we knew we were right, I hated how we were: our nagging, challenging, arguing, complaining. We were caught up in their world, on their terms.

I feel as though I've spent years being "Reality" for men, bringing them back to earth. All the soft, gentle, giving, loving parts of me feel dead or lost, so much of me is spent being a "reaction to." I don't like how I am, but I can't seem to change: I won't be sociable when a man puts down a woman; don't see "jokes" as innocent or swear words as meaningless. I want to live where people know themselves honestly, don't need to play "fastest gun in the West." I want to be gentle, but I have forgotten how. When John calls Marian a dumb twat am I, another "twat," supposed to be deaf as well as dumb? When newspaper cartoons mock the "inherent stupidity" of women, should I be blind? The woman in me is getting killed. I can hardly stand the self I have become.

runs in (the recovery rate) and allow several hours between major usings (garden, baths, etc.) for it to regain its normal level.

Developing your own water system is one of the most empowering (as well as essential) acts you can take on first coming to a piece of land. For me, it was a whole new way to know my land—and intuitively now I feel the pull and movement of hidden streams and pools of water. As I walk, I come to know not just its surface, but its deeper life. Water, which I used to use so lavishly and unthinkingly for years, is now a constant part of my consciousness. I have gained the power of learning in one more way how to provide for my most basic needs. I have dug a well and developed springs: I am no longer dependent on municipal pipelines or expensive contractors.

DEVELOPING A WATER SYSTEM

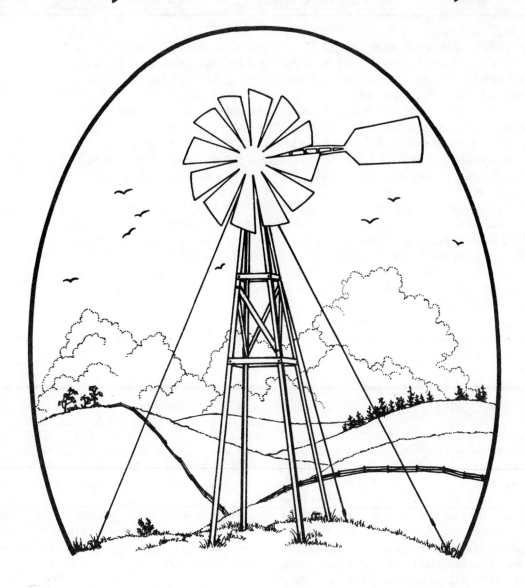

If you have ever developed a source of water—spring or well—you will know it is a monumental accomplishment. If you have also hauled all your water by the bucketful to where it's needed (for two humans, two goats, one horse, eight sheep, and a garden), you will discover, as I did, that the supposed secondary task of setting up a system to move the water is not so secondary after all. But the developing of a water system does not need to be a frightening or intimidating task.

"Water systems" means how you move the water from your well or spring to your faucet. This can be done with natural power—gravity pull from a greater to a lesser height, wind power (a windmill), or water power (a hydraulic ram). Or it can be done with human-made power—electric or gasoline.

Water pressure is what makes water move—pressure is created by some force, either the water's own movement in a stream or spring (caused by gravity) or by the natural external push or pull of a pump. Water pressure is measured in pounds per square inch (psi). The numbers on the pressure gauge of a pump indicate psi ("30" means 30 pounds of pressure per square inch of water). *One psi will lift water two vertical feet in the air.* A normal house needs a minimum of 15 psi in its lines at house level so that the water will come out of the faucets with some force.

Water can only be *sucked up* out of a well or spring twenty-five vertical feet, though it may be pushed *out* of a pump hundreds of feet if necessary. After twenty-five feet, a vacuum is created which causes the water in the lines to boil away. This means that if your well is deeper than twenty-five feet or if your pump is located more than twenty-five feet above the bottom of your well or spring, you must add a jet attachment to the bottom of the water line *or* use a submersible electrical pump in the bottom of the well. With either of these methods, jet or submersible, you are actually sucking only the first few feet from the bottom and then pushing the rest of the way. If you have a gravity system, at no point on the whole line (even if it is hundreds of feet long) can you go more than twenty-five feet above the original source.

Gravity systems have the virtue and beauty of total simplicity. There are no machines or moving parts, only one water pipe from the source to you. The water is moved out by the same force that moves all water, from tiny underground seeps to major rivers: the pull of the earth. When I was fortunate enough to live below a hillside spring, I felt admiration for the forces of nature and enjoyed a virtually infallible system for supplying water. If you are fortunate enough to use gravity power, feel thankful and enjoy your pumpless future!

A gravity-fed system runs on a siphon—a natural sucking action. The water line may run an indefinite distance as long as it ends up below the level of the source. It can even go up and down hills as long as it ultimately ends up below where it began. The water pressure at the end of the line will be ½ psi for every foot of vertical drop from the original source (this is called "static head" by professional mystifiers).

To create a siphon, your line must be completely full of water and the end of your line must be below the source. The easiest way to fill the line initially is to use a hand pump at the spring or well. You should use a fairly large pipe from the spring to its destination—one-and-a-half-inch if you can afford it, one-inch if you cannot. This large diameter provides a greater volume of water when you need it. The pipe can come up vertically out of the spring and then turn at right angles to go down. The only thing to remember is that ultimately it must go down and you must have a way to fill the whole line. If you have a vertical pipe into your

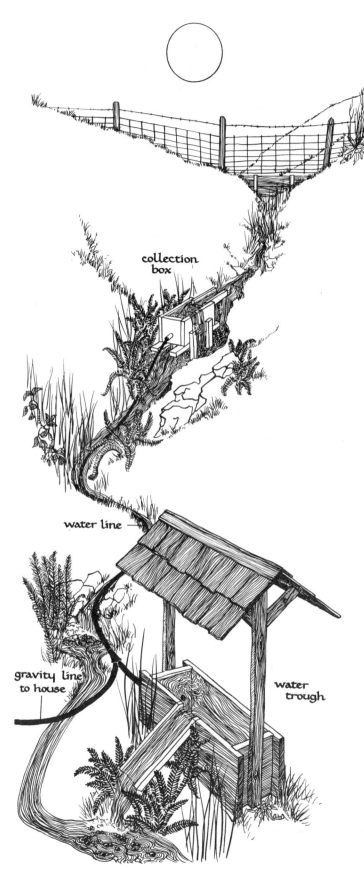

collection box

water line

gravity line to house

water trough

33

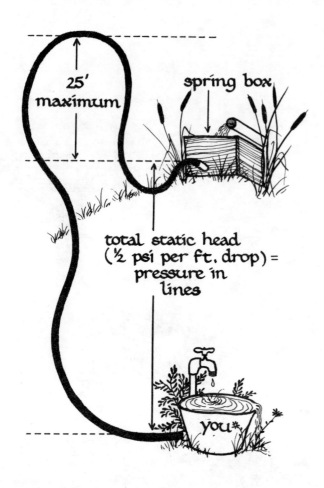

25' maximum

spring box

total static head
(½ psi per ft. drop) =
pressure in
lines

you

of the water in your lines, the water cannot push the air out of the lines nor can it get past the air pockets to you. There are two main causes of being air-bound. The first is that your pipe has too large a diameter for the psi of the water, so that the volume of water does not fill the pipe and air gets sucked back in. This can be cured by narrowing the pipe at the end, as mentioned above. The other cause is that cold water carries a lot of air dissolved in it. When the water heats up in the line (especially in plastic pipe), the air escapes from the water and forms pockets in the lines. The only way to prevent this is to bury your water line, a tedious job over any distance. Otherwise, you will just have to clear the line whenever it happens.

To restore your siphon once it has become air-bound, you have to create more water pressure until the water can push the air through the lines. You can do this by lowering the end of your pipe to raise the psi—this may be twenty or fifty vertical feet below your present location. You can also suck the air and water out of the end of the line, or reprime the entire line by using a hand pump at the source to force water into the line (with your faucet open so the air can escape) until the water flows freely again.

As I said above, power sources can be divided into natural and human-made. Pumps with natural power sources (the hydraulic ram and the windmill) have many of the virtues of the gravity system. They are relatively simple mechanically, do not depend on outside energy sources and often run perfectly for decades. Their major disadvantage is the initial expense, which can be quite high. I have never lived with a windmill or a ram and can only discuss briefly how they work. The sources listed in the Resources section will give detailed information and help you decide if your land is suitable for either type of pump.

Windmills require a fairly reliable supply of wind all year round. When the wind is blowing, the mill pumps water up into a large holding tank (on a tower or uphill from where it will be used). Gravity then supplies water to your house from the tank. This gives you a reserve of 1,500 to 10,000 gallons (depending on the size of your tank). Traditional tanks are made of wood and are quite expensive. Plastic swimming pools will work if you have a hill to set them on. Traditional windmills are expensive (between $1,000 and $3,000 if bought new) because of the strength required to withstand wind pressure and the gears required to use wind from all directions. The size and design of the mill

source, you need a foot valve at the end of it. This lets water flow into the pipe but won't let the water in the pipe drain out again. If you want to insure a steady supply of water, you can put a tank of some sort (a 1,500-gallon water tank to a 55-gallon oil drum) on your line somewhere uphill from your house. This will give you a large volume of water to draw from at any one time. At the far end (the house) of the water pipe you should use an adapter to narrow it down to one-half- or three-quarter-inch-diameter pipe for a foot or so before you connect to your faucets. This prevents air from being sucked back up the pipe when you turn the tap on.

Water lines can be either galvanized metal or flexible plastic. Plastic has the advantage of being cheap and easy to install; it will go uphill and down dale, which you may need to do. Its disadvantage is that it heats up in the sun and will crack after several years if it is not buried.

The only real problem with a siphon system is the possibility of its becoming "air-bound"—having the high parts of the line fill up with pockets of air. When the psi of the compressed air exceeds the psi

I went out to pick the broccoli today and couldn't do it! There are sixteen heads out there, some of them beginning to flower. By next week I'll have to put them in vases! I can take one head at a time (it's hard but I can usually do it) letting the broccoli nourish me as I have nourished it. But to cut the hearts out of sixteen at once, to chop down a dense row to a few spindly stalks, to destroy the perfect symmetry of my garden creation, is so painful that I hesitate, looking for some other solution.

I don't know how to reap what I have sown. I can pick a bit along the way and even cut down a plant here or there as need be. Lettuce and carrots are easy, a little judicious pruning or thinning with no real harm done—my rows as thick as ever. But to harvest, to end a cycle before its end, is infinitely difficult for me. I have a hundred excuses: I want to see an artichoke flower, to know how many pounds a cabbage *can* weigh, to see how large the beets will grow. I don't like frozen vegetables anyway! It's so hard to gather for the future, to take more than my immediate needs.

Who would ever have thought that the price of a simple garden is a moral crisis?

will depend on its location, the normal wind velocity, how deep it has to pump (they can pump wells several hundred feet deep). These mills can also be used to run a home generator to supply electrical current. Plans for less sophisticated and much cheaper homemade windmills are now available. Most of these were developed by the Peace Corps for areas with steady prevailing winds from one direction, which makes them less generally useful in this country. Sources of these plans are in the Resources section. Be sure, if you are considering a windmill, to include the cost of a storage tank in your calculations; it can add considerably to the total cost. If you can get either the windmill or the tank in good used condition, you should of course. But around here old-timers are reluctant to part with theirs, knowing full well their value.

A friend who has a hydraulic ram swears by hers. She says she's been down to look at it twice in five years and has never had a problem with it in all that time. Rams are worked by the flow of the water in a spring, creek, or artesian well. They require a steady volume of water even in the driest of seasons (at least one and a half gallons per minute) and a drop in the water level from its source to where the ram is placed (at least twenty inches). A ram will pump water uphill twenty-five feet for every foot of fall in the water source. With a rapid drop in water level or several rams in a series, they can lift water up to 500 feet. Like windmills, rams are used in conjunction with a storage tank which supplies water under gravity pressure. Rams for normal farm use, without excessive lift, cost between $230 and $500. They are made by only one company in the United States (see Resources) which is old-fashioned enough to supply you with lots of information, help and advice, free.

Human-made power for pumps is traditionally gasoline or electricity. I have a strong prejudice in favor of electric centrifugal pumps when power is available. They are relatively cheap ($125–$200), run without any effort on your part, and are very simple to repair. The main reasons for not using one are the unavailability of power on remote land or a political decision not to be tied to the power companies. In either case, I would investigate windmills or hydraulic rams before a gasoline pump—unless your land is not suitable for either or money is more important.

Having said all those discouraging things, let me describe a gasoline pump briefly. Most gas pumps for farm use are a simple one-cylinder engine (multiple cylinders are used to pump large vol-

I've joined a women's group started by women I've seen but never spoken with. We've had three meetings now. Going to the first one I felt excited, shy, and aggressively defensive—wary of starting all over again with women who have never questioned anything. Each week more women have been coming, even Susan and Kathy from this ridge. When I look around the room I feel surprised to see it filled with so many seemingly home-contented women. The meetings have been terrible—tonight I have my first headache in years and my muscles are cramped tight. We don't know what to say with each other, so we are trying to discuss books most of us haven't read. Each week I try to find something to bring, searching for *anything* by a woman (I never realized how many of my books are by men!). This week I dragged out Sylvia Plath's poems of despair, which I haven't even looked at for years. We sit around the room staring intensely at each other's feet, listening to whatever someone has chosen to read and to the awkward silence afterward. In the car going home, we ask each other, "Did you get anything from it?" as though we were in school again. And though we mumble out frustration and uncomfortableness and talk about only coming one more time, we do still keep coming back as though the contact alone is worth it, no matter how bad the process is.

umes of water). This engine drives a pulley (like the fan belt and pulleys on a car). The other end of the pulley pushes a simple pump, much like the ram in design. From the pump, water goes to a holding tank and then by gravity to the house. Large industrial gas pumps are started from a battery like a car engine. One-cylinder pumps for home use are usually started by a hand crank (like a lawn mower or chain saw). One of the disadvantages of a gas pump is that it is noisy, smelly, and must have fuel carried to it. Another is that it must be started by you each time the water in the tank gets low; it does not operate automatically. Lastly, it takes as much energy to repair as any other gas engine. The advantage is that it is a relatively inexpensive non-electric source of water. A used gas pump around here runs from $50 to $150, depending on size and condition. A plastic swimming pool for a holding tank has about the same price range, depending on size.

I am going to discuss centrifugal electric pumps in more detail than any of the above kinds. This is partly because all my personal experience is with these pumps. They are *the* basic farm pump in America. The only other type of electrical pump for normal use is the submersible pump for deep-

well situations. I have never used one and only know that they exist.

When we went to buy a centrifugal pump for our new well on our new land, we spent hours talking pumps and water with the sixty-year-old owner of a farm supply store. I still remember the conversation clearly, both because of the good advice we received and because of the good time we had getting it. "Get the right pump for your place and set it up right, and that galvanized tank will rust out before you have to pull that pump out for repairs," he promised us. It hasn't worked out quite that way for me—an unexpected freeze cracked my uninsulated pump last winter—but his system is still the best and most practical one I've seen yet.

Centrifugal pumps vary in size and power. You need to find one that fits your water situation. The motors range from one-third horsepower to two horsepower on normal home pumps. The greater horsepower can suck a greater volume of water at once or produce more pressure to lift a greater distance. Variations in size are mostly in the number of stages—each stage in a centrifugal pump contains an impeller that whirls the water through a narrow passage and builds pressure. The more stages, the greater the capacity to produce pressure. Multiple-stage (three and four) pumps generally have larger (one and two) horsepower motors. A one-stage, one-third horse pump costs about $125 right now and a three-stage, one-and-a-quarter horse pump costs about $300. What you need to find is the right combination of stages and horsepower to carry water the height you need and draw out the volume of water you need. If you have to lift more than fifty feet from well to house, you will need a multiple-stage pump. Our friend the pump dealer talked us out of buying a one-half horse pump because its ability to draw a large volume of water would run our relatively shallow, but steady recovery, well dry. Instead, we bought a one-third horse, one-stage pump (we have virtually no vertical lift between the well and the house). This pump takes less water at a time and can pump up to 60 psi safely.

Centrifugal pumps are basically alike no matter what the brand. I have had good experiences with Jacuzzi pumps on three different pieces of land. Barnes pumps have also been recommended to me. Most pumps these days come with plastic innards. You may want to specify brass fittings and impellers if you live under the special conditions we searchers for cheap land often find. Plastic parts get easily chewed up by dirt in the water and crack or round off so a wrench won't fit after about the second time you take them apart. So, brass will

save you money in the long run if you are pumping out of a spring (which may get muddy or fill in) or if you have a limited water supply (so the pump may run dry).

If your well is deeper than twenty-five feet, you need a jet attachment for your pump (or else a submersible pump). A jet fits onto the water pipe near the bottom of the well. The pump pushes water down a one-inch pipe to the bottom and around a sharp curve. The pressure of water whirling at a high speed around a curve creates a vacuum that sucks in more water through an opening in the jet. All of the water is then pushed up again

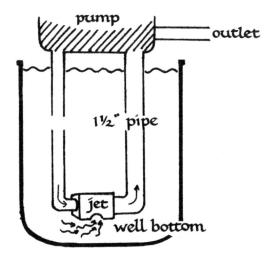

The women's meeting tonight was wonderful! I felt so excited as I sat in the meeting that my hands were shaking. Now it's two A.M. and I can't sleep, hoping to push the clock faster toward next Tuesday.

A new woman came tonight and sat silently listening to our fumbling attempts at discussing *The Dialectics of Sex*. Finally she looked up and said, "What I'd really like to know is why each of us came here and how we're feeling." And we began to talk to each other! So simple! The meeting seemed to explode into a jumble of questions and feelings—about husbands, lovers, friendships, hopes, where our lives have gone. Each of us trying to share part of the history of how we came here—scattered events in individual lives merging together. I spoke of how isolated I've felt, trying to struggle for equality in work with Peter and having no one to talk with about it. I looked up and saw a woman I didn't know smiling and nodding. It was so incredible to hear what we were saying—it's not that we don't feel free to be openly who we are, but most of us don't even know who we are or what we could be!

through a one-and-a-half-inch pipe to the pump where the excess is sent out to supply you with water. The rest recirculates down into the well through the one-inch pipe and past the jet where it picks up more water.

The water system includes not just a water source and pump, but also a pressure tank to store water, all the pipes and fittings to connect the well to the pump, the pump to the tank and everything to the house, and, of course, the electric line to run the pump. The method of joining all these that the pump dealer taught us is the best I've ever seen and not one commonly used.

The essential ingredient in his system is a one-hundred-and-fifty-gallon galvanized pressure tank. This costs about $120, and after three years of use I've decided it was well worth it. Most pumps come without a pressure tank (Jacuzzi's have a three-gallon "hydrocell"), but they are often sold in a package deal with a twenty-gallon tank by the store. A one-hundred-and-fifty-gallon tank stores so much water that you can use fifty gallons before the pump ever turns on. Your pump only runs once or twice a day, saving on electricity and wear on the pump and giving your well a long recovery period between pumpings. With the standard twenty-gallon tank, the pump turns on after every five gallons of water. This constant off and on is the hardest wear you can give a pump, much harder than a long steady pumping.

The other secret to this system is how you place your tank and water lines. The pressure tank contains a large pocket of compressed air which is compressed further by the water being forced into it. This provides constant pressure in your water lines even when the pump is off. When the pressure in the tank gets low enough, the pressure switch turns the pump on and refills the tank. With the traditional arrangement, a pipe connects the pump to the tank and another pipe leaves the far side of the tank for the house. When you are running water and the pump is on, the water is whirled around and through the tank before it enters the lines. Because cold (fresh from the well) water can carry a lot of dissolved air, it tends to absorb the air pocket in the tank as it whirls through. Gradually, over time, it dissolves all the air in the tank and runs it through the lines. The tank is then waterlogged—it has no pressure and the pump runs every time you use water. To fix this, you have to drain the whole tank, use a bicycle pump to replace the air and then refill it with water. I was taught, instead, to simply plug up the outlet on the

tank and take my water lines off from the line *between* the pump and the tank. This means that water flows from the tank into the lines, but when the pump is running it fills the lines first and then the tank. Water never churns through the tank and the tank never becomes waterlogged. It's a simple, logical system that will save you pump repairs and having to drain a waterlogged tank.

Now, hooking it all together: pumps are either mounted horizontally beside the well or vertically above it; they also vary in the placement of intake and outlet pipes. Follow the diagrams that come with your pump and *ask at the store* for specific details about your pump.

I recommend using galvanized metal pipes and fittings between the pump and the pressure tank; it is such a short distance that the expense is minor and they will disconnect easily and last forever. I also recommend using plastic pipe down into the well; it's not exposed to the sun and therefore won't decompose, and it is cheaper (unless you have a deep well jet, in which event it's easier to use galvanized pipe). You should use rigid plastic PVC down into the well so that you are sure how far you are from the bottom. Flexible pipe never completely unbends so you either lose a couple feet of precious water or risk sucking dirt off the bottom (which can ruin a pump). Most pumps use one-and-a-quarter-inch or one-and-a-half-inch pipe for both intake and outlet pipes.

The intake line usually is a threaded hole in the base of the pump. A (male) nipple—a short piece of pipe threaded on both ends—screws into the intake opening on the pump. Tighten this with a pipe wrench (don't forget to use pipe dope first). Onto the nipple screw a union (also galvanized). There are many types of unions, some are simply threaded at both ends and others work through pressure on a rubber gasket. What they do is join two pieces of pipe with a waterproof seal. You need a union here so that you can disconnect the pump from the well when (if!) it needs repairs without having to cut your water line. (I learned this the hard way.) The best type of union to use here is one with a female end to thread onto the nipple and a permanent male fitting at the other end. Onto this male fitting, you thread a metal-to-plastic converter. Your rigid PVC is cemented into this. If your pump is horizontal, you need a right-angle PVC coupling to turn and go down into the well. About a foot from its bottom (two to three feet if you have a sandy or muddy bottom which may fill in), the pipe should stop. At the end of it

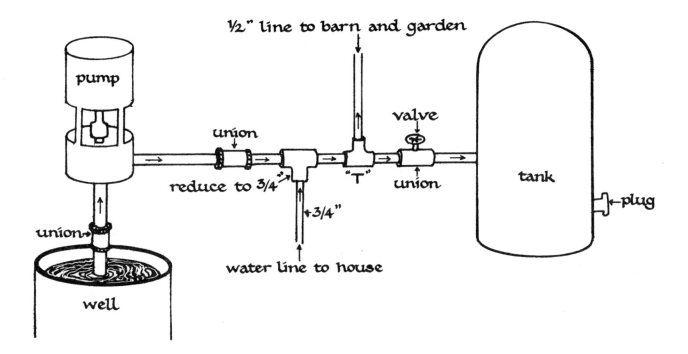

comes another plastic-to-metal converter, onto which threads a galvanized foot valve. This valve allows water to flow in but prevents it from draining out of the pump back into the well. If you feel confused by all these terms and fittings, make a list of what you need. Then go to a hardware store and ask for them. Stop and look at each fitting; you will see how they fit together and will understand the basics of plumbing.

When you connect your pump to the well, make sure the pump is slightly higher than the line from the well (if the pump is mounted horizontally). This prevents the line from becoming air-bound if you run the well dry. When you reprime the pump, the water will flow downhill through the pipe and fill the line.

There are two other openings in a pump besides the intake one. Both are usually in the sides, but some brands have one or both coming out the bottom. One is for the deep well pipe or a shallow well injector (which helps you build pressure); when you buy your pump you will get whichever you need and be shown where it goes. The other opening is the outlet for the pump. Into this opening thread a short length of galvanized pipe with threads at both ends (or a nipple). Onto this goes another union so you can disconnect the pump at this end too if you ever need to repair it. Coming out of the union comes another short length of pipe, also threaded at its protruding end. Next comes a "T" fitting which enables a water line to connect to the system (to the house, garden, etc.). Your main line is probably one and a half inches, so you want a T that adapts down on the outlet side to the size of your water lines. At this point you have to decide whether you want to put two or three T's on your main line from pump to tank (one to your house, one to the barn, etc.) or whether you want to run one major (one inch) line off and then split off of that line. Either way is all right, but it makes a difference in what size T you buy. If you decide to use several T's, you need threaded nipples to connect them to each other. If your water lines are galvanized, they will thread directly into the T. If they are plastic, you will need to thread a metal-to-plastic converter into the T. I recommend using a galvanized converter here (they are usually plastic) because you may want to disconnect it again someday.

After your last T joint comes another nipple and then a gate valve. This valve enables you to shut the tank off if you disconnect the pump or have a break in the water line so that you don't drain the tank of water. After the gate valve comes another short length of pipe, threaded at both ends, and then the tank. Try to place the tank so that the line from it to the pump is level. Make sure you have plugged the outlet opening on the far side of the tank.

Before filling the tank, you need to pressurize it. This is done by attaching a bicycle pump to the air valve on the pump (or an air tank if you're lucky enough to have one); a hundred-and-fifty-gallon tank is pretty heavy to haul to the gas station. Pump forty pounds of air into the tank, measured using a tire pressure gauge. This amount of air lets you use fifty to sixty gallons of water before the pressure in the tank gets low enough to turn the pump on.

Now, all your water lines are connected and all that remains is to hook your pump up to your electricity. The wiring box on the pump has a snap on the cover and is attached to the motor by a thick wire or cord (this is actually three wires). Remove the cover and you will see these three wires: one red or white wire for the current, one black one for neutral, and one ground wire which is green or possibly bare copper. The red (or white) and black wires carry the current and are attached to the electrical switch. The third wire is the ground wire and screws onto the box itself at some point.

Your wire from your fuse box should also be three-strand solid copper. It is important that you buy heavy enough gauge copper (thickness of the wire) to carry twenty amps of current from the fuse box to the pump. Volts decrease over distance unless the wire is heavy enough. The hardware store will be able to tell you what gauge is necessary for what distance. (I used #12 copper wire to carry twenty amps one hundred and fifty feet.) You should connect your pump to its own circuit on the fuse box or to one that has plenty of current. A pump needs only twenty amps when it first turns on, but if it doesn't get enough current then, the motor will eventually burn out. If your wire is to run above ground or be buried directly into the ground (as opposed to running through conduit pipes), buy wire that is especially insulated to withstand weather.

This line from the fuse box enters the electrical box on the pump through a small hole in its side (this may be a plate you will have to knock out with a screw driver). This is usually directly opposite where the wires from the motor enter. Just before the wires enter the box, strip off the outside insulation so that all three wires are showing. At least two of these will have inside insulation around them. The insulation needs to be stripped off the very end of these wires so that they can make contact with the switch. At no point should a bare copper wire ever touch the bare metal of the box, *except* where the two ground wires meet. This will cause a short, the pump won't run and you may get a nasty shock. The black wire from the incoming line goes around the brass screw which is parallel

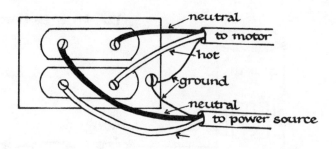

with the black wire from the motor. The red wire goes around the brass screw that is parallel with the red or white wire from the motor. Both screws should then be tightened down with a screw driver. The ground wire goes around the *same* screw as the ground wire from the motor. To make tightening easier, wind your wires around the screw in a clockwise direction so they twist tighter as you turn the screw. It is all *very simple*—if you have remembered to pull the fuse or turn the breaker off before you start; otherwise, you are probably very dead. Now you can snap the cover back on the pump and it's ready to run when you turn the power on.

An alternative to connecting your electric line directly to the pump is to wire an outlet box next to the pump. Then attach a short electric line to the pump as directed above. This line should have a plug wired on the other end. The advantage of this method is that you can turn the pump off if there is a problem without having to walk (or run!) back to the fuse box. You do have to be careful, though, to protect the outlet and the plug from weather or any possible water leaks.

The only other thing you should know about the power box is that it also contains the controls for the pressure switch. These adjust the point at which the psi turns the pump off and on. When you buy your pump you will be told what range it is set for. Our one-stage was set for thirty/sixty, which means the pump turns on when the pressure drops to 30 psi and off when it builds up in the tank to 60 psi. The adjustment technique varies with each brand of pump and instructions come with each one. On my pump, there are two tiny nuts that adjust with a screw driver. One raises or lowers the whole range—it keeps a 30 psi difference, but alters the off and on points. You would make this adjustment, for instance, if you needed a little more lift from your pump (so that it would cut on at 40 psi and off at 70). Find out what the top limit of your pump is and only exceed this at your own risk (what you need is a bigger pump). The other adjustment nut widens or narrows the gap between the off and on points. This makes the pump draw more or less water at each pumping. You would want to make this adjustment, for instance, if your well got very low and you wanted to pump less water at a time, more often (so that you would cut on at 30 and off at 45). *When adjusting these nuts, turn them only very slightly* (one-quarter turn) until you see what difference that has made. Turning too far can cause many problems with your motor and the pressure in your lines.

One step still remains before you have water flowing through your lines. This is to prime the pump, which means fill it with water. This must be done any time the pump has been disconnected or drained (because of a break in the water line, the well going dry, etc.). When you are first setting up your pump, you will have to unscrew a metal plug in the base and fill it with water. It will take quite a bit of water because you have to fill the lines to the bottom of the well too. When water starts bubbling back up out of the pump, it is primed. Make sure you have let all the air work its way out and the line is full of *water* only. Replace the plug (with pipe dope on its threads), and tighten it with a wrench. Now, you are ready to turn the power on, listen to the pump whir quietly, and wait for the tank to fill.

If you ever have trouble with the pump (the motor overheats, not enough water pressure, no water pressure, etc.), the first thing to do is turn off the power (at the fuse box or your plug). The next thing to do is to shut the gate valve by the pressure tank so it will not drain if there is a leak. Once you have fixed the problem you can reprime the pump by simply opening the gate valve at the tank—unless you also drained the tank before you realized there was a problem. Always remember to shut off both the power and the valve and to open the valve before turning the power on again.

The only other general information I can think of is that the pump needs to be protected from weather and freezing. The tank can be exposed to the weather and will last thirty to seventy years, depending on the minerals in the water and the amount of salt in the air. The pump, however, needs protection. In mild (frost but no freezes) climates, a simple wooden box will do. Where there are infrequent freezes, the box should be lined with insulation. A light bulb can be wired onto your power line and placed inside the box to provide extra warmth. Where winters are severe, the pump will need to be placed in an insulated cellar or heated house. The pump may be located some distance from the well, as long as there is not more than twenty-five vertical feet of lift from well bottom to pump, or it has a deep well jet.

All of this exhaustive detail may have made water systems seem as mysterious to you as they used to be to me. It is all very easy, however, once you have the parts in front of you and begin to assemble them. I love plumbing because its simplicity has overcome all my I-can't-do-it fears, and it is one of the few building skills I never hesitate to undertake.

The Growing Season

The first summer
was a funnel; we swept
into the center and around
and around the house we flew
faster and faster working, worrying
our heads confined
to the size of the house
and the trees around it turning
from witches to pudding.

In the garden the squash sucked everything
from the vines like straws and the vines died.
The corn puffed up ears and died from
the ground up. Tomatoes reddened overnight
plopping off the vines. Lettuce never made it.

The weeds won. They outgrew us all.
They grew under roots and pulled them loose.
They grew over the ears, they pulled down vines,
shaded the leaves and finally buried the garden
in a heap: a crop of thin brown stalks.

By August the garden weeds
reached my waist.
The sun was cooler
the color of waterfalls.
The locusts surfaced. Humming
broke the air. Fall.
We gathered wood and apples
and searched the buried
garden for a feast.

Late summer nights when crickets
scratched at the moon we danced
barefoot across the floor
playing music through open
doors, through the hayfield and corn
dancing we cast a spell around
each other to keep us
through the winter.

the BASICS of pLUMBING

There are two really different types of plumbing. The first is the neat and orderly installation of a new system. It may involve a few contortions getting pipes through studs or beneath the floor, but basically it's easy. The pipes and fittings are most likely new. You have enough and they are the right size; they fit together as they are supposed to. And there is perhaps the extra pleasure of soldering copper pipe with a torch. Having long dreamed of being a welder, it was the torch which hooked me into a part-time career as a semi-professional plumber.

My initiation into the second type of plumbing came when a friend asked me to assist her in some repairs on her old house. This type involves crawling on your belly underneath the floor, flashlight held in your teeth, through a muddy slime into which the bathtub drain has been leaking for an untold number of years. It's also the kind of plumbing in which the fitting you're trying to tighten shears off in your hand or the leak turns out to necessitate tearing off four feet of wall boards. Repairing old plumbing satisfies none of my fantasies and is an experience I've learned to avoid; nev-

ertheless, I will include at the end of this section some advice for those of you who do have to deal with it.

Installing Water Pipes: Installing the water lines is usually the first step in setting up a new system. There are four different types of pipe commonly used for home plumbing: flexible or rigid plastic, galvanized iron, and copper. Each is easy to install, the only complication being that each is done differently. Whichever type of pipe you choose, there are some basic things to know or think about. You need to decide what kind of pipe you want to use and to plan your system so that you use only one kind of metal. If you must join two metals, be sure you have plastic-lined couplings to prevent electrolysis—the bonding together of the two metals. Outdoor water lines should be buried to prevent their heating up, getting damaged, freezing, or decomposing. How deep they need to be will depend on the winter climate in your area, but they should be below the frost line. It's easiest to dig your trenches before you lay the pipe—I've put my mattock through several plastic pipes in the

44

process of burying them. You should plan your whole plumbing system to use the least material and make the fewest number of joints possible. Where you can, try to place outlets (sink, tub, etc.) in a straight line or at right angles to each other. Pipes are usually laid beneath the floor between the joists or inside the walls between the studs. If you live where there are no hard freezes, you may want to run most of your pipes underneath the house and come up through the floor directly under the fixture. This is easier than going through studs. However you do it, keep your plan as simple as possible.

For most houses (one bathroom and a kitchen sink), a ¾″ incoming water line is large enough; if you have more than one bathroom or a washing machine, etc., your incoming line should be 1″ or 1½″. Once you get to the house, you should reduce the lines at each outlet down to ½″ so that you'll still have water in the shower when someone else flushes the toilet.

You should plan to put several shut-off valves into your water system—at least one in each major line (house, barn, etc.). I prefer now to put several on the house lines, so that if I get a leak in the kitchen, I still have water in the bathroom. I learned this the hard way after living with only one shut-off valve on my entire water system. You also need one valve on the incoming line to the water heater so that you'll be able to drain the tank if anything goes wrong. These shut-off valves can be connected to any type of pipe—they come threaded for plastic and galvanized pipe or designed to solder onto copper.

Once you've got the basic design of your lines figured out, you need to begin counting up the feet of pipe, number of valves, right-angle fittings, etc. that you'll need. If you're not sure how to connect the copper or plastic or whatever pipe to the sink faucet or the toilet or whatever, *go to the store and look and ask.* This is the most important thing I can say about plumbing. Plumbing is the most obvious and mechanical of country skills: everything fits onto or into something else, and those somethings are sitting out in bins in the hardware store. So, take a diagram of what you need to connect to and a six-inch piece of the size pipe you'll be using to the store and start looking. You can ask for help if you want, especially to check if you found the simplest way to do something (though the answer will probably be patronizing). But I recommend at least an hour of fumbling on your own, picking up parts, threading them together, getting familiar with the sizes. You'll learn more in that hour than you will from a whole plumbing book. Plumbing is absolutely simple and all the parts will be sitting there in front of you.

A few other general hints: There is a gray gooey compound known in the trade as "pipe dope." It is used to seal the threads on pipes so they won't leak. You should use it any time you thread something together—no matter whether it is plastic or iron. Just fill the depressions between the threads; it doesn't take a lot.

Threaded pipes and fittings are classified as "male" (threaded on the outside) and "female" (threaded on the inside). This is because our sexist culture perceives "male" plumbing fixtures as fitting into "female" ones—but anyway, the designation will probably help you when you try to describe to a clerk what you need.

Threaded fittings (except gas ones) always tighten clockwise and loosen counterclockwise. This is easy to forget and someday you will probably find yourself feeling very foolish after thoroughly tightening a joint you thought you were loosening.

I always try to keep an extra stash of basic plumbing parts on hand—most especially the ones for plastic piping. You never know when a line's going to freeze and crack, when a horse will stomp on it, or when a helpful friend will put a mattock through it. Having the basic fittings and clamps on hand to mend a leak means not having to do without water until you can get the part from town—it somehow makes the disaster seem much more minor.

Once you get your whole system put together and water flowing through the lines, go back under the house and carefully check for drips. Drips will waste water and rot floors or foundations. Frustrating though they are when you thought you were done, you should fix them right then. It's a rare plumbing job that doesn't have at least one drip. It's often just a matter of another turn on the clamp or a little more pipe dope on the threads, but before you do anything be sure to turn the water pressure off! Under the old house I mentioned before, we went to give the last fitting one final turn to stop a tiny drip, broke the fitting loose and were flooded with thirty gallons of hot water. Coal-miner fantasies turned to a vision of a coal-mine disaster as we inched along the crawl hole on our bellies. And though it seems funny now, it was awful then—so remember to turn the pressure off, even for the simplest adjustment.

Types of Pipe: Flexible black plastic pipe is the cheapest to buy and the easiest to install or repair of all four types. It comes in any diameter you might want (up to two or three inches). It is designed for cold water, and that's all I've ever used it for. I have watched a friend run hot water from her heater to her sink for four years through flexible pipe, with no problems at all, but I don't recommend it since really hot water will melt or weaken the pipe walls. Plastic pipe between the well and the house needs to be buried since repeated heating and cooling will cause the pipe to crack after several years. The only exception to this is if you want a free source of hot water—the pipe gets quite hot in the sun. We use this principle to run an outdoor shower—hot water by 2:00 P.M. on any sunny day. The only trick is timing your bath just right, as there is no cold to mix in.

In this area, plastic pipe is legal to within six feet of the house; after that you must use metal. I have ignored the laws and used plastic for all my cold-water pipes, but what you do will depend on how vulnerable you are to the building inspectors.

Flexible pipe is really easy to install. You join any two pieces of it together using a plastic connector (which fits inside each piece of pipe) and pipe clamps (which tighten the pipe in place). These connectors may be a simple line one (to connect two sections of pipe or mend a break), a right-angle one, or a T-connector (to join three pieces of pipe). It is easiest to put the pipe clamps

on each line before you join them. Then start the connector into one section of pipe. Sometimes it will go quite easily, with a spiraling motion. Once started, I usually tap the connector into the pipe by placing a board at its end and hitting the board with a hammer. When you go to attach the second piece of pipe onto the connector, it helps to have a partner so you can push against each other; again, a twisting motion helps. If you absolutely can't get the pipe onto the connector, you can heat up the pipe first to make it expand. Hot water works best for this, though you can also use a quick pass with a propane torch. The pipe clamps should be tightened over the connected pipes, about a quarter inch back from the end of each piece of pipe. I use an extra-large screw driver for extra leverage when tightening clamps.

There are also connectors that are fitted for plastic on one end and threaded on the other. These are for joining plastic pipe to metal pipe or faucet fixtures. You tighten the threaded end first with a wrench (carefully) and then fit the flexible pipe onto it. To go from a larger to a smaller pipe, you use a connector that is fitted (for example) for ½″ on one side and for ¾″ on the other. These are called reducers.

There is a new type of plastic pipe now on the market that is for both hot and cold water. It is a rigid pipe that is joined together with a plastic cement. It is not yet approved in this area by the building department, so it's not available in build-

I am coming to love my body—the peculiar, individual smell of sun and sweat on my skin, the feel of strengthening muscles, how I look working half naked and growing skinnier and tougher. I can't remember a time since childhood when I felt comfortable with my body, never being pretty, tall, slender, graceful, desirable, attractive, poised, casual. Never being what a man would want, I've lived as though my body weren't a part of me. I've looked out through eyes that believed that awkward gesture, frizzy hair, self-conscious pose had nothing to do with *me*. Now learning to work with my body, coming to know it as a tool, away from eyes which weigh my value in my shape, I'm discovering a new freedom. It's as though I'm breaking through cellular memories—relearning how to move. Now I can stride, pound, lift, pull, haul, retest all my limits and find them expanding, let myself flow freely through my movements. I want to try to dance, to hug myself. I love the power I'm discovering in this body.

ing supply stores, but it can be ordered from Sears. We used this to replace the hot water pipes in the old house mentioned before, and I am a great believer in it. The rigid plastic pipe is cheaper than either type of metal and easier to install.

The pipes are sold in 8′ lengths, ½″ to 1″ in diameter. They can be cut to any length with a hack saw. The rigid pipe is joined together by a coupling that fits over the end of each piece of pipe. With this type of pipe, too, there are special couplings threaded on one end to join to metal pipe. Again, tighten onto the metal first before fitting the plastic in place.

I've found that it's safest when using this type pipe to fit together all the sections with the appropriate couplings before I glue any of them. Once the glue has set, the pipes can be moved only with a hack saw. Both the pipe and the coupling should be completely clean and dry before the glue is applied. Once they are coated with glue and fitted together, give them a one-quarter turn to form a tight bond. The glue needs to set at least three hours and preferably overnight before water is sent through the lines. Make sure you use enough glue for your first chance is your only chance with this type of pipe; after that, you cannot get it apart, drip though it may.

Galvanized iron pipe was commonly used in the days before plastic and is what is found in most older homes. It's not used as much any more since it is more expensive to install than either plastic or copper. Galvanized pipe is threaded on each end and is joined together by female threaded fittings (fittings that have the threads on the inside). Two pieces of galvanized pipe are joined together by a coupling that is threaded all the way through. Wherever you may want to disconnect the two pipes again easily, you would use a union, which can be loosened from either side. Once the whole system is in place, it is nearly impossible to disconnect it without a union. So use them liberally.

Galvanized fittings come in all sizes and descriptions: bells to reduce from a larger pipe to a smaller, fittings to increase from a smaller to a larger, angle fittings, T fittings, spigots, and caps to end a line. For anything you want to do, you can find a fitting to do it. The real problem with galvanized pipe is that you must have it threaded everywhere you want to make a connection. So your measurements and instructions at the store (where they will cut and thread the pipe for you, unless you have access to a threader) must be very precise or you'll find yourself running to town twice a day to cut this or that. This means you should plan and measure your whole system before you start buying pieces.

To tighten galvanized fittings, you will need pipe wrenches. Occasionally you can get away with a large crescent or open-end wrench, but usually they won't grip well enough or have enough leverage. Pipe wrenches are expensive and should be bought secondhand. They are very handy, if not indispensable, tools for any farm that has a water pump or any galvanized pipes. I have three pipe wrenches, all bought at a flea market. One is very small and lightweight (about eight inches long), useful for small fittings or getting into tight places but without enough leverage to tighten a fitting or break loose a rusted one. The largest (about 18″ long) is very heavy and cumbersome to use, and I have often cursed it. But when I need a good tight seal, or a fitting is really stuck, I'm always glad to have it there. The most generally useful pipe wrench I have is about twelve inches long and of an easy-to-handle weight; it will do any average plumbing job. You need two wrenches to tighten most fittings. One is set to grip the fitting or pipe you're screwing onto; the other goes over the new fitting and tightens it in place. The two wrenches should be set opposite each other—so that you can keep the pipe (or whatever) from turning while you turn the new fitting. The wrench in one hand pulls toward you while the wrench in the other is being pushed away. I've found it helps to put the larger wrench (if you have one) on the piece that is to remain stationary since you'll need the extra leverage to resist your own push. To make sure the fitting is really tight, I switch wrenches at the end and give a final turn with the long one. Most work can be done alone (good training for ambidexterity), though it is a little tricky getting both wrenches in place when you're working flat on your back in a one-foot crawl space. For old pipes that are badly stuck, you may need one person on each wrench and you may want to extend the end of the wrench with a piece of pipe for greater leverage. And don't forget: Always put pipe dope on galvanized fittings before tightening them in place!

Working with copper pipe is my favorite though it is definitely not the easiest or the cheapest. (Copper pipe costs more than hot and cold plastic and slightly more than galvanized, but the fittings are cheaper than those for galvanized.) Most code plumbing is done in ½″ copper these days. Copper comes in two types—flexible tubing and rigid. Rigid is thicker walled, more durable, and is nor-

There are fears in me that go so deep I can't even touch their source, so powerful they paralyze my muscles as well as my will. Some I understand: in my fear of ducks I remember childhood weekends in the city park with ducks biting my legs and fingers as well as the bread. But somehow I can't remember which weekend I became afraid of tools and when I was taught to be afraid of even trying. It's that fear of trying that seems so potent to me now, beyond my control but so in control of me.

Peter and I have finally finished the bathhouse after months of talking about it. It took his being home for me to actually work on it. Somewhere in the middle of it I remembered that two years ago we started as equals, unskilled ex-urbanites. Now when we build, he directs, I follow; he takes the measurement, I hold the tape; he saws, I steady the board; he hammers, I hold. And I agree to this without thinking, feeling insecure and incompetent. While we were framing the third wall, I realized what was happening and asked if we could formally organize it so I hit half the nails, saw half the boards. He said it wouldn't be "efficient" with fishing season about to start: he can build better and quicker. After a long argument he agreed. When I went to nail my studs, he stood right behind me, watching and offering me advice. I was tense and bent twice as many nails as usual.

The plumbing was supposed to be my job, but I put it off until there was nothing left to do. I hate feeling incompetent. When I do, I start collecting lots of information, asking everyone who's ever done it, and reading books until I get enough information and courage to start. Peter just starts, whether he knows how or not. With the plumbing, I finally went to the store to get parts. I was here

mally used for plumbing. Occasionally the flexible type is a real advantage. I have a friend whose water heater consists of coils of flexible copper inside the fire box of a 55-gallon-drum wood stove. My homemade shower attachment is a gracefully curving piece of copper with a shower head on the end. Flexible pipe should be warmed before you try to bend it. Plan on long, flowing curves as it won't make sharp, angular bends without crimping shut. Both flexible and regular copper pipe can be cut to any desired length with a hack saw.

Copper is joined together by "sweat" fittings—couplings that fit over the pipe and are soldered into place. Soldering is not at all difficult, though it takes care and some practice to get a good seal. I use an acetylene and air torch for my plumbing jobs, and I recommend it to anyone who is doing plumbing professionally (it has a hotter flame). For ordinary home plumbing, a small inexpensive propane torch is fine. There are two types of solder,

either of which is fine for copper pipe. Acid core solder takes more heat to melt but forms a harder seal; with a propane torch, I would buy the softer rosin core instead. A roll of solder will last a very long time as it takes only a little bit at each joint. When you buy your solder, get a tin of flux to go with it. Flux, a paste that attracts solder, is put on the inside of joints to draw the solder in.

The first thing to do when you go to solder is to clean thoroughly each piece of copper—the outside of the pipe and the inside and edges of the coupling. Steel wool works best for this—the copper should be shiny and pink all over. Use a clean rag to wipe off any lingering bits of steel wool. Be careful to keep your own hands (or at least one finger) clean too. Apply a thin coat of flux to the outside of the pipe or the inside of the coupling with your clean finger. It doesn't take much; you just want to grease it all over. Then fit the coupling onto the pipe. Make sure it slides on all the way.

about three hours, looking at everything so I wouldn't have to ask for help. Eventually, I found a "this" that screwed into a "that" until I figured out how to connect everything up. I got in an argument with the clerk about the drain pipes; he said they wouldn't drain, but all he's ever seen are the code sizes. I'm sure I bought a size bigger than I need. Dan offered to come teach me how to solder when I put it all together. We were almost finished before I realized all I'd learned was how to stand there and hand him things. He gets so much support from being able to do things for people, he doesn't want to give up the power of his expertise.

I've been doing the shingling all alone. It took me six hours the first day to get out there and start. And still each day, it takes me an hour of thinking of everything else I could do before I get back to it. But once I'm out there, I feel so good. I just move right along, row after row feeling peaceful. Taking time to pick out which shakes go where and think my own thoughts. On the last side now, it feels good to look and realize I did it. Sometimes I miss the extra energy of someone else to work with, when I have to climb down the ladder for each new bundle of shakes. But I think I can only work with someone as unskilled as I, from whom I can get courage, not a sense of my limitations.

I wonder how long it's going to be before I *know* that I can do things? How many times of enjoying the work when I'm doing it before I'm not afraid to start? There's a part of me that doesn't believe in my competence. Fear catches me and for five seconds I cannot pick up the wrench. How many shingles and gates and tune-ups and plumbing jobs will it take before I really learn that I can do what I was taught I couldn't?

As with rigid plastic, I usually fit the whole system together once for test before I begin to clean and flux the parts. When I'm ready to solder, I usually prepare a whole series of joints at one time so I can just go down the line with the torch.

Once each joint has been cleaned, fluxed, and put together, you're ready to solder. Light the torch and turn the flame up part way until you have a cone of blue flame in the center. The tip of this cone is the hottest part and you should touch it to the copper. What you want to do is to heat the copper so thoroughly that the solder will melt and fill the joint as soon as it touches it. The actual soldering takes only a few seconds; it's the heating of the pipe that is critical. If the pipe is not hot enough, the solder won't flow freely and the joint will leak. Heat the pipe right at the joint until you're fairly sure it's hot enough. You can test it by pulling the torch away and touching the joint with the tip of the solder. If it begins to flow, hold it

there until it flows all around; if it sits in a puddle on top, remove the solder and continue heating with the torch. The *pipe, not the torch,* should melt the solder, so don't hold the solder beneath the torch flame. I always unwind two or three inches of solder from the roll, so that I can touch the tip to the joint while holding the roll and not burn myself. This much will last several joints. Be conscious *always* of where you are pointing the torch while your attention is focused on applying the solder to the joint or on unwinding the solder. Solder follows the same laws of gravity we are all subject to, so try to touch the solder to the highest point on a joint and let it run down. The hardest joints to solder are the bottom ones on a vertical pipe—where all the solder must run *up*. On those your best hope is to use a lot of flux and heat the pipe a long time, then touch the solder all the way around the pipe; or you can remove that piece, solder that joint horizontally and then fit it all together again. Once you've put solder into the first joint, move right along to the one nearest to it. This will take less heating than the first one since it will already be warm. When you're first learning to solder, you will probably use a lot more than you need to, dripping it onto the floor and leaving blobs on the pipe. In a perfect solder joint, none shows at all, but I myself have preferred to be safe—even excessively so.

If you're using copper for a whole house, you're bound to have a leak or two. When I put in the bathroom in my house, I was so proud of my handy use of the torch—until I turned the water on. Water sprayed in all directions across the room out of eight pinhole leaks. The first thing to do is to shut off the water pressure and drain as much water as you can out of the lines. Two things make repairing leaks hard—the solder is slower to heat than the pipe (harder to solder onto), and the water in the pipe cools it down. Sometimes you can just boil the water out of the pipe, using the torch longer than usual. You can also stuff a bit of bread into the pipe to absorb the water (it will dissolve later). If you're lucky, just heating the pipe may be enough to make the solder run and fill in the hole (which is usually tiny). Just to be sure, you should also apply more solder. As I said, the solder doesn't get as hot as the pipe, so this can be hard, especially on vertical joints. If you just can't get the new solder to adhere, you may have to heat the joint until the solder melts. Then pull the pipe out, clean and reflux it, and start again. If worse comes to worst, you may have to cut the joint out with your saw and replace it with a new piece of pipe— but this is a real hassle, involving several more joints, so try to avoid it. As you get more proficient, you'll find you have few, if any, leaks.

Drain Pipes: I've seen wide variation in drain systems—from elaborate ones with clean outs and air vents to very funky ones made of beat-up flexible pipe. And they've all worked. What you need to consider in a drain system is your particular situation: are you draining onto the ground or into a septic tank; are you subject to building inspection; are you draining one sink or a sink-shower-toilet combination? Code drain systems are made of rigid pipe (clay, iron, or plastic), 3″ in diameter. This system includes air vents, P-traps, and other elaborate contrivances. It is basically a good system if you have a flush toilet and septic tank, though I believe in modifying it somewhat when you can.

If you are not going to have a toilet or septic tank, much simpler drainage systems work fine.

The best way to drain waste from a shower or sink is to run your drain pipes several feet (at least ten) from your building and drain them into a gravel-filled hole. This keeps bacteria and waste from accumulating at the surface. Drain pipes for this system can be either rigid or flexible plastic. I recommend rigid since it is easy to install and can't pinch shut, but it is slightly more expensive. If you choose to use flexible instead, be sure to install right-angle couplings each time you change direction, so the pipe won't get crimped shut. With either type, you will need a fitting for the bottom of the sink or tub that adapts from the metal of the drain to the plastic of the pipe. Then attach the pipe with either clamps or glue, depending on whether it is flexible or rigid. Because there's no easy way to clean out a clogged pipe with this system, always be sure there's a screen or trap over the top of the drain to keep out debris.

The basic code drain system is pictured here, with the exception that the drain pipes shown for sink and tub are two inches in diameter. Code requirement is an unnecessarily large three-inch-diameter pipe for all drains. The drains for sink, tub, and shower are all essentially the same. Beneath the fixture is a P-trap—a curved piece of metal or plastic pipe. This is to keep air bubbles from blocking drain lines. Below that a vent pipe is attached so that air in the lines can escape. Conventionally, there is an air vent for each fixture. These are sometimes all joined together and vented through one roof stack. I suspect that one vent—at the toilet—for the whole system would be adequate. Below the air vent comes a Y-fitting to join this fixture to the main drain pipes. Y's can be adapted to fit different-sized pipes together. In this diagram, two-inch pipes have been used to drain the sink and tub; these attach at a Y to the three-inch line from the toilet to the septic tank. Where only two pipes come together at a Y, the third arm is covered with a removeable cap. This is called a "clean-out." It allows access to the lines if they ever become clogged.

The toilet connects to a plastic phlange which goes through the floor. Toilet drains are all one standard size, which the phlange fits. The phlange is bolted or screwed to the floor. Between the toilet and the phlange is a wax seal that prevents leaks— it fits over the bottom of the toilet and into the phlange. This seal needs to be really warm (about 80°) to fit into place without cracking. A three-inch drain pipe extends below the phlange. Next comes a T-fitting for an air vent. Then there is a Y with a

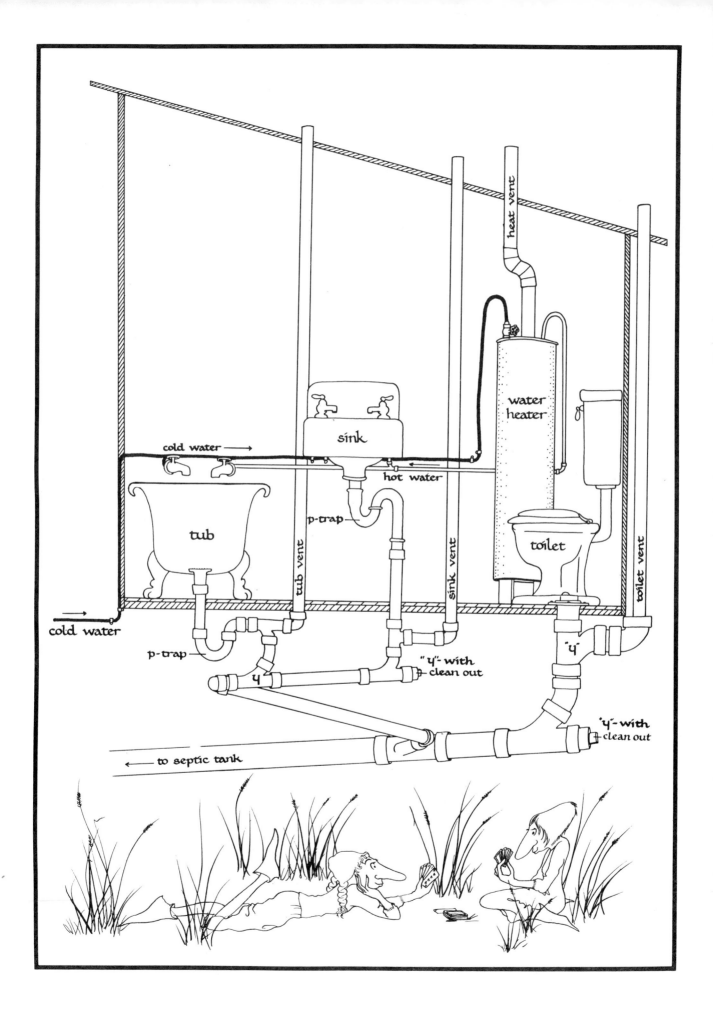

clean-out, joining the toilet to the main three-inch drain line.

Whenever possible, the toilet should be in a straight line with the septic tank. The more bends in the line, the more chance that it will clog. It's also easier to use a snake to clear out a straight line if it does clog. The lines also need to slope steadily downhill, more sharply on the toilet line than the sink, shower, or tub. Plumbers' tape—a metal strap with holes in it for nails—is the best way to support the pipes beneath the floor and insure a steady slope. As with rigid water pipes, you should fit the whole system together before you start to cement anything. Once it's cemented, it's there for good.

Repairing Old Plumbing: The principles and techniques are the same for dealing with old plumbing as those I've just covered for installing new pipe. The way you connect galvanized, copper, or plastic pipe is the same whether it's brand new or twenty years old. I'd like to cover here a few of the basic problems you are likely to encounter in an old house.

The most common problem is low water pressure in the house. This can either be caused by a leak in the water lines or by clogged pipes, and it is often difficult to tell which. If the pressure is equally low (doesn't come in spurts), then most likely there's a leak in the incoming water line. Unfortunately, it's also hard to track down the leak once you've decided that there is one. Check underneath your house, around any exposed pipes, and along walls or baseboards for signs of drips or dampness. Leaks in underground lines are hard to detect since water flows downward and won't necessarily show on the surface. You may have to dig up the whole line, but before you do that, check the pipes for corrosion causing low pressure inside the house.

To check for clogging in water pipes, you simply have to disconnect them and look at the inside of the pipe. With galvanized pipe, you can do this at the end of a pipe (a faucet, for instance) or at a union. Galvanized pipe usually cannot be disconnected at any other place. Copper pipes can only be easily disconnected by sawing apart and resoldering the joint back together when you're done. Copper is less likely to stop up, however. Old galvanized pipes are sometimes simply clogged by rust (the tin coating on the pipe has worn away, allowing the iron to corrode). A high iron content in the water will also clog both copper and galvanized pipe over a long period of time. In this type of corrosion, your water is usually rust-colored, and

the enamel of sinks and tubs gets stained. The other common cause of both clogging and pinhole leaks is electrolysis—the bonding together of two metals. This happens when an electric current flows through the water lines and touches a joint between two different metals. The current causes electrons to fly back and forth between the two metals, eating them away or filling the pipes with flakes of this new compound. The best solution for an electrolysis problem is to separate the two pipes and replace all damaged pieces. If you must still connect the different metals, use a special plastic-lined union so they don't touch each other directly. As I mentioned before, it's best to use only one metal in the whole system if you can. Tracking down the source of the current in an electrolysis problem is usually more difficult than finding the damaged parts. It may come from the electric wires being grounded to the water pipes. This is easy to spot and can be corrected by attaching the ground wire to a copper rod driven into the ground. The current may also be coming from an electrical short (usually in the pump or water heater). These are harder to find, requiring an ohm meter and experience in testing circuits. You can stop the electrolysis, however, without stopping the flow of current by simply making sure no two different metal pipes ever touch.

Whenever you have pervasive clogging or multiple leaks in an old house, you should carefully evaluate whether it will be easier to replumb or to repair. Both galvanized and, to a lesser extent, copper are difficult to remove and therefore to repair. It's often possible to replace sections of a water system (the bathroom, for instance) with new pipe, without having to replace the whole system. This often saves tearing out walls or floor boards, which is a big plus as far as I'm concerned.

The other common problem in old houses is clogged drains. This may either be the drains themselves or the septic tank. To check and clean the drains, you need to buy or rent a "snake"—a long metal line with a hook on one end to clear obstructions. Everywhere on your drain system where there is a Y with a clean-out, you can remove the cap and snake out the lines. If only the toilet is clogged, for example, you would crawl under the house until you found the clean-out on the toilet drain and then try to clear the line. If you can't find any obstructions in the lines themselves and the water still won't drain, the septic tank probably needs to be pumped out. The most usual cause of clogging in a septic tank is the use of detergents or chemicals which kill the bacteria necessary for decomposition in the septic tank.

If all of this exhaustive detail makes plumbing seem anything but simple, let me reassure you once again. There are many types of piping and fittings, but they all fit together as easily as Tinkertoys. Plumbing is a refreshingly unintimidating skill, though far too few people realize that. Because of this, in fact, you can probably find plenty of work as the neighborhood plumber if you wish.

We talked at the women's meeting tonight about passivity, and I thought I had nothing to say. I'm such an open, aggressive woman. Then I heard Kathy and Janet describing themselves that way, women who seem to me quiet and gentle, and I wondered what makes us see ourselves as so aggressive. It's as though just being ourselves is stepping so far out of line that we've come to accept it as being aggressive and dominating.

Hannah started talking about when she wanted a chicken house. She spent days talking to Jeff about "its needing to be done"—hoping that he would "offer" to do it or that she would get the energy from him to not be afraid of trying it herself. Sarah talked about trying to draw in her tiny cabin with three children talking and playing—how she gets more and more anxious and irritable but isn't conscious of it until she starts screaming. She never simply acknowledges her need for privacy and demands the kids respect it. I thought of when Peter wakes up in the morning all full of energy and wanting to move right then. I feel pressured and tense, needing to wake up gradually. When we work together he likes to work hard for five or six hours then stop for the day. By then I'm tired and strung out like a wire about to snap. Alone, I work in spurts taking breaks to rest or think, but I still get as much done in a day. Why can't I control my space or energy except when I'm alone? Our pace is his, our methods his, our priorities his. It's not that he obviously tries to control me; it's that I abdicate. I don't think I know how to live with other people and stay on center. I'm always passively responding to whoever's energy is around me.

As we were talking at the meeting, we realized that instead of directing our lives, we spend them providing for others. That we not only let others' energy prevail, we always subordinate ourselves to the needs of those around us. Sarah says she's been craving a drawing pad for months, but the children need socks. A $.98 drawing pad! Judy talked about when her family goes out to restaurants; her mother always orders the cheapest thing on the menu, hamburgers and chicken wings. Janet said that she had worked all afternoon splitting wood with Joe and a male visitor. Then when she sat down to dinner, she ate less than she wanted, feeling they somehow deserved it more. When I go to the grocery store, I automatically buy the kind of cheese Peter likes.

As we talked, incident piled on incident. It's amazing in how many minute and subtle ways we've voluntarily subordinated our needs to everyone, anyone else around us. We don't see our needs as valid or equal, don't function as mature and deserving people. Ah yes, those aggressive women lost in the middle of the grocery store wondering what he would like the most!

53

outhouses, compost privies, and septic tanks

I live in what is becoming an increasingly common situation: although our house is equipped with a flush toilet and septic tank, we walk two hundred feet to our outhouse each time we feel the call of nature. The toilet uses four gallons of badly needed water with each flushing, water that becomes polluted by the sewage it carries. The septic tank cost us $800 to install five years ago, and I now look on it as a rather costly monument to my city naïveté. I have forgotten that the outhouse is an inconvenience and am only really glad of the con-ventional bathroom on the wettest of winter nights, when abundant water allows its use anyway.

Whether you've just moved onto your land and need a temporary outhouse, or you don't have or want to put money into a septic tank, a pleasant sanitary outhouse is easy to build. To begin with, it can be a very simple structure, or even almost no structure. Basically you have a hole in the ground x feet deep, and you need to cover it to keep it sanitary. Our structure is three-sided, 4′×4′, with a roof sloped from seven down to five feet. It's

1″×6″ redwood plank and batten construction over a frame of two by fours. Half of the inside is a box built against the back wall with a hole for a seat, and open underneath. We live in California where the weather is relatively mild, so we dug a hole on the edge of a thick grove of trees and faced the open side of the outhouse toward the direction of least wind and rain. An adjacent sapling redwood gives some privacy but leaves a spectacular view of the ocean. I really prefer this open design to the dark box of a four-sided, sealed with a door, outhouse.

The simplest structure I know belongs to a friend who has a plywood or plank platform built

I've been restless and unable to work all morning, thinking about the women's meeting last night and what passivity means to me. Suddenly it has become clear to me—it's so obvious, but I have never seen it before. I've abdicated all control over my life, have centered it so completely on Peter that I've become paralyzed. I spend my life listening to the wind so I'll know when his boat is coming in, making sure he has everything he wants, doing everything "we" planned. I do lots of things—work on the garden, split wood, work as a waitress, fix the pump —but those things, even the job which consumes so much of my life, are all peripheral. I don't do them because I choose to or want to but because they sustain our relationship, because "we need to." It's so incredible—I don't exist outside our relationship any more. I don't know who I am or what I want to do.

I'm not even sure how I got here. I always felt so directed and in control. There was never any simple point at which I decided this is what I'm going to do. It all just seemed to happen. Looking back, I guess it started when I quit my job in Pennsylvania and moved to the farm—but then it seemed like a pause, not a break, until I could figure out what to do next. Everything since has just happened without my ever questioning or choosing: being with Peter more than the others; moving to California because he wanted to and I wasn't doing anything else; his getting a fishing job he loved and me doing more and more of the housework because I just happened to have more time; me dreaming of babies and land of our own. A perfectly natural evolution—I've never paused to look at, much less question, this path we're on. What could be more beautiful than to devote a lifetime to building a relationship?

But it hasn't been working. We aren't devoting our lifetime to building a relationship. I'm devoting my life to taking care of him, and he's devoting his to doing what satisfies and pleases him. I never chose to run this place alone—it was "our homestead." But it never has been except in fantasy.

Suddenly today, thinking about it all, I've realized that the solution isn't for Peter to quit his job and stay home. It isn't for us to build the perfect family homestead when he doesn't even want to be on land! I have to begin to find some meaning in my life that's not dependent on him, to begin to choose again what I want to do and be. It seems so simple and so obvious, but I've been so immersed in being a wife that I don't even know if "our" directions are really mine.

So simple and so obvious. But I'm feeling panicked. What am I going to do? It's been three years since I made choices for myself, and I don't know where to begin. What I'm doing doesn't satisfy me, but I don't know what will. I'm so scared I won't be able to do anything meaningful or creative. I'm scared, but somehow I've got to start. Clearly, something has got to change.

over the hole, with a trap door on leather hinges, that you squat to use. Another neighbor's old family homestead had a two-seater outhouse, which has to be moved less often and leads to who knows what sorts of communion?

When you think of building an outhouse, there are several basic things to consider. First, be sure to locate it *below* your water source to avoid any contamination. Try to pick an area with good drainage. Next, the hole you dig *will* fill up, so think of the outhouse as an occasionally movable structure, and build light. Make your design simple, and then use the lightest materials that will work. For the outhouse to sit sturdily through the seasons, it may take two or three people to move it, but do consider relocation when you build. For instance, don't make sunken posts your main corner pieces. The building doesn't have to be completely watertight (especially if it's only three-sided), so it shouldn't be too intimidating for new carpenters. It's nice not to have puddles on the seat or snow coming through the walls, but it's not like a bedroom, either.

The seat can be any height, or nonexistent. There is convincing evidence that squatting is the most healthy position, or barring that, to have your seat so low that your knees come above your waist. I have a bad knee, so I like the conventional armchair height, but set your seat to suit yourself. If you have a lot of people in your family, a good design is to dig a long trench for your hole. Then build a box over the trench for your seat. As you fill up one section of the trench, pry up the next board on the box and move your seat down. Make sure that your seat has a well-fitting cover, to keep out flies or bees that carry germs.

If an outhouse is working properly, waste matter will be decomposing and won't smell at all. Decomposition works best in an alkaline environment. To develop that you should dump wood ashes (from your stove) down the hole periodically. Urine is very acidic and tends to neutralize the alkali, so whenever possible, don't urinate in the outhouse. Our general rule is Piss Anywhere, as long as you're at least fifteen feet from the house and not in the flower bed. We keep a roll of toilet paper by the front door. Paper, even toilet paper, fills the hole more quickly and slows down decomposition: use as little as possible, or better yet, keep a can in the outhouse to collect used toilet paper and then burn it in your stove. Tampons and sanitary napkins will also burn in a good fire, so don't put them in the outhouse either. I think it's a

good idea to question some of our fetishes against bowels, but at the same time keep your outhouse sanitary. You may want to keep some disinfectant there for rinsing hands. Decomposing waste matter also gives off heat, and all last winter we had a racoon who made a neat nest on the outhouse seat every night.

When your hole gets full (don't wait until it runneth over), pick a new location, move the structure and fill in the rest of the hole with dirt, covering it completely. A $3' \times 2' \times 3'$ hole lasted a year for two people on this farm.

Lastly, try thinking of your outhouse as a real structure in your life, and build or decorate it to please you. We really do have Renaissance tapestry art on the walls of our outhouse, and we buy new prints to change the decor whenever our whim or mildew moves us. Next to my present one, my favorite outhouse was a seat with a roof set on poles, no walls, in the woods on the edge of a basin overlooking the Cascade mountains. One commune built theirs in the center of a huge old burned redwood stump—high charcoaled walls with a circle of sky above. Another friend bent thin sapling trunks with plastic to form a domed hut. Be as fancy or as plain functional as you want, just remember that it's pretty easy and inexpensive to take care of this basic need in the country.

If you don't want to brave winter storms to reach the outhouse, or if building inspectors frown on such a solution to a universal need, there are two other choices: the compost privy and the conventional septic tank. The compost privy has recently been approved in this area (after a long struggle) as a legal means of sewage disposal; it is quite commonly used in Sweden and Norway. The compost privy is both sanitary and ecological; it can be built as an outhouse or can be built into a house, with an outside door to remove the final product—clean, fresh humus. The privy uses basic composting techniques to convert waste products (sewage only; it cannot handle sink or tub water) into humus. It is an aerobic decomposition process (bacteria in the presence of oxygen). This process gives off only humus and carbon dioxide as byproducts, no odiferous gases; its high temperatures destroy any dangerous bacteria in the feces. There are two basic designs of compost privies—the Van der Ryn and the Clivus systems. The Van der Ryn privy is somewhat cheaper and less complicated to build, but requires maintenance (turning the pile) each month. The Clivus system is the one used in Scandinavia and needs cleaning only once every year. Sources of plans for both types are listed in the Resources section.

Basically, each type privy is a cement or cinderblock box with two chambers. In the collection chamber, soil and straw or leaves cover the cement floor, supplying bacteria for decomposition. Air vents supply oxygen to speed the process, and a vent pipe carries away carbon dioxide. Depending on the system, the pile must be periodically moved to the aging chamber (Van der Ryn) or it automatically moves itself down a sloping floor (Clivus). A trap door allows for the removal of the humus end product which is disease-free and safe for garden fertilizer. The cost of building a compost privy is between one and five hundred dollars, depending on the design and materials used.

A friend who has built one says she had no difficulties once she learned to work with cement and cinder block. It is a bit unsightly though, she says, and should be designed into the house whenever possible. I would guess that compost privies are going to become an increasingly common and legal method of waste disposal. They waste and pollute no water, produce fertilizer, and cost less to install than the conventional septic tank.

Twice now while writing "septic tank," I've written "sceptic tank" instead—which betrays a little of my attitude. I've put septic tanks in on three different farms, but I don't recommend them to anyone. As I have said, they are expensive even if you do the (arduous) work yourself. Having said that, let me share what I know of installing one. A septic tank is simply a large box (usually concrete) with a partition wall reaching not quite all the way to the top. Sewage enters the tank on one side of the partition. Liquid wastes float up over the partition, into the second section and out the drain. Solid waste settles in the first section until bacteria decompose it into a liquid; then it too drains. Because of this bacterial action, it's important not to drain detergents or other pollutants into the septic tank. Washing machines should not be connected to a septic tank. If the tank clogs because of a lack of bacterial action (this sometimes happens after years of use), it can be pumped out and started again. Some books recommend regular pumping once every five years. There are companies with pump trucks to do this job in most rural areas.

Outflow from a septic tank goes into "leach" lines —pipes with drain holes in the bottom. These lines slope downhill in gravel-filled trenches to distribute the waste over a large area. The pipes are then covered with a foot or two or earth. Most pipes now are plastic (PVC) which is cheaper and easier to install than the old iron ones. Between the septic tank and the leach lines is a small cement "distribution box" which is merely a junction point for the several lines. The length and number of the leach lines, the depth and slope of the trenches, etc. is regulated by the local health department and varies depending on soil conditions and the number of people in the house. Consult your local inspector for the rules in your area. One problem with septic tanks is that they drain badly in heavy clay soils and may require extra leach lines.

At our first country home, we dug the septic tank hole and trenches ourselves with shovels. It was a far more difficult job than digging our well. The septic tank hole was 12′×6′×6′ deep and the three leach lines were each forty feet long and two and a half feet deep. We built the septic tank out of cinder blocks lined with cement; we hauled the rock for the trenches ourselves; we bought and laid the pipe. And still it cost us several hundred dollars for materials plus days and days of digging.

Even if you're short on money, I would recommend hiring a backhoe to do the digging for you. A skilled operator, used to septic work, can do the job in an afternoon (at about $20 an hour), getting the proper slope and depth to the trenches, etc. Then you can build your own septic tank in the hole (if local regulations allow) or pay someone to deliver one. There are new fiber glass septic tanks being sold now at about one half the cost of cement ones and can be hauled on a pickup truck. The cement ones are expensive, weigh several tons and require special equipment to install. The drain pipes are available at any building supply store. The rock you may have to have delivered (it takes at least a dump truck load, depending on the length of your lines). As with any other project, you shouldn't settle for a package deal ("We'll install your septic tank for $1,000") unless you have money to waste or until you've checked to see if you can do it more cheaply yourself. Usually you will save money doing it yourself, even if you contract out most of the work (to a backhoe, rock hauler, etc.). As one final word on septic tanks, I want to relate my experience on this farm. Having decided to make our house strictly legal, working hard for the $800 for a self-installed tank, spending two days shoveling gravel and laying pipe, I confidently greeted the health inspector. His first words were, "Well, it looks fine, but what did you do it for? I'd have let you get by with an outhouse."

ELECTRIC PUMP REPAIR

The first time I ever lived with a pump, it was one that broke down often. I had only the vaguest conception of what a pump was and no idea of how one worked, so each time it broke down I would haul it in to a pump dealer to get it fixed. This meant hauling the entire eighty-pound pump up a steep canyon slope, as I was afraid that touching a single bolt might do irreparable harm. After one such trip, my (by now) friend the repairman said, "You know, you really could do this yourself." And with that I began to learn how really simple a centrifugal pump is. Looking back now, it's hard to remember that state of being afraid to touch, but then for most people, a pump is more alien than a car engine. It's something you've never seen before and have no understanding of.

As with any other curative process, there are two stages to pump repair: learning all the parts and how they fit together, and learning how to diagnose the trouble. I am still caught sometimes tearing a pump apart only to find I had a vapor lock in the water line. It's not a tragedy to do so; pumps come apart and fit back together easily. But it's a lot of trouble for no reason.

All my experience has been with Jacuzzi pumps, so what I will describe is based on their design. Centrifugal pumps are basically the same, though, and you should be able to figure yours out even if it's a different brand. If you can, try to get an owner's manual from the manufacturer—that will at least give you a diagram of the insides of your specific pump.

A pump has three main sections—the motor, the mounting frame, and the base. The motor is a simple electrical one, composed of a large coil of copper wire. Out of it comes a steel shaft that drives the impellers. This shaft joins another shaft that comes up through the base. When the pump is working properly, the shaft should spin easily with finger pressure. The base may have several stages (from one to four on a normal house pump), each containing an impeller, which whirls the water around through a narrow opening to create pressure. This is what does the pumping. A pump must be filled with water to create a siphon (a natural sucking action); the motor then turns the impeller(s) which pushes the water out through a small opening under pressure. The vacuum that would be left by the water being pushed out is filled by

more water being sucked in. Pumps are just about that simple.

To take a pump apart, you begin by disconnecting the incoming and outgoing water lines; then you take off the motor. You may have to disconnect the electrical wires to the motor or you may not (depending on whether you have a place to set the motor next to the base or not). I usually disconnect them since its only a matter of turning three screws; then there is no danger of breaking the wires by dangling the motor from them. To do this, you follow the electrical cable from the motor to the electrical box on the pump. Snap the cover off the box and trace the wires that go to the motor. There should be three: a white, a black, and a ground wire. Before you touch anything, be sure you have unplugged the pump or turned off the fuse. Loosen the screws and remove the wires, then

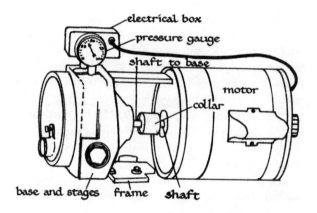

pull them free of the box. If you also have to disconnect your incoming electrical wires (if they are on a long line instead of a short one with a plug), make sure you remember which wires go where. Draw a diagram if you need to. Rewiring incorrectly can short out your motor.

Next, loosen the four bolts that go through the frame into the motor. This should be done with an ordinary open-end wrench or a socket set. If you have to use a crescent (adjustable) wrench, do so very carefully. They do not fit perfectly on a bolt and tend to round the corners. Once these bolts are removed, the motor will spin around but still won't pull free, since it is still attached by the collar at the end of its shaft. This collar attaches by set screws that tighten down into a groove on the shaft. Set screws have concave heads and are loosened by hexagonal allen wrenches which fit down into them. Allen wrenches can often be found in those $.99-each tool bins in hardware stores, and it's nice to have a set on hand since nothing else will do their job. Turn the screws counterclockwise with the wrench until they come out, then slide the motor free from the collar and set it aside. (Be careful not to lose the set screws—they are small. Either put them in a tin can with the bolts and main seal or, better yet, screw them back into the collar for safekeeping.)

The next step in disassembling the pump is to remove the collar from the other shaft, again using allen wrenches. You may have to hold the collar with a pair of vise grips to keep it from spinning while you turn the allen wrench. If your pump has been sitting a long time, these set screws may be rusted in place. If so, spray them repeatedly with WD 40 or Liquid Wrench and work them gently with the allen wrench. To get added leverage, you may need to clamp another smaller pair of vise grips onto the allen wrench. Once the screw has broken free (started to turn), use the allen wrench alone. Remove the collar when the screws are loose and set it aside. (I should have mentioned before this that experience has taught me to disassemble the pump in some handy place, like on a sheet of plywood or the living-room floor. Hunting for set screws in two and a half feet of grass is no fun.)

Next, use your open-end wrench again to remove the four bolts that run clear through the stages and thread into the bottom of the base. Pull out the studs when they turn freely (don't force them; they will slide out when they're completely unthreaded). Again, these may be rusted in place. You can spray WD 40 around the head but not on the threads, so all you can do is slide a pipe over your wrench for more leverage and try some more. If, by bad luck, the head of the bolt shears off, you should still be able to slide off the stages and then remove the stud using vise grips. Once the studs are out, the frame section will slide right off. Try to hold and control it as there is a gasket between it and the first stage that you want to remove intact if possible. So, pull the frame loose gently, trying to either bring all the gasket with you or leave it all behind. Whichever side the gasket sticks to, you can then remove it by careful use of a knife. Try not to tear the gasket—it is replaceable and inexpensive, but sometimes hard to find.

Once the frame is off, the inside workings of the pump will be exposed to view. In fact, the main seal may literally pop up at you. It is held in place against the bottom of the frame by a compressed spring which is released when the frame is removed. The main seal has two pieces: (1) a spring which has a carbon and rubber ring on its upper (motor side) edge and the seal itself, and (2) a rubber and hard plastic ring. The rubber edge of the ring goes up against the bottom of the frame; the carbon and rubber end of the spring presses against the plastic. The repairman warned me to be careful with the seal, as the carbon can crack, but the only time I've ever had trouble with a seal was when I put it in upside down (plastic first). This was a week after I taught a class in pump repair and my pump shot off like Old Faithful four times a day. I waited to fix it until I had a new seal, but when I took the pump apart again, I found the old seal in fine shape despite its ordeal. So, be careful with main seals, but don't be intimidated by them.

On general principle, I keep several gaskets and a new main seal in the house as a standby kit. These are the things most likely to need replacing

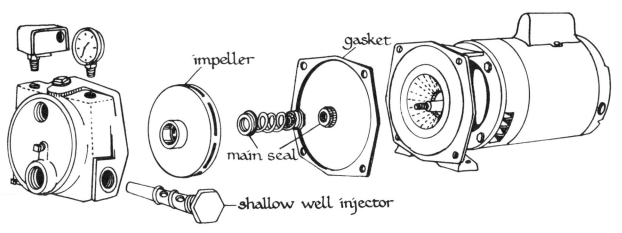

impeller · gasket · main seal · shallow well injector

when you take your pump apart. It's handy to have them around when the nearest store is twenty miles away and there's often a three-day wait for parts on a new model pump. Gaskets cost about $.60 each. If you ever do get stuck, you can cut your own out of a sheet of cork gasket material, obtainable at any auto supply store. Just be sure to buy the same thickness. I have never yet needed a new seal for this pump, but my old one blew seals regularly (we ran it at too high a pressure). I guess I just keep one around as a good luck omen (they cost $3.00 to $5.00).

Below the seal you'll see the impeller threaded on the shaft. This is a round, thin (⅜" approximately) disc of plastic (most likely). It's not very impressive. I kept looking for airplane propellers. The impeller threads onto the shaft with a nut that is cast into its center. A socket wrench is the best tool for loosening the impeller since you have to get down in at the nut. Hold the shaft with the vise grips so it won't turn. If your impeller is plastic, this nut will begin to round at the edges after several times of being taken off and put back on. Be as gentle as you can. Plastic impellers also get chewed up really easily by sand or dirt. If for some reason you have to work on your pump often (twice or more a year) or your well or spring tends to silt in, you may want to replace the plastic with brass impellers. They last a lifetime.

Once the impeller is off, you can lift off that stage of the pump to expose the next impeller. Between every stage is a gasket, so follow the same precautions as above. When you have removed the last impeller (if you have a one-stage pump, *the* impeller; there is nothing to lift off), the pump is apart. The only other thing you may want to loosen is the shallow well injector (if you have one) which threads into the side of the base section near the bottom. It has a rubber gasket, is made of plastic and also gets easily chewed up or clogged by dirt. It, too, can be replaced with brass. The rubber gasket occasionally wears out. They cost about $.20.

The pump goes back together the same way it came apart, in reverse order. The shaft is the only tricky part. First the impeller goes back on. Tighten it down with the socket under easy hand pressure; don't strong-arm it as this can crack the plastic. Then put the gasket onto the edge of that stage using *non-hardening* gasket sealer (Form-A-Gasket is a good brand). Get the right stuff; Permatex in a tube *says* it's non-hardening—but it

works like cement. Line up the holes in the gasket with the bolt holes in the stage and make sure there are no tears or wrinkles in the gasket. Repeat this process until you have put on the last impeller. Slide the main seal onto the shaft (right way, please!) and then set the frame and the last gasket into place.

Some pumps are reassembled in the other way: main seal, impeller, first stage, impeller, etc., until the last stage goes on the shaft. Follow your owner's manual or look to see how yours fits together. It will be obvious. Slide the studs into position and start them turning by hand (always do that with a bolt to make sure it's threaded straight). They should turn easily. Tighten with a wrench, using a good strong turn but no super effort. Gaskets are made to be snug but not smashed, and loosening that too-tight bolt will take four times the effort that putting it there did.

Now, the shafts. The set screw goes through the collar into a notch that runs the length of each shaft. They tighten up against this notch but do not thread into it. You need to have them lined up right so you don't damage the threads. Either do this by eye (peering through the holes in the collar) or by setting the screws part way, sliding the collar around the shaft until the screws slip into the notch, then tightening them down. To make it easier to take apart next time, I always smear a little grease on the notch and on each set screw. The collar should sit as far down (toward the base) on the lower shaft as it will go easily; don't push. Next, set the upper set screws part way into the collar. The impellers and main seal work by being *suspended* in the pump by the upward pull of the shaft. To do this, you have to pry up on the collar until you get the shaft from the motor fastened; after that, the motor will hold it up. This is best done with two people. One takes two large screw drivers and pries up on the collar (about ¾"). The other turns the motor shaft until its notch lines up with the set screws, slides it into place, and tightens down. Make sure the collar spins easily beneath your fingers when all is tight and that you don't hear an impeller scraping. If you have a leak in the main seal or hear an impeller rubbing when you have the pump back together, you can usually fix it by raising or lowering the position of the lower shaft.

Before tightening down the bolts for the motor, notice which direction the frame is facing. One side of the frame is flat for the pump to rest on. Make sure neither the air vents nor any protrusions from the motor are toward this flat side. The shaft will spin so you can turn the motor to any position. Tighten the bolts to the motor, reconnect the electrical wires, and the pump is back together. Reconnect the water lines, prime the pump by unscrewing the priming plug and filling the pump with water, and it is ready to run.

Now, how do you figure out what's gone wrong and if you really do need to tear it all apart? First, get to know your pump when it's in good working order. Know at what pressure it normally turns off and on. You can tell if it is within the normal range by looking at the pressure gauge. Know how it sounds when it's running. Know what a free-moving shaft feels like. Know how warm or cool the motor feels. Make friends with your pump and it will serve you well.

Pump problems come in two types—sudden catastrophes and chronic problems. Chronic problems (too low a water pressure, gravelly noises in the pump every time it runs, not pumping enough water, etc.) happen when you don't connect the pump properly in the first place. In general, make sure you have enough electrical current to the pump, that you are not sucking over twenty-five feet in depth, and that your pipes are the right diameter for the pump. (The water systems section in this book describes in detail how to set up a pump. This section restricts itself to sudden catastrophes.)

Decreasing water pressure is a common danger signal. You walk outside and see a geyser shooting up from your water line (the horse stepped on it). You turn on your faucet and get no water at all. You hear your pump running constantly. You do a routine check on your pump and notice the motor is overheating. And there must be more—the kinds of things that can happen to pumps or any other machines defy imagination. Anyway, as soon as you notice trouble, do something about it. If you have the slightest suspicion that the pressure is low, *check it out.* You know what pressure turns your pump on. If the gauge is below that, something is wrong.

The first thing to do once you think something is wrong is to close the valve from your pressure tank to your water lines. That way, you can't drain all the water out of your tank if you have a leak. If there is a valve between your tank and your pump, shut it too. Then turn off the electricity to the pump. Now, you can calmly proceed to isolate the problem. If your pressure is dropping or you have no pressure at all, check for two subconditions:

A. *The motor runs, but there is low or no pressure.* This means that the motor and the pump sound normal but you're not getting pressure into your lines. Most likely there is a leak in the system somewhere. If it is at the pump, you will probably have found it already. The main seal or a gasket may be spraying water. This usually happens if you are running the pump under too high a pressure. Whatever is leaking will have to be replaced. (The main seal sprays around the shaft, a gasket out the sides.) Or you may have a leak in the pump itself. This is caused by ice inside the pump cracking it. If you're lucky, it will be just the frame section, which is relatively cheap; if not, you may have to replace one or more cracked stages and impellers. You can buy new parts from a dealer. An unusual freeze caught most of northern California unprepared last year and there was a month backlog for parts from the factory. My frame cracked then and I tried welding it together, but it never really sealed properly. When parts were available again, I had to replace it. To prevent a cracked pump: insulate your pump house or box; keep a light bulb or other heat source near the pump during a freeze; leave a faucet running at night; and change the pressure setting so the pump runs more frequently and cannot freeze. If you can't find a water leak on or near the pump, start tracing the lines. If they are buried underground, this will take some time. You may have to leave the pump running so that enough water will flow out the leak to soak the ground above the leak.

Most other causes of *low* pressure are chronic ones. If the motor has loose connections or isn't getting enough voltage, it won't be up to speed and pressure will be low. If air is getting into the suction (intake) pipe, pressure may also drop. Check for cracks in the pipe and make sure the bottom of the pipe is completely submerged in water. If the impeller is partly clogged (dirt, twigs, leaves, tadpoles, etc.), pressure may drop, but usually it's so clogged the pump isn't working at all. Try to eliminate all other possibilities before you take the pump apart for a supposedly *partially* clogged impeller. (One good but not foolproof check is to open up the priming plug and look at the water: is it muddy or cloudy? And to take out the shallow well injector: is it dirty or clogged? This will help you to decide whether to take the pump apart or

We've been talking for weeks in the group about a women's fair—a place to share with other women here the experience of women together, teaching each other our crafts and skills and about ourselves. There are so many women scattered along the coast who never see or get to know each other. Last night we were sitting around, talking about how fine it would be to have one and where we should have it and what it would be like. And I said very quietly, "Well, I will be on a co-ordinating committee if others will too." And Jane said, "Oh, farout, I'd like to do it to." So now there's a women's festival.

After I volunteered, I realized that no one else had even thought of a structure or the mechanics of how to do it. And suddenly I remembered that I used to do things like that! I had forgotten—that part of my life seems so long past. Now I am excited, drawing up lists and adding on possibilities. It will be so good to be doing something of my own, to try to use my skills again. It's like remembering a whole other self, another personality, another set of actions. But it's coming back to me now: the awareness of process, the following ideas through to their practical conclusions, the feeling competent . . .

not.) A fine film of rust is normal on interior pump parts; don't be confused by it.

B. *The motor tries to run and can't or is hot and not running, and there is low or no pressure.* Usually this subcategory means you can hear the motor whirring and shutting off immediately or clicking off and on. The hot motor situation is included here because it is usually a variation on a clogged pump, not a broken motor.

If the pump is trying to run and cannot, something is stopping it. If the temperature is below freezing and your pump is inadequately insulated, suspect frozen impellers and hope you caught it before it cracked. Turn everything off and expose your pump to sunlight or some other heat. When the pump and lines have thawed, turn it back on and see if your problem is solved. If you cannot thaw the pump, open the priming plug so the ice has an undestructive passage, keep the pump as warm as possible and turned off.

If the weather is above freezing, you have most likely clogged your pump or run it dry. First check your well or spring for signs of either. If the bottom of your pipe is suspended in mid-air, the pump is dry and you must wait for the well to recover and reprime the pump (making sure to get all the air out). If the bottom of the pipe is resting on the bottom of the well, *or* if you feel that it is in watery sand or silt, the pump is probably clogged. You can also check the priming hole and the injector as mentioned before. If the pump is clogged, take it apart and clean each impeller, each stage, and the injector. If any are chipped or damaged, replace them. Remove the suction pipe and clean it out. Then either shorten its length or dig out the spring or well, or both. Usually with both these problems, the motor will be trying to run. However, sometimes it has overheated so badly that the safety mechanism is keeping it from running. If it is hot *and* not running, you usually have a clogging or airlock problem. A hot and running motor and a cold and not running motor are quite different problems.

Unfortunately, diagnosing the problem isn't always easy. You may have checked and found no sign of clogging and a low but not dry water level and still the pump won't run. It is possible that the well has run dry without your noticing it (while you were at work or because of the stored water in the pressure tank), had time to refill itself before you realized something was wrong, and left the pump or water line air-bound. Last summer I had a pump that wasn't running but there was no sign of a dry well. Deciding that it must be a clogged pump, I tore it apart and found nothing wrong. I put the pump back together but put the seal in upside down. Then I tore the pump apart again and put the seal in properly. It still ran erratically. Then I realized there was air blocking the suction line! I should never have taken it apart in the first place. From such experiences do we learn!

If your well is low enough to have gone dry, check this out *before* you take the pump apart. Open up the priming plug and pour water into the pump. Keep doing this with the plug open to try to let air bubbles escape. Then close the plug, open the pressure tank valve(s), and turn the pump back on. Leave the nearest faucet wide open. Manipulate the switch manually so it goes on and off—on (two minutes), off (one minute), etc. This works air bubbles into and through the pump to clear the line. To operate the switch, open and close the points with a small screw driver. The

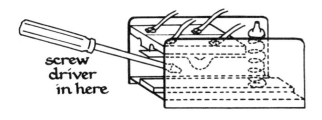

points (located in the electrical box below the contact plate for wires) are at the end of a thin strip of metal resting on a spring. You will hear them click open and shut and the motor will start and stop. If water sputters out of your faucet, that is a good sign. It may take a long time to clear the line. Continue until the pump will pump in one long, normal, quiet run. If this doesn't seem to help, you'll probably have to take the pump apart to look for other problems, but don't forget the possibility that there is still air in the lines when you have it all back together.

Now other possible problems:

C. *The motor won't run at all and is cold.* Usually this is a simple electrical problem. Check first to see if a fuse has been blown or if there are any loose or dirty connections in the electrical box (turn off the power before you touch the wires). Then check the points. Make sure they are clicking open and shut, look to see if they are dirty or blocked. (A friend once found an earwig between her points!) Your points may need to be replaced; on some pumps this means a whole new electrical box.

Check to see if the spring is exerting pressure on the points. It is rare, except on an old pump, but occasionally the pressure switch goes out and the points cannot close. The switch consists of a rubber diaphragm which expands and contracts with the water pressure in the pump. When the pressure gets high, the diaphragm expands, compressing the air above it which pushes the spring up and opens the points (stopping the motor). When the diaphragm contracts, the spring drops, closing the points and starting the motor. If the spring breaks or the diaphragm has a hole, the switch quits working and the pump either won't run at all or runs until it blows a gasket from excess pressure. It's easy to check a broken spring—look at it. To get at the diaphragm, you unscrew the pipe that holds the pressure switch. But try not to do that unless you have eliminated all other possibilities.

The other reason for the motor not running is a problem inside the motor itself. Turn the power off and then remove the top of the motor (it screws or bolts on). Check to see if the wires to the electrical box are securely connected. Look at the large coil of copper wire inside the motor. It should be shiny and metallic. If it is black or burned, you may have shorted out the motor. Don't be confused by paint sprayed over the whole pump at the factory. At this point, you've either found your problem, or you need someone who understands electrical motors.

D. *The motor and pump run, but the motor overheats.* Again, the problem is most likely electrical. Are you running at the right volts (110 or 220)? Are there enough amps available for the pump on that circuit? Most pumps need 20 to 30 amps, depending on the motor size; a pump should be on its own circuit. Is your electric line heavy enough to carry the current? Volts are lost over distance if the wire is too small (twelve-gauge copper for fifty to a hundred feet, ten gauge for one to two hundred feet, if you have 110 volts; fourteen gauge for 220 volts up to two hundred feet). Are all your electrical connections tight and clean?

Also, check to make sure the air vents in the motor are not blocked by insulation or clogged. If you have mounted the motor wrong, the vents may be blocked. Take the lid off the motor, if necessary, to make sure they are clean—I once had a family of field mice build a home in my air vents.

And last of all:

E. *The pump runs all the time.* If it's running and the pressure is low or never enough to shut it off, you have a break in the system somewhere. If the pump literally runs all the time and pressure continues to build, shut it off immediately before a gasket blows. Either your pressure switch is not working, the points are stuck shut (possibly jammed by some foreign object), or you wired the pump wrong. Check the points and the wiring first.

If your pump runs every time you use water or keeps turning off and on, off and on, your pressure tank is waterlogged. This means the air cushion inside it has dissolved into the water and run out of the tank. To restore pressure, you have to drain the tank completely, pump more air into it with a bicycle pump and then refill it. For a permanent solution, see the section on setting up your pump. This situation is very hard on your pump and will lead to other problems, so fix it.

These are all the things that can happen to a centrifugal pump. Of course, there are variations (what can you get between your points?), but the basic problems are simple, mostly requiring logic and common sense to trace. Repairing any part of the pump is also very simple, though it helps if you're not under an immediate need for water pressure or if it's not freezing cold. It helps to be prepared too: have basic wrenches and gaskets on hand, check your pump regularly, and watch your water level. Don't take your water system for granted. Some sunny day when you're feeling good and have some extra gaskets, take your pump apart. Get comfortable with it so fear won't stop you when you are trying to cope with a water crisis. And, who knows, after a few minor catastrophes, you too may be the neighborhood pump repairwoman!

SPLITTING WOOD

Splitting wood is not grabbing a piece of wood, laying it on its side, and pounding it with a heavy axe until it falls apart.

To me, splitting wood is knowing wood, knowing the trees and how they grow. Knowing to use fallen or deadstanding wood. Knowing that Douglas fir and redwood have wide straight grain and tanoak and madrone are close grained and solid. Knowing seasoned wood and how the axe just needs to touch it right and it falls open. And knowing green wood and how the axe has to push its way through.

It's having an axe that is light enough for me to guide with hand and wrists and heavy enough to make an impact on the wood.

And it's putting this piece of wood and this axe together in a way that uses least of my energy to break it open. I always look at the particular piece on the chopping block. I look at the placement of all the knots and plan to hit parallel to the biggest. I look to see if the wood is naturally splitting anywhere. When I've found the place (not always the first try) I lift up my axe and look at the place until

a circle is formed between the place on the log
through the chopping block, through the ground,
up my feet, legs, into my arms, out the axe, and I

let the circle close

splitting wood

Types of Wood: If you have a choice, try to balance the types of wood you cut and split. Different types of wood burn hotter, quicker, slower, or cooler; each type serves a certain purpose. Soft woods (redwood and fir) burn quickly but don't give off much heat, making good kindling for starting slow-burning woods. Sappy wood (pine and some fir) burns hot and fast, warming up your room and oven temperature quickly. Hard, close-grained woods (oak, birch, alder, madrone, etc.) burn slowly and very hot, giving you sustained heat and using less wood. Most types of wood split more easily when they are dry and aged. This makes already dead trees or downed wood your first choice. If you have to fell live trees, it is best to do so in the spring, leaving the rounds to dry all summer before you split them. There are some kinds of wood, however, that are virtually impossible to split once they are dry. Bull pine and eucalyptus are among those which must be split green or not at all. Learn about the wood in your area and try to collect some of each type for your wood pile. This effort will save you much wood when you begin burning through the stack: a stack of fir will vanish four times as fast as one of mixed hard and soft wood. Learn too, which woods make poor firewood. For example, burning green madrone will eat out a stove. Eucalyptus burns with a lot of smoke. Too much redwood causes a pitch coating

on the inside of the stovepipe that can catch fire. Ask old-timers and people who cook and heat with wood to share what they know.

If you cannot cut your own wood and are forced to buy it, these differences in wood types will be mirrored in the cost. Wood is measured and sold in cords. A cord is a stack measuring eight feet by four feet by four feet. A cord of hard wood will cost up to twice what a cord of soft wood does.

It is important to stack your wood dry—and if you live in a rainy area, to put it under cover before winter. Wood that is split and stacked wet will not get enough air to dry out and will rot. A woodshed only needs to have three sides, if it can be faced away from bad weather. If your shed doesn't have a floor, you can lay saplings on the ground and stack your wood across them. This keeps the wood dry and lets air circulate around it. When stacking wood in your shed, mix the different types in each layer so they will be easily accessible when you build your fires.

Axes: Axes vary a lot in size and type. They are tools and like any good tool, having the right one for the right job means less effort and work for you. Using the right-sized ax lets the ax split the wood for you. We have a hatchet, two axes, a splitting maul, and two wedges. The hatchet is used for splitting kindling and trimming branches from

The festival has taken over my life; I feel anxious and excited, perpetually tense, tentatively hopeful . . . it's not so much the meetings, as the following up of every little detail, making sure each thing is covered. I feel as though if I relax for one moment, all the details I am remembering will slide away . . . I can't do anything any more without it leading to remembering something else that must be done, even chance words in the newspaper set me off.

This is usually the hardest time of year for me—fishing season—yet I haven't even been home enough to notice Peter isn't there. I work in the garden between going to work and going to meetings. Sometimes I don't even see the land from morning to midnight. But it feels good. I feel frantic and overworked and tense and so *alive*.

trees. The lighter (3 pound) ax will split up to a 12-inch-diameter log and almost any dry, straight-grained wood. It is easy to lift, handle, and aim. The larger ax (seven pounds) is something we grew into and still use less often, but its heavier weight will split larger rounds or closely grained wood with one swift blow, and sometimes it is a joy to feel it at work. The splitting maul could be substituted for the large ax in most situations. A maul is a combination of an ax and a wedge—it can be swung or it can be set in a log and driven with a sledge hammer. Ordinary axes should not be driven; their steel is insufficiently tempered and can splinter, causing a serious accident—even blindness. We use the two steel wedges to split very large rounds and knotty wood or to free an ax stuck in a difficult piece. We set them in the wood with a small hand sledge—the size of a hammer with a sledge head, which is easy to handle and does the job. We bought good-quality axes ($7.00 to $15) and have been glad we did. An ax is a lifetime tool. Good quality in an ax means a well-balanced head and handle, which is a pleasure to hold and swing, and good steel in the head, which will hold its edge through the years.

If the head gets loose on an ax, you can buy tiny wedges for a few cents apiece. They come in different widths and are hammered into the ax handle where it comes through the head, perpendicular to the head. One or two should tighten up your handle for another year's use. Handles are replaceable if need be, for three or four dollars. If your handle is getting chipped and worn just below the head, practice your swinging and aiming so you split the wood with the head, not the handle (this is where a well-designed head makes all the difference). If you're careful not to hit the ax blade into the ground (dirt dulls it pretty quickly) or to hit it on nails or rocks, it should last forever. A blade that has become dull or slightly nicked can be returned to new by sharpening it on a grinding wheel.

Using Wedges: Wedges are an easy tool for splitting very large or knotty rounds of wood. Always have two wedges, so that if one sets in the round without splitting it, you can get it out again (and split the round) by using the second wedge. Take a look at the round and see if there are any likely cracks in which to set the wedge. Always set the wedge across the grain of the wood and never across or through a knot. Knots are the strongest part of the wood and will not split, though you can

pound through them if you want to work that hard. Parallel to a knot is a good place to set a wedge, as the wood is curving around the knot naturally and is likely to split there. If the first wedge doesn't split the round, look for cracks beginning to form a pie-shaped piece within the round. Drive the second wedge into one of the new cracks and the round will split. Once you've reduced a round to halves or quarters, use an ax to get it down to stove size. Wedges are efficient tools, but driving them takes more energy than splitting with an ax does.

If your ax gets stuck in a round that won't split, you can free it by setting a wedge parallel and close to the ax blade. As you drive the wedge in, the grain will loosen and you will be able to pull the ax back out again.

Another technique for splitting rounds with wedges is adapted from the "checking" method used when you split posts. I learned this technique watching a friend split posts and use it almost every time I'm faced with a large or very knotty round. Draw an imaginary line across the round, avoiding knots as well as you can. Beginning at one end, drive your wedge in about an inch. Then pull it out (rocking it back and forth is a good way to work it loose) and set it further along on the "line." Drive it in again and reset. Continue until you've actually created the line you imagined. Set your wedge on the line a few inches from either end and drive it in. As soon as the wedge is well driven in, there will be a popping crack and the wood will split along the line. Even the most difficult rounds will usually yield to this method. It's a little slow but wonderfully effective, and it will give you the feeling that you can split anything!

Chopping Block: When splitting stove wood use a good, solid piece of wood as a chopping block and you will save both your back and your ax. In addition, you'll make splitting easier. The block should be big enough in diameter to hold your wood comfortably. It should be level on top and bottom so that it doesn't rock. The type of wood you use will vary according to your area—try to choose a block that won't split the first time you use it! The height of your block should be whatever feels most comfortable to you, depending upon your height, the size of your ax, and the lengths of wood you're splitting. If you are splitting in an area with sloping ground, always stand *uphill* from your block. This minimizes the chance of an inaccurate swing following on down to your leg. All of this may sound obvious, but I've watched more than one novice try to split with *her* wood on the ground. The ground is soft and gives, making the piece harder to split or letting it fall sideways when struck. If the ax misses and hits dirt, or passes cleanly through the round and hits the ground, it will be dulled and even nicked. A good solid block should be part of your wood-splitting equipment, as important as your ax or your wedge.

The last day of the women's festival—it has been everything I ever dreamed of and better—something I've never experienced before. To my surprise it hasn't been the activities that matter to me, not a specific workshop even if I learn something new. It's the feeling of this place that's so important to me, the simple fact of being here. It's a village, a culture, a world of women; for a few days we can live out our fantasies, make our own rules. I can't get over how beautiful women are. Even the plainest face seems beautiful as we look each other in the eye, meet as friends. There is such openness here, touching, hugging, laughing, dancing. Bodies move freely without fear of being adjectified or alienated. There are no leaders; everyone leads. Most of all I am amazed by these incredibly strong, confident women: Ann who has a garage in Oregon; Liz who has a cattle farm in the valley; Laura who helps run a five-acre truck garden. Their faces glow with self-assurance. They are tough and strong and gentle and real, and I want to know them well. Women together; women framing walls, fixing pumps, and chain-sawing wood; women cooking gourmet feasts for two hundred and washing pots; women singing and playing fiddle and dancing the Virginia reel. Women loving each other, loving ourselves. I wonder if this is just a magic world flourishing briefly in redwood groves, or if we can carry the magic home with us into ordinary lives. I hate to leave.

Staying home is so hard; I no longer know what to do with myself. The festival has left me raw—full of hopes, questions, possibilities, not knowing what to do. I wonder what I'm doing with my life. Why I'm living with a man who knows so little of my essential self and doesn't like most of what he does see. Why I feel so tied and bound, as though I have become his Siamese twin and there is no life without him. The world, of moderately happy couples—with animals and gardens they've never even wondered if they want—no longer feels the best of all worlds. Rural homestead land doesn't seem all that different from suburbia except one works harder. Do I really want to live this way, for years and years, being the least that I might be?

In Indiana
the corn comes up sharp
green
like shards of glass
broken glass
broken fields

this spring the whole country is
splintered

my sister walks beside me
grow up strong I tell her
lift weights
push up
have strong arms
strong legs
hands that can crush things

don't be like me
another weak woman
fingers like lace
a paper doll

my sister does not hear me
already she is dancing
barefoot in the corn
her feet turn the earth
her body is straight
hard as a cob
her breasts are small
unripe kernels
her feet turn the earth
my sister is dancing
her feet turn the earth
like the blades of a plow.

CREE FELLING

In my first years here, I never even considered cutting down a live tree. The trees were magical presences surrounding us. Everywhere I looked, they were majestically and simply *there*. I spent hours in their company, relearning a sense of quiet, recovering my sense of the holy. Giant, fire-blackened stumps stood everywhere on the land, remnants of the incredible redwood forests that once stood here. There is nothing that will ever replace those forests and nothing as eloquent as those stumps. They made me realize how much we can harm the land, sometimes in irreversible ways. In years of farming, trying to make a productive life on this land, I have never lost that initial sense of reverence and understanding. I hope that in sharing the following skill I can also share my belief that we must use such skills wisely and sparingly. If we must sometimes cut down trees to open space for growing food, for pasturing our animals, if we learn to harvest wood from the forests, if we can learn to cut and mill wood for building our structures—let it all be done with a great deal of careful thought. We *can* have the foresight never to destroy what will take lifetimes to replace or to change radically the natural balance of the land.

A few trees cut purposefully and used carefully will probably fill your needs without harming the land. Whenever you can, though, harvest dead and fallen wood for your stoves. Make use of suckers (small trees growing up beside and stunted by larger ones) for pole framing or fences. Recycle lumber whenever you can. Familiarize yourself with faster growing and more prolific species and use them rather than trees that take centuries to grow or have been made practically extinct. Consider replanting what you take. And remember that felling a tree is taking a life—an act that is serious and worthy of respect.

The trees that I have cut have been mostly fir and bull pine. On the average, they have been one to two feet in diameter at their base, and fifty to eighty feet tall. These trees have a lot of limbs and usually grow in fairly dense stands, often sharing their space with thick brush and smaller trees. For tools, I've used a chain saw, a large bow saw, a two-woman saw, and a hatchet in various combinations. The hatchet is for clearing away small limbs and brush; the saws are used for the actual cutting of the tree. You may want to use a large ax rather than a saw—this is a traditional tree-felling tool requiring a lot more arm strength and energy.

The first step is to clear away brush and lower limbs so you have a good clear work space. Then assess the surrounding trees and find the clearest space where you might fell the tree. You can pretty well judge the height of your tree and the amount of space it will need to fall clear. Do this carefully —a falling tree that gets caught up in surrounding branches or jammed against another tree can be quite a project to unlodge. When you have decided which direction you want the tree to fall in, you are ready to make your first cut.

The cuts you make are called kerfs. You make two cuts directly opposite each other. The tree will usually fall in the direction of the deeper and lower kerf. I say "usually" because there are a few rare cases in which a tree might fall backward—I have seen this happen only once, but it is a *danger* you should be conscious of. I always make my deeper kerf first on the side I want the tree to fall. The height at which you make your kerfs should be whatever feels comfortable for you. I like to work at about waist height. The tool you use is also up to you. A chain saw or bow saw works well if you are alone. A two-woman saw will be easier than a bow saw if you are working with a partner.

First cut a horizontal line at least one third through the diameter of the tree; then cut another line two to five inches higher, forming a forty-five degree angle with the first cut. The wedge piece between the cut will usually fall out when the two cuts meet. You may have to give it a chop or two with your hatchet to unlodge it. This first kerf may be deeper than shown in the diagram, but a cut one third the diameter of the tree usually works well.

Next, go to the side opposite your first kerf. Be sure you are careful to make your cuts *directly* opposite each other or the tree may twist when it falls. The bottom of your second kerf should be at least two to three inches above the bottom plane of your first. The larger the tree, the greater this distance should be. You will have to do some experimenting to get a feel for this. The bottom of this second kerf can be at a slight angle. I like to cut the notch a few inches deep first. (This keeps the tree from binding the saw.) Then continue the lower cut.

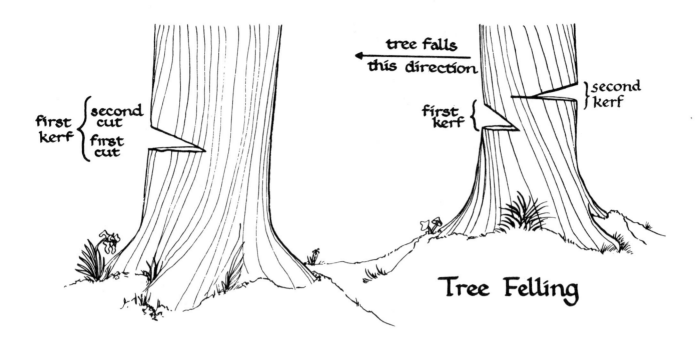

first kerf { second cut / first cut

tree falls this direction

first kerf {

second kerf

Tree Felling

If you have been using a chain saw, I strongly advise switching to a handsaw for the final cutting on this second kerf. You will want to be able to hear the first cracking noise indicating that the tree is beginning to fall. With a chain saw running, this is almost impossible. That cracking noise is the tearing of the fibers of the tree as it loses balance and begins to fall. You may have to cut a few more minutes, but then you can stand back as the tree tilts slowly and then topples. Always be aware that a tree might not act quite the way you expect—it might twist, bend, spring, or not fall quite the way you intended. Be ready to get out of the way if it begins to do any of these things. (Behind a good-sized tree might be your safest place.) A tree falls very slowly, majestically, and with a final resounding crack and thud. There is a moment for quiet contemplation before you begin the job of sawing off the branches and cutting up your wood.

CHAINSAWS

Chain saws have always been an approach-avoidance trip with me. I love their efficiency; I hate their loud and angry nature.

The first year I lived here, we cleared land and cut our firewood with handsaws. It was hard work and it took a lot of time, but we had that time and the work felt good. As winter approached, though, we realized we had in no way cut enough wood to last us very long, and we panicked. We asked a friend who had a chain saw to come over and cut up some of the big logs that we had laying around. These would have taken hours to saw through with a handsaw.

He came with his monster machine which fascinated and terrified us all. Our country peace was broken. The noise was deafening and overpowering. I was eating an orange and couldn't taste it; the chain saw demanded all my consciousness. The animals fled. We stood at a distance, fingers in ears, and watched the chain saw rip through the wood in no time at all, leaving us a three-month supply of warmth and comfort and hot meals.

Our pioneering spirit was corrupted; we wanted one of those machines. So, after much discussion, we decided to cash in some of our peace and quiet for this faster and easier method of getting in our wood supply and clearing our land. We (two women and two men) went together to pick out the chain saw. From watching how the men in the area related to their chain saws, we two women realized that it was extra important that we become adept at using one. The chain saw is a sacred male tool. Women are in no way expected to be able to use one. And many women have accepted that place out of lack of confidence—no one to teach them supportively, or no real desire to deal with such an ugly, intimidating machine. "The men cut the big wood and the women collect the squaw wood (kindling), get the fire going, and start dinner," one man glowingly told me how they did it in the old days and how it should be (and pretty much still is) done today. And everyone stays in their place.

This system seemed even more outrageous when I found out how easy it was to use a chain saw. At the store we all tried out carrying and maneuvering various chain saws. It is really important to have a saw that isn't too heavy or bulky to control easily. There's no reason to get a huge saw either, unless you're going to be cutting down and/or cutting up huge trees. A lightweight (9½ pounds, 20″ bar) saw can take care of most or all of your wood-cutting needs. A new small chain saw costs around $200 but it's possible to get a reliably rebuilt saw for half that, and it will work just as well. Ask in the store about used and rebuilt saws.

Chain saws run on a mixture of gasoline and two-cycle motor oil. The oil-to-gas proportions depend on the saw type and model. It is important to have the oil/gas mixture exact, or you can burn up the engine. Most saws have the correct ratio printed on the cap of the gas tank. They also need

oil to lubricate the chain; a light (not over SAE 30), cheap, non-detergent oil will do for that. When you pour in the gas or oil, clean off the saw and use a funnel so that you don't get sawdust in your engine. We're not always that neat, and I'm sure we've gotten sawdust in with the gas or the oil at times. It's not something to panic about, but it's not taking good care of the machine, which is something we're learning about the hard way. Take care of your machine, and it will last longer and run better. After you fill the gas tank, it's a good idea to move away from the fueling area to start the engine.

The starting mechanisms on all chain saws are basically, but not exactly, alike. On our chain saw, you flip the ignition switch to "on," push a little knob to the side of the throttle trigger that holds

the throttle open while you are starting, pull out the choke if the engine is cold, and then pull the starter cord to fire the engine. Once the engine starts, you press the throttle trigger which releases the knob holding it open. By then you should be able to also push in the choke. The throttle trigger controls the chain revolving around the bar. When sawing, hold the throttle all the way open. To stop the engine you turn the ignition switch to "off."

Now to cut wood . . . My first reaction to the chain saw was to hold it as far from my body as I could, to hold the throttle loosely, and to sort of lean over and lay the bar on the wood. If you are in control of the saw, there is nothing to fear. Stand sturdily, feet apart, with both hands firmly gripping the saw, one on the handle on top and one

holding the throttle trigger and its handle. Cut with the body of the saw close to the log rather than cutting with the tip of the bar. That way you have a firmer hold, pushing the saw against the log instead of suspending it. The saw can buck if the chain hits an obstruction or when you're cutting small branches, so always hold it firmly. Don't cut over your head or in any situation where you don't have complete control of the saw.

When cutting, the chain needs to be lubricated. Some saws have automatic oilers, others have manual ones. Ours is manual—there is a knob at thumb's reach of the throttle trigger that should be pushed every five seconds while the chain is cutting. When the knob isn't giving any resistance, that

means it is low on oil and time to add more. Chain saws with automatic oilers usually have a manual oil button. This is used when you need extra oiling —as when cutting very sappy or hard wood. If the teeth on the chain look brownish and burned, it means you are not oiling enough. Mixing kerosene with the oil will help when cutting really sappy wood.

When cutting a log, note how it is positioned and supported. If it might roll, stand on the uphill side. To cut a log lying on the ground, cut it two thirds of the way through, then roll it over and cut to meet your first cut, to avoid hitting the dirt— which will dull the chain. If the log you're cutting is too heavy or too long to roll over, cut a few rounds most of the way through and then cut one all the way through, carefully avoiding touching the ground. Then roll the whole section over. If a log is supported on both ends but free in the middle, it will close on the saw blade if you cut straight from the top. So, cut one third through from the top and then finish by cutting up from the bottom, using the top of the saw blade.

Cutting from the bottom like this is harder because you're pulling up on the saw rather than pushing. I don't feel I have as much control that way. If possible, I wedge something under the log near where I am cutting to push it up so that I can cut down from the top without binding. Binding the saw can mean having to chop it out of a big log with an ax, so figure out what the log is going to do before you cut it. Logs supported just on one end are the easiest for me to cut. Start from the unsupported end and just cut from the top. No binding

or hitting the dirt problems. In the books, they say to cut one third from the bottom and then finish from the top so that the log falls away from the saw blade. I've never had a problem with it falling into the blade by cutting straight down. Just watch where the log is going to fall and be ready to pull the saw out or kick the log away.

Chain saw bars come in two types: the roller bar and the hard-nose. The roller bar has a gear in its tip that must be greased every time you use the saw. I like the roller tip on a small saw as it takes some of the strain off the engine for easier cutting. The chain needs to be kept in adjustment too. On hard-nose bars, it should hang down 1/8" maximum from the bottom of the bar. On roller bars, the chain should fit snugly, but you should be able to snap it easily. The adjustment mechanism is different on each saw (on ours it's a matter of loosening two nuts and tightening a screw), so check your owner's manual. Always check the tightness of your chain before you start to saw. As you cut, the chain heats up and gets looser. Don't adjust the tension when hot or it may snap as it cools. If you're doing really prolonged cutting and the chain is too loose, you can tighten it, but be sure to loosen it again when you finish.

Taking care of your chain saw means your saw will last for years and really serve you as a tool. You should keep on hand a chain saw tool (combination wrench and screw driver), a soft bottle brush, a wire brush, a 7/32"-diameter round file, a file guide, and a grease gun, if you need one.

Most of the maintenance we do on our saw comes under the heading of cleaning. The air filter needs to be removed and cleaned after every few cuttings. Be careful as you lift it off to keep dirt from getting into the carburetor below it. The filter can be brushed clean or soaked in solvent (kerosene is best) and left to dry before reusing. The bottle brush is to keep the chain sprocket and oil holes clean. On our saw, we remove two nuts and lift the blade cover off. The end of the bar and the sprocket for the chain can then be brushed clean of wood chips and twigs. Make sure the holes in the base of the bar are clean. These let the oil flow onto the chain.

We use our wire brush to scrape sap or pitch off the teeth of the chain. A really dirty chain can be soaked in kerosene first. The wire brush is also handy to clean out the groove in the bar into which the chain fits. This gets clogged occasionally. While the chain is off in order to clean the bar, we usually take the bar off too and turn it upside down so it doesn't wear more heavily on the bottom side.

The most difficult and tedious chore connected with the chain saw is keeping the teeth sharp. For a long time we just casually filed away at them and occasionally took the saw to a local shop for a professional sharpening. Finally a friend taught us what to look for in a tooth, and we discovered we were only cutting with half our teeth! Since then, we've bought a precision filing guide to get the right angle and depth on each tooth. The cost equals about four professional sharpenings and means that *we* can keep our chain in perfect shape ourselves. The edge of the teeth should feel as sharp as a kitchen knife. When you cut, the teeth will throw out wood chips when they're sharp, fine sawdust when they're dull. The file guide will hold

your file in place to sharpen each tooth at the same angle and the same length (so each one cuts). It will also prevent your filing the tooth away. With the guide, you do all the teeth on one side of the chain, then switch the guide to do the ones on the other side. The guide is tedious to set up the first time, but ours has needed only minor adjustments since then. When you file your saw, tighten the chain more than you would for cutting so that it can't wobble or slip. (Always remember to readjust it when you're done.) Use the file in firm, smooth strokes—three or four will usually be enough. Don't force the file or keep carving away at the tooth. You're finished when the tooth shape looks right, has a shiny edge, and feels sharp to the touch.

Besides the teeth, the depth guides also need to be filed. They are small points that stick up between each two teeth and determine how deeply into the wood the teeth can bite. They should be kept longer (.35″) if you cut mostly soft woods; shorter (.25″) if you cut mostly hard woods. There is a cheap tool for measuring these accurately. At all times, these guides should be shorter than the teeth (watch this as your teeth wear away). The depth guides should be filed with a flat file and should have a slightly rounded shape.

Occasionally, you will have trouble starting your saw. This is usually just a dirty or worn-out spark plug, and we always keep an extra one on hand. If the carburetor or idle speed needs adjusting, you should consult the owner's manual for your model. These are really useful books that will help you with basic maintenance and simple repairs.

In general, take care of your chain saw. Have a long talk with whoever you buy it from, or get a manual on your model so that you know how to maintain it.

I feel really good about knowing how to run a chain saw. It's a powerful, intimidating machine that I am able to control and use to get the wood that I need to survive. And that's a new experience for sure, controlling a machine basic to my survival. In the city, some unknown entity controlled the heat, the water, the subway. But in the country the control is clearer—either you get a man to do it at the price of losing some freedom and self-respect, or you do it yourself.

Chain Saw Teeth

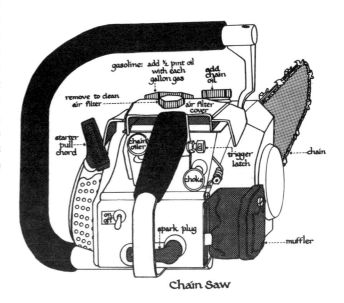

Chain Saw

the two-woman saw

A wonderful alternative to the chain saw or simple handsaw is the two-woman saw. This is a tool that is inexpensive, efficient, and a delight to work with. It is a lesson in harmony and co-operation as you work with a friend to cut your winter wood. Combining your strength and energy, you can cut through the biggest and hardest of logs. You can work in the forest without disturbing or polluting it. Most two-woman saws are unearthed in junk shops or antique stores. They are usually priced under ten dollars, but you'll probably have to rum-

mage a bit.

These saws come in various lengths, ranging from three to six or eight feet. The size of the logs you'll be cutting should guide you in choosing your saw. The steel of the blades varies considerably—an old Simmons Company blade, clearly and proudly stamped, is the best one we've found. Look for a fairly stiff blade as this will generally be the best steel. A rusty blade can be scoured down with steel wool and oil if it's not in too bad shape. The wood handles may be replaced if they are missing

or cracked. Your saw will probably need sharpening, and the teeth must be set at slight angles so that it can pull the sawdust out of the cut properly. Sharpening, and especially setting, should be done by a professional until you learn these skills—a sharp saw incorrectly set won't cut.

If you cannot find someone who will teach you this skill, get the book, *Old Ways of Working with Wood* by Alex Bealer (see Resources). The first chapter has pretty specific directions for sharpening and setting your saw.

When you go to work with your saw, you'll discover a rhythm with your partner. You should each, in turn, *pull* the blade toward you, then relax as your partner, in her turn, *pulls* it away. Trying to push the saw may cause it to buckle and bind. Work at an easy and non-tiring pace. You'll find that the energy and muscle for this work will come from your whole upper body, not just your arms. If you keep your saw well-sharpened (and oiled to prevent rusting), it will probably become one of your favorite and most-used tools.

WOOD HEAT STOVES

Ashley: An Ashley or good Ashley-type wood heat stove provides thermostatically controlled heat that will keep you consistently toasty through the bitterest cold. The Ashley burns slowly, can take large chunks of wood, and conserves fuel. It will pay back your initial investment by using much less wood, and it needs less attention to keep the fire going than the average heat stove. Properly banked, your Ashley will burn all night or all day if you are out. It cannot be used for cooking, though. The larger Ashley can heat a house with many rooms.

Log Burner: A log burner is a rectangular heat stove usually made of cast iron. Hence, it is durable and heat retaining. It will take a hefty chunk of wood and burn for hours once it has a good bed of coals. Logs may be loaded in through the front door or through the hinged top. This type of stove can be used for cooking, too, as the top heats up well and is spacious enough for a couple of pans and a coffeepot. The log burner will heat a good-sized room or a small cabin.

Homemade Log Burner: This homemade stove is second best only to an Ashley. It starts with little kindling and relatively large pieces of split wood. The damper and draft can be closed almost immediately after the fire is lit. It will warm a large, drafty room in fifteen to twenty minutes and will burn whole logs, unsplit, once it's going well. It is made of heavy steel and is long lasting. With a knowledge of welding, it can be made very cheaply. The inside of a hot-water heater that forms the body of the stove, the angle iron legs, the 1¼″-diameter pipe for the bottom draft, and the scrap iron for the stove pipe connection were all scavenged for free on this stove.

Pot-Bellied Stove: For the small room or cabin, a pot-bellied stove is a durable and safe heater. Made of cast iron, it has a good life span and holds its heat well. It will tuck safely into a corner and keep you cozy if you're willing to tend it. The small size makes it necessary to cut and split quite small firewood.

Tin Stoves: The main advantage of these stoves is their cost—they are by far the cheapest wood stove sold commercially, even though the cost has doubled in the last four years. Because they are made of fairly thin tin, they heat up very quickly to warm up a room. Their drafts close down tightly and they burn wood efficiently. They come in a range of sizes. The smallest requires much cutting and splitting of wood; the largest will burn whole logs with a good fire. The real disadvantage of these stoves is that they don't seem to last more than about four years, and possibly less depending on climate and conditions. They rust and corrode easily—internally from pitch and dampness in the wood, externally from dampness in the air. Small holes can be patched with asbestos "stove cement," but you can't build a whole stove top out of stove cement!

Franklin: The Franklin combines the heating efficiency of a cast-iron stove with the aesthetic advantages of the open fireplace. It may be used for cooking as well as heating. Front doors open to allow insertion of large logs and odd-shaped chunks of wood. Some Franklins also have side or top loading doors, and some expensive models have thermostatically controlled vents.

Roof Jacks: Important in installing a wood stove is having a safe way to vent the stove pipe through the ceiling and roof. The code-approved method is with an insulated, double-walled roof jack. These are virtually fireproof, but they are also very expensive. If you have a wall-board ceiling and a tin roof, or if your roof is exposed on the interior (so that you will notice smoldering before a fire starts), you may want to use a simple, single-walled roof jack. The safest way to use one of the inexpensive single-walled jacks is to buy one at least one inch larger in diameter than your stove pipe. That way you create your own double-wall insulation. Cover the stove pipe with a wide brimmed metal hat to prevent rain leaking through the air space. Friends of ours have also designed a homemade jack that is probably even safer than the single-walled tin one. It uses a piece of ceramic drain pipe wider in diameter than the stove pipe. The drain pipe is held in place on the inside of the roof by a homemade metal collar that is suspended from the rafters by three wires. With all three types of jacks, you must be careful to place metal flashing between the jack and the roof, providing as much insulation as you can between the hot stove pipe and the roof itself. Leaks around the jack may be stopped with "Wet Patch"-type roofing cement.

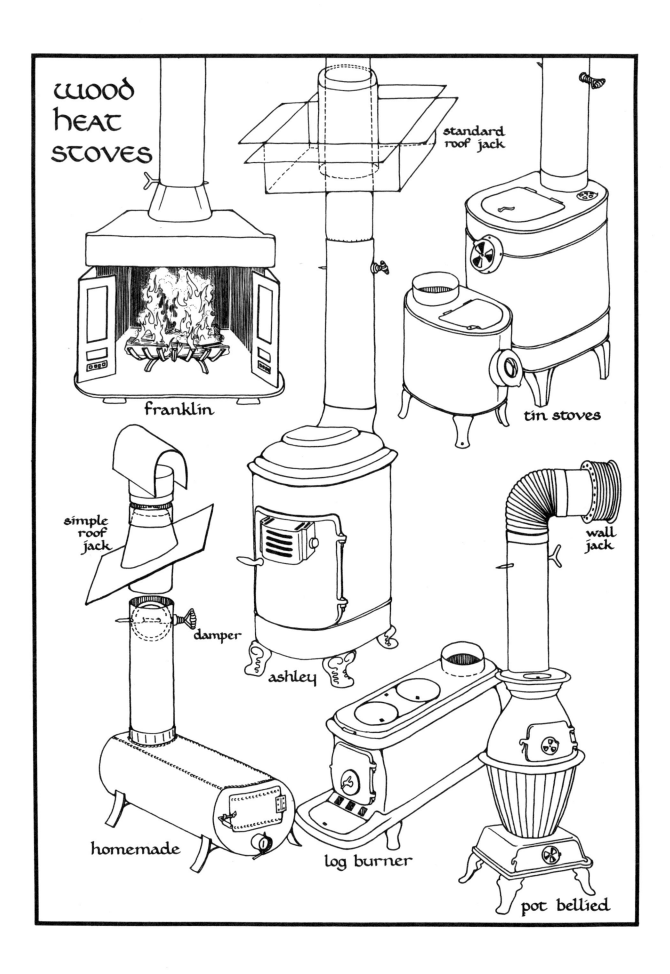

WOOD HEAT STOVES

standard roof jack

franklin

tin stoves

simple roof jack

damper

ashley

wall jack

homemade

log burner

pot bellied

CHOOSING A WOOD COOKSTOVE:

🍂 Make sure you have two dampers, one (called the draft) under the firebox (this one may be on the front or the side of the stove) and one at the base of the stove-pipe to circulate smoke and heat around the oven. This last is really important, make sure it is in working condition.

🍂 Check for rust to see if any parts are rusted through. Surface rust can be cleaned, rusted through cannot.

🍂 If your stove had water pipes in it which have rusted out, plug the holes with plumbing fittings and pipe dope so smoke doesn't escape.

CARE OF YOUR STOVE:

🍂 Rusty stoves: use steel wool and wire brush to clean the rust, then stoveblack to protect it. We tried grease but it was like living in a pancake house. Grease spilled as you cook will help protect the stove top.

🍂 To keep stovepipes free from accumulated soot and creosole (can cause chimney fires) put a few teaspoons of "red devil" in a hot fire each week. Also tap your pipes weekly to make soot fall down (elbows cake up the worst).

🍂 Use a spatula to clean out soot and ashes under plates on stove top and down along the side of the oven and in "clean out" below oven. Keeps the stove from smoking and makes the oven heat well. Do it monthly!

USING YOUR WOOD COOKSTOVE:

🌿Have a wood box for kindling and firewood near the stove. Dry wood in the oven with door open.

🌿If you need help getting fires started, use kerosene, never white gas or gasoline which explode and can burn you badly.

🌿Start the fire with fir or redwood (softwoods) to build heat quickly, then for baking use oak, madrone, apple or other hardwoods with damper shut.

🌿Begin your fire with dry kindling and work up. As it gets hot, shut the draft. When it's really drawing well, shut the damper for the oven. Then play with dampers - sometimes reopening the draft part way causes a better draw, sometimes all shut are fine (fine means no smoke, the more that are shut the more heat and the less wood). With a really good fire I also shut the damper or stovepipe.

🌿Stoves in good condition may have water pipes in back. these can be attached to a water tank for free hot water.

🌿The greatest joy of woodstove cooking is the total range of heats available to you all the time you're cooking. Just slide your pan across the top from hot to warm. Use warming ovens up top for rising bread, making yogurt and warming food. The oven heats evenly from all four sides and bakes better than any I've ever used.

🌿Simple, beautiful, ecological: food, heat, warm water always there. What other work of art nurtures me so well?

🌿The small and large plates on top can be removed and pots put directly over fire for faster heating.

🌿Many stoves have a water reservoir hanging from one side which keeps water at perfect dish washing temperature, whenever you want it.

NON-ELECTRIC LIGHTS

Living without electricity turned out to be much more pleasant than I expected. Though I missed having an electric water pump and was glad when we finally got the power for one, adjusting to an ice box and non-electric lights was easy. The first lights we bought were ordinary, old-fashioned kerosene lamps. They were inexpensive and cast a soft, glowing light. The light was so soft, in fact, that I began to understand the old dawn-to-dusk life-style. I had trouble at first with the globes getting all blacked up. One night watching a cowboy movie, I had a revelation. In the movies, they light the lamp, then gradually the room gets brighter.

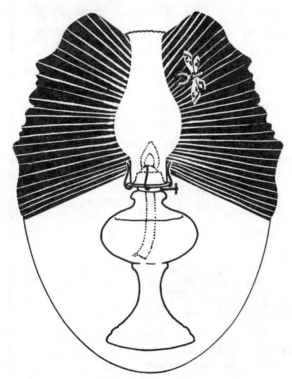

Kerosene

Aha! If the wick is *down* when it's lit, there's no black smoke when you replace the globe! Yes? Yes. Later, still, I learned that the wick needs to be trimmed every week to give maximum light. The shape of the wick also determines the shape of the flame, affecting the brightness of the light and blacking on the globe. I've heard tales of old-fashioned butterfly trims and such, but a gently rounded peak in the center of the wick has worked

best for me. This gives a high flame for bright light that burns straight up the chimney. To trim a wick, you remove the globe and flip open the hinged section of the wick holder to expose the wick. Cut off all the charred parts with scissors and then clip the wick to the shape you want. The wick should always be hanging in kerosene and should be replaced (just wind in a new one) when it gets too short. Placing rocks in the bottom of your lamp means it will take less kerosene to fill it when the wick gets short.

Much as I liked the aesthetics of those simple lamps, I found that I really did need more light at night for any sort of work. So I bought an Aladdin kerosene lamp, which uses a mantle to reflect as much light as a hundred-watt electric bulb. These lamps come in all kinds of models—brass and aluminum, hanging and table, glass- or paper-shaded. They range in cost from $25 to $40, the difference being in beauty, not quality. I bought the hanging type and prefer it to the table type as a light source since the light shines down on what you're doing and the mantle is above your line of vision (no glare). The problem with the hanging model is

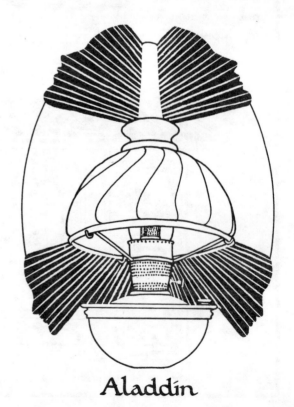

Aladdin

that it hangs down two feet—and from my seven-foot ceiling that means that just about everyone bumps his head on it—regularly shattering the mantle and occasionally the chimney. A friend who has a table lamp says her mantle lasts about four months.

Basically the Aladdin is just a kerosene lamp with a round wick, a tall chimney, and a mantle to reflect the wick's light. It is simple to assemble and requires little care, except wick cleaning and mantle replacement. The mantles ($1 each) are very fragile and shatter easily if poked or knocked. Because all Aladdin parts are imported, they're sometimes very hard to get. I now keep a stockpile of three or four mantles on hand against such emergencies. The Aladdins are very economical to run—mine goes about twenty burning hours on one-third gallon of kerosene. And they do give off a lot of light, more than you can imagine from a kerosene lamp. They're brightest if left unshaded, but the mantles glare without the shade. The Aladdins also give off heat and on very lazy and not too cold nights, I have kept myself warm by sitting near it. This is just about their only potential danger, as they can burn anything just above them. The hanging models come with a metal reflector which must be hung above them to prevent the ceiling from catching fire.

Other than that, Aladdins are relatively safe, much more so than ordinary kerosene lamps since they cannot break and are designed not to spill when tipped over. Kerosene is also less flammable than white gas. A kerosene fire, if you ever have one, should be put out with flour, cornmeal, dirt, or milk, but never with water, which causes an explosion.

The Aladdin mantle will turn black with carbon when the lamp is turned too high (the wick should be turned up till the whole mantle glows white, but no points of flame stick through). The carbon can be removed by turning the lamp down very low and waiting until it burns off. It will also burn off while the lamp is still turned up if salt is sprinkled down the chimney. However, I've found the accumulated salt is badly corroding my burner, so be sure to clean it off as soon as you blow the light out. After a while, you will learn how high the lamp can go and will rarely have trouble with carbon.

I loved my Aladdin from the time I got it and never begrudged the initial investment. But after several years, I have several complaints too. The mantles seem unnecessarily fragile to me and I

question the planned obsolescence of them. The winding mechanism on mine slips, and I know others with the same problem. My shade has come unglued and fallen apart. They don't seem made with the kind of care I would expect in a thirty-dollar lamp. They are also just tricky enough to use so that unsuspecting visitors shatter mantles and sometimes chimneys (which crack if not firmly in place). It gets costly replacing mantles in addition to buying kerosene. But despite the complaints, I can't begin to express what joy it is to have adequate light that is easy to install, can be moved to whatever room you're working in, and is comparatively inexpensive to run.

After a year of just kerosene lighting, we installed propane gas lamps in the kitchen. These are relatively expensive and take some effort to install, but once done, the lamps are in place forever and need no further work. They give off a very good, bright, non-glaring light, and they are cheap to run

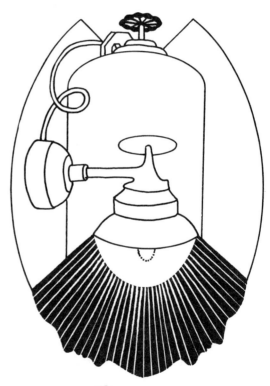

Propane

(two lamps burning about five hours a day use fifty pounds of gas a month—a half tank). Propane lighting is also very safe since the lamps are not moved for refilling (you just connect a new tank to the whole system) and they cannot be tipped over. I recommend them very strongly to anyone who plans on maintaining a non-electric life-style.

The gas lamp fixtures are usually mounted on the wall and have glass shades. They also use a mantle (much less fragile than the Aladdin) that costs $.75 each and lasts for months. New fixtures cost about $15. Used ones are hard to find but are sometimes available at trailer parks or where used truck campers are sold. We bought both a new and a used one and have found the new one (a slightly different design) gives off more light than the old one.

The directions for installing propane lamps may sound difficult, but it is really only unfamiliar, not hard. To connect the lamps, you need a gas tank and a regulator. You can either buy a small one or rent a 100-pound tank (with regulator) from a gas company. They may deliver refills or you may have to go to town to get them (depending on the condition of your dirt road and how far up it you live). The tank is connected to the lamps and anything else (stove, gas refrigerator, water heater) by flexible copper tubing. The lamps need ¼″ tubing and everything else needs ⅜″—so the main line (to major appliances) should be ⅜″. Try to plan the placement of your appliances and lights for the most economical use of copper tubing since it is expensive.

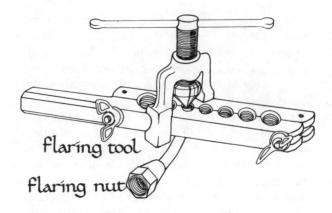

flaring tool

flaring nut

pipe cutter

The copper tubing attaches to everything else (the regulator, fixtures, itself) with what are called "flare nuts." These are threaded on one end and open on the other. The nut slides onto the tubing with the threaded end toward the end of the tubing. Then you use a flaring tool (a simple hand vise) to flare the end of the tubing into a trumpet shape. (I borrowed the flaring tool from a plumber friend, but they may also be rented.) Then bring the nut down to the end of the tubing. The flaring has made for a tight, gas-proof seal between the tubing and the nut; it also prevents the nut from slipping off the tubing. The threaded end of the nut can then be screwed onto the tank regulator, light fixture, or other appliance. Be sure when you buy flare nuts to get the right size for your tubing; both come in ¼″, ⅜″, and ½″ sizes. Gas fittings thread the opposite of all others, so they tighten counterclockwise and loosen clockwise. Use a wrench to tighten the flare nuts, but be careful not to mash the soft copper.

If you have a ⅜″ main line, you will have to narrow down to ¼″ for the lamps, using a T adapter. To make the connection, cut the ⅜″ line and fit flare nuts on each end. These nuts screw onto threads on the T. The ¼″ line attaches to the stem of the T, also using a flare nut.

After you have made all the connections, turn the tank on using the valve on top. Light a match at each connection to make sure there are no leaks. If there is a leak, retighten that connection. If that does not work, turn the tank off and disconnect that nut. Put pipe joint compound ("pipe dope") on its threads for a better seal and make sure the flare nut is tight against the tubing when you reconnect it.

The only other type of bright, efficient non-electric light is the white-gas lantern (most are made by Coleman). They are more dangerous than kerosene or propane lamps, as white gas is explosive, and these lamps will leak if they fall over.

Coleman lanterns come in single and double (more light) mantle types. All models will either stand on a table or hang from above; special wall brackets are also available. They use about two gallons of white gas a month with constant use. Most makes of mantles can be used with these lanterns, but the Coleman brand, silk ones, seem to last the longest. They are smaller, more compact,

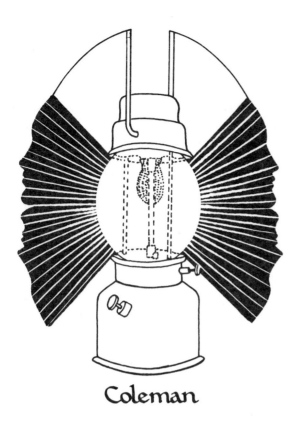

Coleman

and much less fragile than the Aladdins. The top shade tends to throw light sideways and downward at a 45° angle, so the lamp gives the best light when hung overhead. About every six months it's advisable to clean the fuel injection mechanism on the lamp and occasionally (every year or so), the generator will need to be replaced.

I've never owned one myself, but a friend who has one wrote: "I like to use the Coleman when I need to see very well (intricate sewing, reading, etc.) at night, but eventually the noise of the generator makes me turn it off. Because a Coleman operates on white gas with a pressure-pump system, there has to be a build up of pressure created before it will ignite. This pressure has to be maintained (by pumping) if the lamp is to stay at its brightest. Naturally, this creates a constant hissing, somewhat like a small blowtorch. The Coleman's bright light can be compared to a 30- or 40-watt electric bulb; its noise to a miniature steam engine. If you're the sort of person who prefers the quiet life, this noise may eventually get to you."

FARM TOOLS

Heavy-duty Stapler: This is also known as a staple gun. It differs from the office type in that the staples shoot into a surface (like a nail) rather than fold and clip things together. This stapler is invaluable around the farm for tacking up lightweight tar paper, attaching plastic to wood, putting up fiberglass insulation, and so on. The gun must be pressed flat against the surface for the staple to set firmly. It's difficult to use this tool overhead. Sears sells a good, inexpensive model.

Fence Pliers: This tool is absolutely indispensable for fencing projects. It can be used to cut the wire, hammer or remove fencing staples, and even to do some light fence stretching. In the barn, the fencing pliers are useful for cutting baling wire and pinch-hitting for the conventional hammer.

Draw Knife: A tool designed for shaving wood, the draw knife can be used for planing and carving, as well as stripping bark from poles. It consists of a one-edged blade about ten inches long with a curved handle at each end. Use it by taking each handle and drawing the blade toward you with a gentle downward pressure. The amount of pressure will control the amount of wood or bark shaved. Keep your stroke smooth, short, and controlled to avoid cutting yourself. The most elegant draw knife I've seen advertised is in the Woodcraft Catalog (see Resources), selling for around twelve dollars.

Rope: Rope comes in a multitude of grades, sizes, and types, and is the one thing you seem never to have enough of. Nylon and plastic ropes are difficult to tie well—they slip, stretch, and are stiff. Once a knot sets in them, you will have a time getting it undone (especially if your horse is at the other end of the rope!). I advise soft hemp ropes, or good-quality cotton ropes (keep these off the ground—they rot).

Tin Snips: With a pair of tin snips you can cut flashing for your roofing projects, cut aluminum gutter pipe, and cut roofing paper without tearing it. They look like a pair of gigantic scissors with wide, flat blades.

Spading Fork: For turning the soil in your garden, use a spading fork. It is basically a pitchfork with wide, sturdy tines. It will break up the soil as you work because the soil falls between the tines in small pieces. If you use a shovel for this same work, the soil stays in large clumps, and if you just turn these clumps, you are likely to end up with your topsoil underneath your subsoil. The spading fork is also a useful stable and chicken or goat house cleaning tool if you use straw bedding. The lighter pitchfork is properly used for pitching dry hay and straw.

Pick Ax: This one is for breaking up rock, hard clay, or packed sand and gravel. It was absolutely necessary when we hand dug our well and put in water lines, but I've used it very little since then.

Mattock: There are two types of mattocks: the pick mattock and the cutter mattock. The pick mattock has a broad blade at one end and a pick-type blade at the other. The cutter mattock has two broad blades, but one is horizontal and the other vertical. The mattock is invaluable for breaking up hard earth, turning sod, and clearing new ground.

Round-point Shovel: A short-handled, pointed shovel is perfect for digging in narrow spaces, like new outhouse holes and wells. I use a long-handled, pointed shovel for digging and filling in holes and trenches.

Scraper and Square-point Shovel: These tools are also useful in cleaning out your animal houses. The scraper is a blade about a foot and a half across and ten inches high with a long wooden handle attached. It is rather like a cross between a snowplow and a hoe. It is used to scrape the floor clean once you've removed the dirty bedding or litter with your square-point shovel and spading fork. The square-point shovel is basically for scooping up things rather than digging in the earth.

Wheelbarrow: There are two types of conventional wheelbarrows; the contractor's type and the garden type. The contractor's wheelbarrow is built for construction work, so it is deep, sturdy and heavy. It has enough capacity and strength for carrying great loads of compost, split firewood, and feed sacks. The smaller, shallow garden wheelbarrow must make two or three trips for every one the contractor's wheelbarrow makes. It is lighter and

FARM TOOLS

hay hooks

fence pliers

draw knife

tin snips

rope

pitch fork

scraper

staple gun

hoe

rake

pick axe

crowbar

pry bar

mattock

post-hole auger

post-hole digger

spading fork

round-point shovel

square-point shovel

conventional wheelbarrow

Vermont cart

cheaper, but seems to make the work endless. My favorite wheelbarrow is an old handmade wooden one that came with the farm. It is old and crotchety and lives in the garden, where it does light work hauling weeds or transplanting soil. Its simply crafted wooden presence is a reminder that we *could* get along without Sears . . . two-wheeled garden carts of the type found in most hardware stores are, in the experienced cautions of gardening friends, not worth buying. They are difficult to maneuver, have a small carrying capacity, and are not very rugged. The handsome Vermont Cart discussed below is an exceptional species of garden cart.

The Vermont Cart: This is a large, sturdy, perfectly designed cart that has a million uses around a farm or homestead. It is widely advertised and, astonishingly enough, every bit as good as it claims to be! It will take an incredible load of garden compost, stable cleanings, firewood, baled hay—or what have you—and roll along merrily with its heavy load. Because it is so well balanced, it doesn't take a lot of effort or muscle to push or pull (no strained backs or weary arms). It tips up for easy loading or storing. This cart is initially quite expensive, but worth saving up for. It is one tool that will earn its keep and be around for years.

Hay Hooks: With a pair of hay hooks you can move and stack heavy bales most efficiently. The hooks are short, thick, curved metal attached to a wooden handle. You can move a bale by sinking one hook deeply into the hay and pulling or tipping the bale.

Don't try to use your hooks on the baling wire—the wire will most likely snap at an inopportune time, and you'll land on your back with hay spilling everywhere. If you have a strong back and lift properly, you can use a pair of hooks to lift and roll a bale into place . . . or you can work with a friend, each handling one hook.

Posthole Diggers: The clamshell-type posthole digger is really an elaborate special shovel. To use this tool, take it with both handles closed and slam it into the ground. Pull the handles apart and the shovels will close around the earth and lift the dirt up and out of the hole. As you dig the hole deeper, enlarge the area of the hole so that you can still pull apart the handles when the shovels are in a foot or deeper. Rotate the shovel now and then to dig a uniform hole. The auger-type digger should be used in loose, loamy soils. It bores into the earth and spirals soil up from the bottom of the hole. Very efficient in the right type of soil, this digger is nevertheless limited in use.

Pry Bar: This is a four- or five-foot-long solid iron bar with one end flattened to a point. The pointed end is used for lifting (using the bar as a lever) and supporting heavy objects like logs or well rings. It is very useful for turning big logs when sawing. The rounded end of the bar is used for tamping (packing down) dirt in postholes. A crowbar is a shorter, curved bar which is used for taking old boards off of walls, prying out large nails, and digging trenches in tightly confined places (under stable walls, for example) for drainage.

SECTION II
TAKING ROOT

hammer and nails

A three-year-old friend of mine was visiting for a few days recently, and while trying to put wood in the stove he realized he couldn't reach the nail where we keep the stove handle. So, without a second thought, he picked up his hammer (which he had brought with him along with his other toys) and a nail, and put up the hook for the stove handle that was at the right height for him. I watched quietly, amazed.

I thought back to a little over a year ago when I moved out to the country and, with much more hesitation than my friend, set out to take care of my needs. My first experience with hammer and nails was a disaster. I was trying to build a water tower out of oak branches, and probably bent four out of every five nails trying to pound them through the wood. I felt incompetent and inept. And I was. At twenty years old, I had handled a hammer less successfully than my friend.

Now, a year's worth of hammering (and sawing and measuring and chiseling and screw driving) later, I feel more able to build what I need. It took practice and watching people around me and picking up useful hints. There's a lot more to building than just hammering nails, but it's a good place to start. These are some useful things to know:

To drive a nail, hold the nail in your left hand (if you're right-handed), and tap directly on the head of it a few time with the hammer to set it. Hold the nail at a slight angle—it is less likely to bend and it makes a stronger joint. Then, take away your left hand and hammer it the rest of the way in. It's most important to hammer directly on the nail head, so that the nail doesn't bend. This sounds simple and it is, but it requires developing co-ordination and a sense of the hammer; its weight and its force. It is best to hold the hammer near the end of the handle. This gives you more leverage and, therefore, more force. I still feel I can aim better, however, holding it closer to the middle. Always hammer with the face, not the cheek of the hammer. The face is hardened for the purpose of hammering; the cheek is much weaker.

To hold nails too small to hold with your fingers, put them through a piece of cardboard, set them with the hammer, and then tear the cardboard away.

When nailing into hard wood, it helps to dip the end of the nail in paraffin or grease. Also, driving the nail slowly, with a series of lighter-than-normal taps, will keep it from bending in hard wood. Under certain conditions, you may have to pre-drill your nail holes in hard wood.

If you're doing fine work, use a bell-faced hammer. It has a convex face so that the nail can be driven flush to the wood without leaving hammer marks. Or, you can use a nail set. Drive the nail in as usual, then place the nail set directly on the head of the nail and hammer down on it, driving the nail about $\frac{1}{16}''$ under the surface.

To pull a nail out, use the claw of the hammer. Curved claws are best for pulling nails; straight claws for ripping boards apart. Slip the claw under the nail head, claw pointed away from you. Then pull toward you. When drawing a long nail out, put a block of wood under the hammer head to give you better leverage. Try to pull a nail out with, not against, the grain of the wood; going against the grain rips the wood. If you have trouble pulling a nail and you are not worried about doing fine work, you can bend the nail back and forth, using the claw, and break it off.

Another alternative with a difficult nail is to bend the hammer side to side, gripping the shaft, rather than pulling the head of the nail. The bending will draw the nail up bit by bit. This is especially useful with really long nails that are set deep into the wood. If you have to remove roofing nails, a small lightweight hammer will pull the nails out without damaging the broad heads. A heavy normal hammer usually tears the heads off.

The strongest kind of nailing you can do is clinch nailing. Hammer a long nail through both boards, bend the end over and hammer it into the surface.

Anchor nailing is also good when you need strength. Drive nails into board at opposing angles. Any time you are using more than one nail, in order to make a stronger joint, anchor nail them.

Whenever you cannot hammer straight through two boards to connect them, as when you join an upright to the floor, hammer the nail at an angle through one board and into the other. This is called toe-nailing.

Use the right nails for whatever you are doing. Different types of wood receive nails differently. Hard or close-grained woods will be more apt to cause nails to bend. You can try using heavier (stronger) nails or lighter (may pass through the grain easier) ones. Old, recycled wood may split a lot and must be handled gently.

If the wood splits when you drive a nail into it, you need a thinner nail. Also, the blunter the nail, the less chance of splitting. Blunt nails bore through the wood; sharp ones drive like a wedge between the grain. You can turn your nails upside down and blunt the ends with a hammer or clip the tips with a pair of pliers. However, blunt nails won't hold as well as sharp ones.

If you are driving a nail in and it begins to bend, you have two alternatives. The first, which often works, is to straighten the nail as well as you can and then hammer it *very softly*. For some reason even a bent and reluctant nail will be persuaded by this treatment. The second alternative is to pull the nail and set a second one in a slightly different spot. Usually when a nail bends it means that there is some difficult grain or a knot in your board. Driving a nail into a knot is almost impossible—the grain of the wood is so close, it becomes like iron and is often brittle.

Nail sizes are measured on the "penny" system.

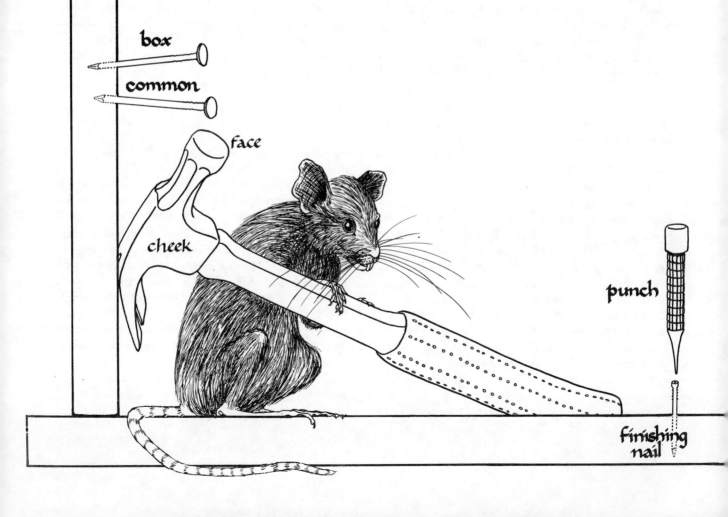

box

common

face

cheek

punch

finishing
nail

Penny is written as "d". They are also divided into box and common types. A box nail has a thinner shaft but the same length as a common nail of the same size ("d").

You can use common flathead nails for rough, small-headed finishing nails for fine work.

Galvanized nails are coated to make them rust resistant and should be used on all outside work. They look dull rather than shiny. Don't put them in your mouth; they're poisonous.

Ring-shank nails, which have a threaded shaft, are good when you want something to hold well. After hammering together our horse's feed box for the second time, I realized that ring-shank nails would be ideal for building animal feeders which take a lot of wear, tear, and pressure. They're really hard to pull out, though, so you can't make mistakes.

Learning to use hammers and nails, as well as all other woodworking tools, is not difficult; it mainly takes practice and demystifying for yourself tools and the process of building. Play with tools; get used to how they feel; get comfortable with them. And, please, let's give tools to our daughters to play with too.

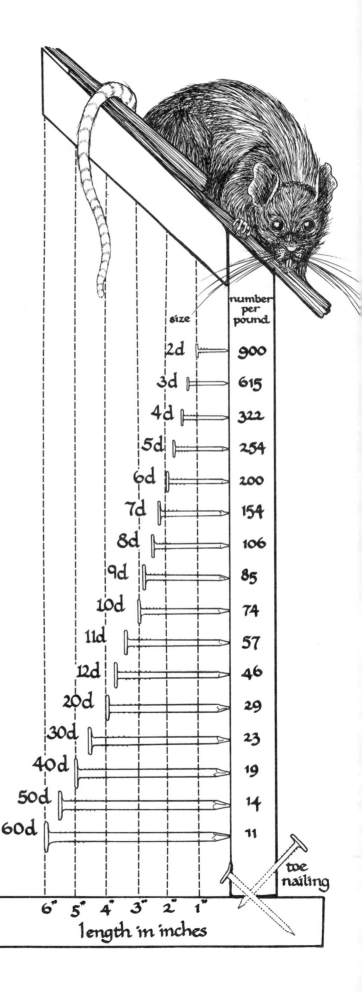

size		number per pound
2d		900
3d		615
4d		322
5d		254
6d		200
7d		154
8d		106
9d		85
10d		74
11d		57
12d		46
20d		29
30d		23
40d		19
50d		14
60d		11

toe nailing

6" 5" 4" 3" 2" 1"
length in inches

CARPENTRY HAND TOOLS

It's difficult to describe even a set of tools for the most basic carpentry job. There are a number of books that could be a better guide. Sometimes the first chapter of a carpentry book will list major tools and illustrate them. And there are books devoted exclusively to tools; the Resources section lists some, but check your library or bookmobile for them. But it's a jump from the story book picture of an ideal tool box to your own needs for your own job, to yourself standing in the hardware store where there are fifteen different kinds of hammers, different prices, *et cetera*. I was thinking about the tools with which I did my first carpentry work—how I could carry them all with me at once (now I have three tool boxes). I used an old leather shoulder purse of mine as the tool purse for small tools and carried my one saw and two crowbars in the other hand. There are a few indispensable tools with which you really do most jobs. After that, you can always use more tools, but they veer toward convenience and then toward luxury—unless you are drawn into carpentry as your full- or part-time work. This short list attempts to describe the essentials, and what to know and to look out for when you're buying or acquiring them; also, when to invest more money and when you might as well get a cheaper model. Along with the consumer-guide slant, I have mixed some information about using these tools.

Hammers come in curved-claw or straight-claw styles, different weights (sixteen and twenty ounce, most common), and with handles made of varying materials (wood, steel, fiber glass). Trying out different hammers and getting your own sense of how they handle is a good idea—it is surprising how much difference all these variations can make in how well this tool works for you. The curved-claw, sixteen ounce is most ordinarily used as an all-purpose hammer. Some people prefer the straight-claw hammer for ordinary use: it's not quite as efficient as a nail puller, but the angle of the claw is handier for prying, for reaching into corners or spots where the curved claw cannot reach. As far as weight is concerned, for sixteen penny nails or · larger, a twenty-ounce hammer should be used. It does more of the work for you,

although at first it may seem heavier to lift. Usually, grasp the hammer at the end of the handle, not in the middle: this chokes the arc of the hammer and gives less force to the blow. Practice this way will give you accuracy—don't expect to do perfectly at first. It is a very good habit to acquire, though, well worth some diligence. Some carpenters use twenty-two-ounce hammers, which are even better at sinking sixteen-penny nails. But for finish work, small nails, and small hammering spaces, a sixteen ounce does the job better than a bulkier, larger hammer. Wooden handles often feel good to the hand, seem to mold themselves to the hand. They can break, but are replaceable. Rubber and steel shank, leather and steel shank, or other such combinations are more durable. Wooden handles do seem to absorb more shock.

The Estwing brand (rubber and steel shank) is particularly good (my favorite so far in several years of carpentry experience). The main thing be-fore investing is to try out a few different kinds of hammers first, and then decide the best partner for your own hand.

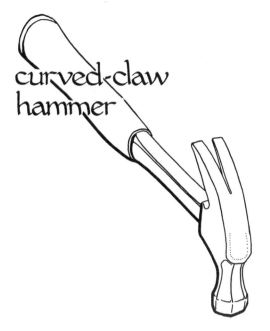

curved-claw hammer

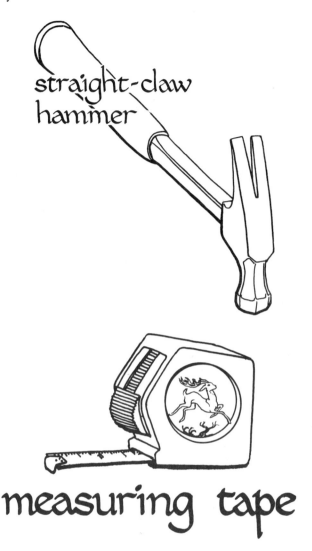

straight-claw hammer

Measuring Tools: A good sixteen-foot tape is worth investing in and will last you for several years with care. Tapes come with or without power lock—a button or other mechanism to hold the tape out when you don't want it to wind up imme-diately. Instead, you lock it and this frees your hand to do something else; the tape stays extended. This lock can also be used as a gradual stop when you are letting it wind up—preventing the spring from making the whole tape wind up too fast, from snapping and breaking, or from snapping on your finger. Some of the best brands in tapes are Stanley

measuring tape

Walking to the bathhouse today, holding my new twenty-ounce hammer, I suddenly understood the *Whole Earth Catalogue* meaning of "tool." I always thought tools were objects, *things:* screw drivers, wrenches, axes, hoes. Now I re-alize that tools are process: using the right-sized and -shaped object in the most effective way to get a job done. It's having a well-balanced hammer *and* knowing how to hold and swing it that makes it a *tool;* changes the whole work process from a struggle to a pleasure.

and Disston. Luftkin too, but their power-lock mechanism tends to stick or to be hard to press down. It's little things like this that add up to really enjoying working with a tool and feeling smooth about it. My current preference is for the Disston.

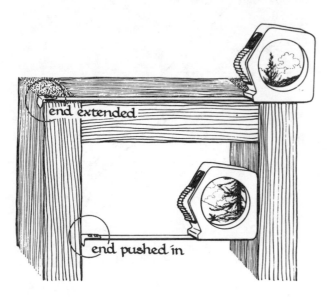

Get at least a twelve-foot tape, if not a sixteen—others are simply too short for measurements of whole boards, whole walls, etc. A twenty-five-foot tape is sometimes handy, but I find it too bulky for everyday use. Also, it's not needed that often.

Tapes usually have a *movable tip* at the end (the beginning, actually—anyhow, where it starts at "O"). It's important to notice how this functions in order to get accurate and consistent measurements. There are inside and outside measurements, terms which refer to the manner in which a measurement is taken. The tip of the tape gets pulled out for outside measurements (e.g., a board of a wall), and stays pushed in for inside measurements (e.g., between two studs, between two walls, or between any two points).

Combination Square is the only other tool I can think of that is used as often as a hammer, tape, or saw. Its main use is to get a straight line drawn across the stock (piece of wood) where you intend to saw. The square is held firmly up against the edge of the stock, marking a perpendicular line across the wood. If there is a knot or other irregularity right at the spot where you are holding your square, this may throw the 90° off. In such a case you can flip your square over or move it to a different spot altogether. If you want to be absolutely sure about a square cut, such as on a large

piece of wood ($4'' \times 12''$ or even a $4'' \times 4''$), lines should be drawn on all four planes and should meet at the corners. The combination square can also be adjusted to different lengths (it's twelve inches long).

combination square

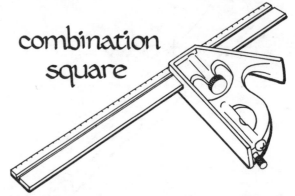

Note: always draw your line on the outside edge of the blade, it is the most accurate. Also, when sawing off the board, its helpful to mark an X on the side of the wood which is to be discarded (the sawed end). Usually a measurement is taken to *in-*

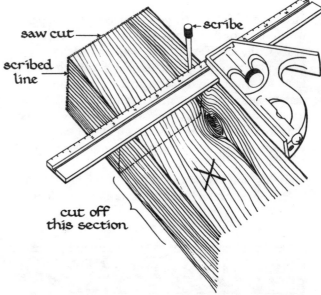

clude the scribed line (as a way of being precise and uniform). Thus, when sawing, the line is left and the X is sawed off. If the line were sawed off, the board would turn out less than the exact length.

If the X piece is at all long or heavy, it's wise to hold it up near the sawing point (being careful not to have your, or anyone else's hand anywhere near the saw). Otherwise the piece, because of its own weight, may tend to split off, leaving you with two ragged pieces of wood or causing your saw to bind.

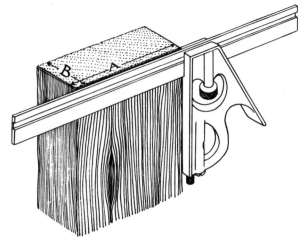

After you have made the cut, the combination square can be used to check the squareness of the cut by simply holding it up to the wood, in both planes A and B.

The Stanley combination square is the best on the market; it is made out of sturdy metal and adjusts smoothly. There are various cheaper versions that look much the same, but the danger in a square of flimsier metal is that it bends, gets smashed and out of shape much sooner and easier. Once bent, even slightly inaccurate, it is a nearly worthless tool. It is a tool to be handled carefully, never thrown down or left where it could get walked on. Unlike hammers, there are no aesthetics involved in choosing a square. The way it feels is not much of an issue. It should be a strong, precision tool. With hammers, it could happen that you'd choose an inexpensive one and use it with ease or that a more expensive model would feel clumsy and stiff for you. In the case of the combination square, however, the more expensive tool is unquestionably worth its price.

Handsaws: There are two different types of saws and if you look at their teeth you'll see the difference. The cross cut saw is used to cut *across* the grain of the wood and is an indispensable tool. The teeth are shaped like knife points and cut in both directions. It is also used for making diagonal cuts. The rip saw is used to cut *with* the grain of the wood; it is used less often but is the only tool for the job. It has larger teeth placed at right angles to rip the board as well as cut it. Such a saw produces a coarser cut. Hardware people talk about "points" when referring to saws. The size of a saw is measured by the number of such points per inch. There is always one less tooth to the inch than there are points. Fewer teeth per inch makes it easier to cut and makes a coarser cut. More teeth per inch makes a smoother cut but takes more work. If you look closely at the teeth on a saw again, you'll see that they aren't perfectly flat. They are, in fact, set or bent alternately to one side and then the other. The channel or cut that the teeth make is called the kerf, and is as wide as the set of the teeth.

Like any other cutting tool, saws get dull and need to be sharpened to do a good job with a reasonable amount of effort. It's a lot of work to sharpen a saw correctly and involves knowing how to set the teeth properly. You can have it done professionally or you can read up on how to do it and purchase or borrow a saw-setting tool. If you feel a hard drag when using your saw, it needs sharpening. Also, don't let it get rusty. Take off any rust with steel wool and some oil. Wipe your saw with light oil after each use to keep it in good condition.

As far as handsaws go, an eight-point cross cut saw is the most all-around tool. Rip saws are usually five-and-a-half-point. For finer work, cutting small wood, or finish work, a ten- or eleven-point cross cut saw is used. Sandvik is an excellent brand—Swedish import, expensive. Disston, Stanley, and other U.S. firms make a whole line of saws, varying in price and quality. It's a good idea to invest in a good-quality saw—made of good steel—because it will last through many resharpenings and will hold its sharpness longer. Use a cheaper saw, or one you do not value as much for cutting through old wood, where you might encounter nails and damage your

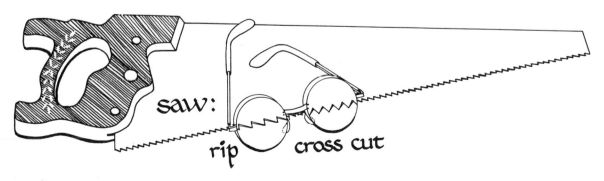

saw: rip   cross cut

saw. When cutting through old or knotty wood, be extra careful that the saw doesn't jump back on you.

When sawing, start your cut with a few light nicks in the proper place. Follow these with slow, deliberate strokes. Don't hurry. These first few strokes will determine if you get a square cut or not. This accurate initial groove will keep the saw from jumping off the board and cutting your hand. When sawing, a lot of force and muscle should not be necessary. Let the saw do the cutting and use long strokes so all the teeth can work for you. A good precaution: keep your left hand from getting too close to the saw (if you're sawing right-handed). This precaution is one to remember when using practically any tool where one hand is active and the other is holding, resting, or whatever. It's one of the most common accidents to bang, cut, poke, smash, or wallop one hand because you are concentrating on the other.

Two other useful handsaws are the coping saw and the keyhole saw. A coping saw can cut circles, arcs, and curly edges. It is an ingenious tool with a thin movable blade. Loosen the tension on the

coping saw

blade and turn it in the direction of the curve. It is a simple saw that gives professional, artistic finishes to square-board carpentry. The keyhole saw is a small, narrow saw that cuts holes in the center of boards, inside curves, and windows in walls. Drill a hole in the place to be cut, insert the keyhole saw and use it to cut out a hole big enough to get another saw into. (In most cases, keyhole saws are too tedious to use to cut the entire hole.)

keyhole saw

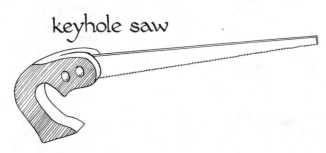

Leveling Tools deserve a whole discussion to themselves. Levels are used to get things level horizontally and vertically (i.e., plumb). Plumb bobs are used to arrive at vertical level only. If your have to choose, a two-foot level and a plumb bob are the first leveling tools to acquire. Together, they can do most jobs and prove to be versatile tools.

Levels are made of different materials (aluminum is common) valued for being lightweight and easy to handle. Some are of magnesium, even lighter weight. This is especially good for long levels, as it allows a bulky tool to be handled easily in one hand—especially convenient if you're on a ladder. The vials in levels are encased in either glass or plastic. The plastic tends to get scratched and for this reason is a poor idea, although it is not breakable. Glass will give you a clear, unscratcha-

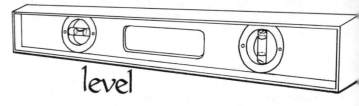

level

ble view and, if one does ever break, the vials are replaceable. Levels have different body styles—kind of skeletal or more fully rectangular and boxlike. I usually prefer the boxlike style (for two-foot levels) because of the versatility as a straight edge—it can be used to mark or draw straight lines.

Long levels (six feet) are unbeatable for precision and handiness, but they are expensive. Also delicate. Do not leave them on the floor to be stepped on, or leaning against a wall or anywhere they could get bumped and fall down. The two-foot level is a convenient size to handle and is not as expensive as a larger size. Its shortness allows for getting into tighter spots. And for long spans, if you find a very straight long board, use it as an impromptu extension of your level. Lay the long board on top of the two points to be leveled (or hold it in position), then put the two-foot level on top of the straight board. This method may not be as accurate as a longer level, but it is a good substitute.

The torpedo level is a small (one foot or smaller) and very sensitive level. It is used mostly in cabinetmaking, shelves, etc.

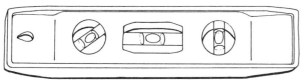

torpedo level

String levels are very small, only several inches long. They can be used over a long distance since they are attached to a string of whatever length. They are inexpensive, but that's one of their few advantages—the string sags and the measurement is inaccurate.

string level

Plumb bobs come in various shapes and are fabricated out of dense metals. A sharp point, which is hard and durable, is preferred. The heavier the bob, the less it will sway and the more accurately it will stay in position, making it easier to mark or measure. A string of whatever length is attached to the top of the bob. The bob is suspended (by hand or from a nail) and used to get points that are vertically in line with each other.

plumb bob

Wrecking Tools also deserve a whole discussion. For one thing, they are useful in ways besides tearing things apart or down. For example, when you are working alone, pry bars are handy as shims or for leverage or for holding something up until you get it nailed. I'd like to mention two other wrecking tools. The Wonderbar is indeed a wonder, with many applications and advantages. Wonderbar is a trade name (it goes by other names, too) for a small, flat prying tool. It is easy to handle, not burdensome to drag around like a crowbar and it fits

on my tool belt. Nevertheless, it is good at prying, nail pulling, even crude chiseling. It's my favorite and most often-used wrecking tool.

The cat's paw is especially designed for digging out nails sunk too far into the wood to reach with a hammer claw or crowbar.

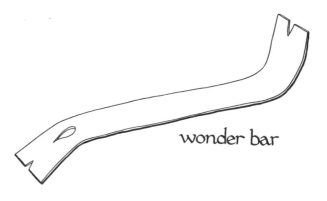

wonder bar

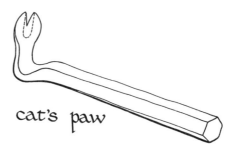

cat's paw

There are three tools that are fairly inexpensive and definitely handy to have in your tool collection; they are three of my favorite and most often-used tools (and I couldn't resist adding them to this list). When you are thinking of building a basic tool collection, these are some of the first to consider beyond the basic hammer, saw, etc., essentials.

The *bevel* is used to measure angles. You can set it at any angle and transfer the mark to another piece of wood. Great for working with rafters, doing diagonal bracing, etc.

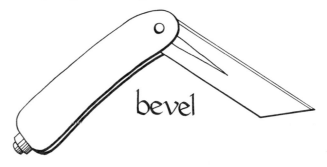

bevel

The *scratch awl* is just a pocket tool that I automatically include in my tool belt. Its very thin point is good for scraping, poking, making marks (instead of a pencil), picking the paint out of old screw slots, etc. Often I use it to make a small indentation before drilling—just a tap with the hammer on the awl makes a small hole just right for guiding the drill bit at the start.

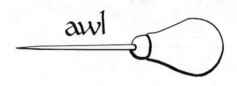

The *chalk line* is another tool of unlimited uses—as many as you can dream up. It's a string wound up in a container of chalk (refillable). Pull the string out, hold it tight at both ends, and snap it in the middle—automatically, a straight line is marked for you. Good for marking a line to cut plywood, for example, especially if you have to cut at an angle; or to put a mark to line up the bottom plate of your wall.

POLE FRAMING

I began my goat barn with my husband, Chris, late in the fall, under the pressures of the winter rains. We spent one day cutting, stripping and trimming poles and the next framing the building. The third day, we began to put up the walls. Chris was in a bad mood that day, impossible to work with. I told him I would rather work alone than have my pleasure in the process spoiled by his anger and frustration. I began to do the work he had always done: measuring, cutting, nailing. A six-year-old friend did the work that was usually mine: holding, handing, helping. Suddenly I realized that I had merely substituted the child for the woman; that he would learn no more than I had "helping" build my first house. So we began to share the work equally—each measuring, holding plywood, nailing, handing boards. I resigned myself to the wasted lumber and inexact cuts of a person learning. And in all that day, neither he nor I made any real mistakes. He measured as well as I did, cut as straight as I, and toe-nailed poles without bending nails. By the end of the day, we—a woman and a six-year-old child—had put up four walls, two windows, two doors, and a roof. We were flying with the power of what

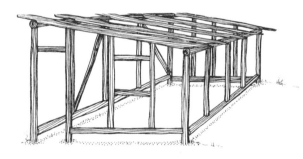

we had done—for the first time in each of our lives, we had actually built something. Never again would we stand and "help."

I stood in the middle of that dirt-floored, low-ceilinged, plywood shed and knew that wherever I went after that, I would always be able to provide myself with shelter. And since then, I have had a new sense of possibility. Fear does not paralyze me any more when I know I need a lean-to for milking or a cabinet to hold my tools. The principles of framing are the same no matter what the job, and I am no longer afraid of creating with wood.

Pole framing is the simplest of all construction since there is no foundation or floor. It's the best place for a beginner to start; more elaborate buildings follow the same basic plan and will come easily once you are familiar with the principles of framing. Pole-framed buildings are perfect for woodsheds, goat barns, chicken houses, tool sheds, etc. They are framed with young trees both for simplicity and to save on lumber costs. The walls and roof can be made of boards of plywood or even plastic, depending on the building's use. My goat barn is plywood with mineral-coated roll roofing over plywood for the roof. My milking shed, which gets no wear and tear, is just plastic stretched between poles.

I try to plan all my buildings to be in units of four feet, since that's the dimension of plywood sheets and a standard size for lumber. My goat barn, for instance, was 12′×12′, with vertical supports every four feet. The end posts on each wall are sunk in the ground (in two-foot-deep postholes) to anchor the whole building since there is no foundation. These should be soaked in creosote or other wood preservative before use. The interior posts are nailed between top and bottom "plates" (poles that run horizontally the length of the wall at ground and ceiling height). These plates are needed to anchor plywood top and bottom and save digging more postholes for the vertical supports. A twelve-foot front or back wall will look like this: Since your floor will be uneven, it's

Front or Back Wall

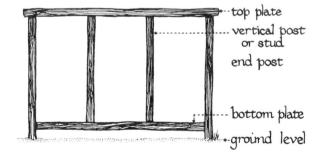

not important that your bottom plate be on the ground. In fact, keeping the plate a few inches above the ground will keep it from rotting.

In order for the roof to drain, it must have some slope (at *least* one foot in twelve feet of wall). So, the front wall is framed the same way as the back, except the vertical posts (studs) are at least one foot shorter than those on the back wall. (My goat barn was eight feet on the back wall and four in front to save on plywood, since the goats do not need as much height as I do).

Side walls are slightly, but not much, more complicated. Rafters connect the front and back walls, running from top plate on the back to top plate on the front. You need your end rafters in place in order to determine the height of the studs on the side walls. Cut the rafters longer than the length of the wall to allow extra length for slope and overhang (which helps protect your low wall). The rafter has a notch cut in it, so it will fit tightly over the top plate, like this:

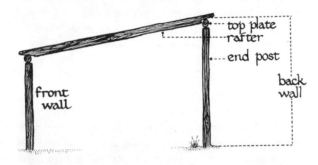

I have found that just sawing the rafter at an angle (as shown) works fine at the low end, but the high side needs to be notched. This notch is the hardest thing of all, but it's not very hard. Just spend a little time visualizing how the rafter needs to fit. Saw down into the rafter the depth of the diameter of the top plate. Then hold a hatchet or wood chisel across the end of the rafter and tap it with a hammer. The piece should pop out and you have your notch.

Once the rafters and end posts are in place, the bottom plate should be nailed in. Then you drop a line from the rafter perpendicular to the bottom plate every four feet. This line tells you the height of your studs (they differ since the roof slopes). A framed side wall twelve feet long would look like this:

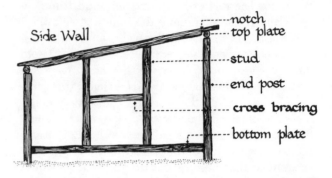

To frame a doorway, you must sink end posts on either side of the door. Essentially, you end the wall at the door, and these posts must be set in the ground. It is easiest to put the door at a corner, where there is already one end post. Studs in that wall must still be every four feet from the end, so there is something to nail plywood to. Windows require a stud on either side of them and a plate at the top and bottom. This is how to frame a door and a window on a side wall:

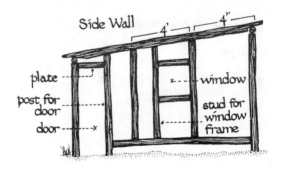

Doors and windows are framed the same way on front or back walls.

Planning Your Building: The first thing to do is to decide how large a building you want, which direction it should face, and where the windows and doors should be placed. Put a solid wall or corner (no doors or windows) toward the direction from which most bad weather comes and slope your roof away from bad weather. Here, for example, most storms and all sun come from the south, so roofs that slope north (low wall to the north) last much longer than those facing south. We get almost no storms from the northeast so I placed my door in that corner. Try to plan your building so that it uses the least amount of plywood or boards possible by staying on four-foot modules and lowering the ceiling on animal barns or storage sheds. Once you have your building in mind, sketch a simple framing plan for it. Then count the number and size of poles that you will need.

Poles: They should be young trees, as straignt as possible and of hard wood if it is available. In this area, redwood "suckers" are perfect since they grow very straight and need to be thinned anyway. But they are a soft wood and do not last as long as fir or oak. Choose trees that do not taper too rapidly; your poles should be three inches in diameter at the ends, though I have used some rafters that tapered off to two inches since there is no snow here. End posts should be at least four inches in diameter, preferably more, at their base. All trees should have the branches stripped off, as close to the pole as possible, so the knobs won't interfere with boards or plywood. Use a hatchet or a draw knife to do this. Bark should be stripped off to prevent rotting. Redwood bark will peel off by hand; most other kinds need a draw knife. (Goats will strip the bark off for you too.) End posts should be soaked to above ground level overnight in a wood preservative (Penta diluted with used motor oil or creosote may be used for this). If left unsoaked, the end posts are liable to rot and your house may topple in several years.

Framing: Clear your building site and get it as level as possible. Mark off your corners and check to make sure they are reasonably square by measuring the diagonals, which should be equal. This is important or your siding will not fit at one corner. Once you have squared your corners (it took me over an hour to do that), dig the postholes for the

end posts and door frame. Posts should be two feet in the ground for an eight-foot-high wall, one and a half feet for a four-foot wall. Put your end posts in place and tamp the ground as you fill the hole. (Tamping is packing the dirt down with a pry bar or piece of iron pipe capped at one end.) In wet weather you will need to add gravel or rocks to get the hole to pack down. The posts must be firmly in the ground.

Frame your end (not sloping) walls first. Cut your bottom plate a fraction of an inch too long so that it has to be pounded into place with a hammer and has a tight fit. Toe-nail it into place. Toe-nailing is nailing at an angle through one piece of wood and into another. It is used when the post is too thick to nail through directly.

Use sixteen-penny or twenty-penny galvanized nails, depending on the size of the pole. Avoid nailing into knots or where branches were; the wood is so hard it will bend nails, and it's easy for a beginner to think she is at fault. The top plate goes on after the bottom plate. It sits on the top of the end posts. Nail directly through the top plate into the end posts. The studs are cut next—again, cut them just a little longer than your measurement so that it is a tight fit between the plates. The studs should be at "four-foot centers"; this means that the four-foot point is at the middle of the stud, top and bottom, so that both sheets of plywood can nail to the same stud. The base of the studs are toe-nailed

into the bottom plate; the top is nailed directly through the top plate into the stud. Since your bottom plate is off the ground to prevent rotting, brace under it with a block of wood so it will not give as you nail. Side walls are framed in this same basic way, except that the top plate is a sloping rafter and the studs are of varying heights.

Rafters must be notched at one end and cut at an angle at the other so they will fit over the top plates of the end walls. They should overhang the low wall by at least six inches with a trench dug below the overhang to drain water away from the building. Rafters can be at four-foot centers if you use a plywood roof and have no snow. Otherwise,

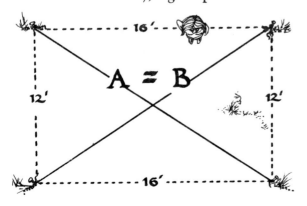

they should be at two-foot centers. Short poles nailed between the rafters (in the same way as studs) will provide bracing and strengthen your roof. Place the rafters directly above end posts and studs for greater strength and nail directly through the rafter into the top plate with a twenty-penny nail.

Walls may be made of plywood or boards, depending on what is available and cheap in your area. I have described framing presuming you are using plywood. There is no real difference with boards, except you might want to put your studs at two-foot centers and brace between them at the corners. Plywood is very strong and will act as bracing on a building. Boards, run vertically, will not stiffen up your building or act as bracing. Horizontally run boards, milled shiplap style, for example, *will* help to brace.

SIMPLE SIDING

Finishing the walls is an even easier task than framing them, and a very satisfying one since the walls go up pretty quickly and turn a collection of poles into a "real" building! The first thing to decide is what materials you want to use for siding. This will be determined primarily by cost and what is available in your area, though you may want to consider aesthetics too. The basic choices are plywood (left bare, painted, or shingled) and one-inch-thick boards of various dimensions (1"×6", 1"×8", etc.). Boards may be new lumber or old siding salvaged from some former building. They may be anything from pine to oak, depending on your area. But it is important to know what kind of wood you are using and what its properties are. My first year of carpentry experience was in Pennsylvania and I was using seasoned oak boards. After six months, I could still set only one nail out of three, and try as I might, I still could not "feel" when I was hitting it wrong. It was a shock to send my first nail through a piece of California fir and discover that I could, indeed, nail quite well! Hard wood (like oak) and poor quality nails are a bad beginning for a carpenter—it is too easy for a woman to assume that she and not the materials are at fault.

Plywood is the easiest of all siding to use, and it is a good choice if you live near lumber mills where "blows" are available. Blows are sheets of plywood having some fault, and are sold at half price or less. If they are carefully chosen (don't buy blows that are coming unglued between the layers), they will work just as well as regular plywood. Plywood for outside walls must be exterior glue (most blows are, but you should check this).

Siding should be at least ⅜″ thick and preferably ½″ or ⅝″ thick. Roofs should be ¾″ thick, though ⅝″ will do where there is no snow.

If you are going to use plywood, plan your building so that you can use whole 4′×8′ sheets (either horizontally or vertically) wherever possible. Frame your building with your studs on two- or four-foot centers so that the sheets of plywood fit the frame. Plywood should be nailed every ten inches to the studs and the top and bottom plates. If you are using uneven poles for your frame, place your nails quite close together in the bumpy spaces. Plywood has the advantage over boards of being fairly flexible and a well-set nail can force it to fit tightly on an uneven surface. It springs back when hit and the nail will have trouble penetrating unless you keep the plywood pressed against the stud. If you cannot press the plywood with your body, put a wooden shim between it and the stud until your nail has penetrated the plywood.

Exterior plywood of good quality may be left exposed to the winter for several years—but it is not indestructible when left unprotected. The simplest and probably cheapest method to finish plywood is paint and sealer. "Wet Patch" is a black plastic roofing cement that may be used along all top edges, especially where the plywood is coming unglued or is nailed over uneven surfaces. Wet Patch is a sticky, gooey mess that is easily applied with a stick. It forms a good, flexible seam and never hardens completely. Painting is not really necessary for protection, but it will help and it certainly does improve the looks of plywood and Wet Patch!

The more expensive, more permanent and most beautiful way to finish plywood is to shingle or shake it. This process makes a permanent, durable wall that insulates as well as protects.

The greatest problem with using boards as siding is to be sure you get a water- and wind-tight seal between them. Before board siding or plywood is applied, tar paper can be stapled to the bare studs. This additional step will provide an adequate weather barrier, as in "tar-paper shacks," that can protect you from the harsher elements like cold and rain. Tar paper is not expensive, and I recommend its use in animal shelters as well. If used on the inner walls of most animal houses, the tar paper will have to be protected somehow. Horses, sheep, and goats may rip the paper from the walls if feeling bored or mischievous.

The easiest and most expensive way to insure a tight seam is to use milled siding–*shiplap, V-rus-*

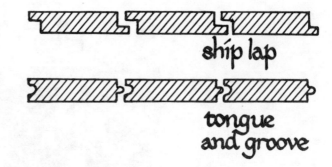

ship lap

tongue and groove

tic, tongue and groove, etc. These are boards that have been especially milled to form a crackless exterior. Viewed from the top, the boards fit together like interlocking jigsaw pieces. Milled sidings can be used horizontally or vertically, nailed directly to the studs. If you are nailing any of this siding in place, allow a small space between boards to allow the wood to expand and contract during damp weather. The thickness of an eight-penny nail is one convenient measure to use for this spacing.

Unless you have a lot of money or have made a very lucky score, you are unlikely to be using milled siding. Regular boards have the problem that they do not fit tightly together when simply nailed up side by side. There are two ways to deal with this—using board-and-board, or board-and-batten techniques.

In using *board-and-board* construction, two

Board'n'Board

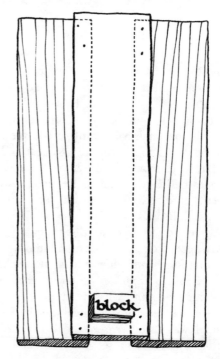

block

Something Peter said yesterday keeps bouncing through my head. He talked about his job being in a perpetual state of flux, change; that if you gave it all your energy, you could ride with it, but if you didn't it was chaos. He said our relationship was like that now too—always changing, never static, not to be counted on. He said he needed a source of stability, a home to come back to. I interrupted him, became angry. I realize now that I was very afraid. That source of stability used to be me, my home, the world I created and maintained for him. Dimly I had realized that and this was part of why I co-ordinated the women's festival, began a life of my own. I wanted him to want *me*—not what I stood for, not the security of wife and home. I am terrified of being trapped forever as wife and home. Loving him, not wanting to be separated but never being wanted for myself. Now when I hear him speak of building a separate cabin of his own, of wanting a home that is not me, I panic. I am sure that once he has that, he won't want any part of me. I can't stand the possibility of having to live with that pain next door to me all the time. Already I see him with a wife and children saying, "Oh, yes, Jenny and I are good friends now." The wife has curly hair, smiles serenely, and doesn't make demands. I know her well.

boards are nailed vertically alongside each other with a space left between and a third board is nailed over with an equal amount of overlap on either side of the space. It is important to use well-seasoned lumber or your wall may shrink, buckle, or gap as it ages. If you have to use unseasoned lumber, add several extra inches of overlap to allow for shrinkage. Old boards taken off existing buildings make the best walls as they will not shrink or warp more than they already have. They do, however, tend to split easily when nailed.

The top board should be overlapped at least an inch and preferably an inch and a half on the bottom boards on either side. The closer to the edge of a board you nail, the more likely the board is to split. The overlap must be sufficient to allow you to nail through top and bottom boards without being so close to the edge as to split either one. Your nails should be thin (box nails have thinner shafts than common nails) and just long enough to go through both boards and set well into the top or bottom plate. Hitting each nail once on the tip to blunt it helps prevent splitting.

It is also important to make sure that your boards are put up straight and parallel to each other. A slight crookedness in one board will quickly become a gap that cannot be covered. The best way to keep your boards straight is to cut a two by four to act as a spacer or template. This template should be the length of the space between the boards. For example, if you are using 1″×8″ boards overlapped 1½″ on each side, the space between boards should be 5″ (8″ minus 3″ for overlap). The template would be cut 5″ long.

Because the nails for the top boards also go through the bottom boards, helping hold them in place, underneath boards need only one nail placed at top and bottom centers.

Nail the first board on your wall, keeping it straight on the end stud. A level might be useful at this point. Then hold your template up against the first board at the top, butt your second board up against the template, and nail it in the center to hold it in place. Then lower the template to determine the spacing of the boards at the bottom and nail the second board in place too. Always use the template top and bottom; do not rely on your eye to determine when the boards are parallel. Continue all across the wall, placing bottom boards. Then go back and nail up the top boards, overlapping them carefully and equally on both sides.

If you are using boards of varying widths, make several different templates for each size gap. Work across your wall in groups of three, two bottom boards then one over the gap, to be sure you have enough boards of the right size for each gap. This is slightly more time-consuming, but worth it.

Board-and-batten walls are made by nailing boards side by side across the wall, fitting them as well as you can. Then, thin boards or "battens" are nailed over each seam. These may be literally batten or lath (very thin strips of wood about 2" wide) or 1"×2" or 1"×4" lumber.

An unorthodox but proven way to use siding, especially if you are going to have an interior wall, is to nail boards up horizontally each one overlapping the one below it. Tar paper can first be stapled to the studs. Starting at the bottom, nail the first board flush to your vertical studs. Lap the next board at least an inch over the first and nail it only at the top into the studs, not at the bottom. The pressure from being forced out should hold the top board firmly against the lower one. There will be air gaps at either end of the wall, so leave space on the end studs to hammer a 1"×2" or a 2"×2" board vertically flush up against the siding.

Board'n'Bat

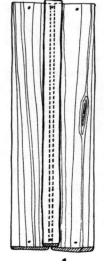 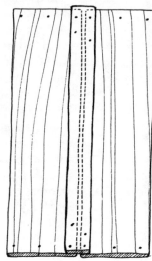

narrow boards   wide boards

We finally bought a truck! An old red '56 Ford, a wonderful machine—the realization of another fantasy as once my work boots were. I remember how those first boots felt. Symbols of my countriness, I wore them even on hard city pavements. I watched each step I took for the pleasure of seeing work boot prints in the mud. I kicked trees, forged through wet leaves, stamped and pounded, all to test the new-found power of feet ready for work. Then came the levi jacket and finally, now, as the jacket is looking faded and well-used, the truck. Today I went out on the back meadow with truck, jacket, and boots, to pick up old fence posts. I found myself transformed into every cowhand I had ever seen in a movie. I drove the truck through the field with keen eyes and stern jaw, "riding fence." I worked my small, thin, "wiry" body to exhaustion without even noticing it—I was so enraptured with the theater of being a rancher.

When out of forests, flatlands
women came
they laid fields down in seed
grain grew, women grew it
Knew what sacrifices, offerings to make
with wonderful names
Ashtarte, Isisar, Isis ruled.
Capsicum, cayenne, coriander, fenugreek
herbs of the earth relieved fevers
healed wounds
cured.

Cyclical life
life rhythm
moonlife
womanlife

When women planted, raised animals
made shelters
as they built fires, gathered wood
they loved one another, this was understood
women loved women
blood women   womb women
tasters of the moon

Weavers, spinners
they dyed laughter into their colors
shared sorrow on the threads of their looms
a line of cloth on their backs
and rings

They danced
chanted earthsongs with their children
voices of their land, their work
their mothers gone a thousand years,
remembered
o Ashtarte, Ishtar, Isis
goddesses, shoulder to shoulder
back upon back
women were lovers: earth lovers
woman lovers
star gazers.

ROLL ROOFING

One of the first projects we undertook when we began restoring an old farm to working order was reroofing our barn. An old, redwood-sided building about twenty by thirty-four feet, it has a fairly steep-sloped roof and a stable added to one side. When we first had looked at our land, we'd thought the barn was beautiful—functional—perfect. A closer look once we'd actually moved to the place revealed a lot of sky through our roof! The old shingles had rotted, split, and blown away. After a little research into the cost of galvanized and aluminum roofing, shingles (wood and asbestos), we decided to roof with what looked like the cheapest and easiest material to work with: roll roofing. "We" were three women and two men, all city suburb raised with almost no experience or knowledge about anything we were undertaking. In our eyes, roofing the barn would be a simple matter of unrolling the paper and nailing it down—a few

hours work and we'd have a rain-tight roof "Good for at least five years." We learned some things in the process.

It *was* simple . . . relatively. We bought heavy-grade mineral-surfaced roofing. This roofing is available at most hardware or building supply stores and must be put over an existing wooden roof. It will cover old shingles or protect plywood from excessive exposure to the weather. Heavy-grade roll roofing is coated on one side with mineral granules. It comes in different weights—the heavier, the more durable the roofing. The heavier grades are a little more difficult to work with as the roofing is thicker, therefore stiffer, and more likely to crack and buckle. One roll of any grade will cover about a hundred square feet of roof. It weighs sixty-five to one hundred pounds per roll and comes in assorted colors—red, orange, green, blue. Lightweight tarred felt (with no mineral coating) is also available. This is usually used as a moisture barrier between outer and inner walls, helping protect the stud framing, under shakes and shingles and so on. It is not really waterproof, in that it allows a certain amount of breathing. If you are pressed for money, however, it will make a fairly good temporary roofing. A roll costs about the same as a roll of heavier roofing, but covers much more square footage. It will protect your roof through at least one winter season and may be covered the following year with shakes or better roofing.

With the roofing, we bought large-headed galvanized roofing nails and tar (tar comes in gallons or small cans; by the gallon it is a lot cheaper). We learned early about the tools you need: something with which to cut the roofing (we used tin snips and later a mat knife—the knife did a smoother and easier job); some light hammers; a few flat sticks with which to smear on the tar; and some sort of scaffolding if your roof is steep. A carpenter's apron (store bought or easily made) with pockets for nails and a slot for your hammer is indispensable for freeing your hands.

We began the barn roof on the stable side. Because each roll must overlap the one below it, you begin at the bottom and work up. At first we took the whole roll and unfurled it across the roof (we did *most* of this roofing project this way until we

realized we could measure out what we needed on the ground, cut it, and roll it up into a smaller, more manageable size). One person held the strip in place, a second nailed it down along the top edge. We spaced our nails a hammer's length apart. A third person then came along behind the nailer and smeared a two- to three-inch strip of tar along the upper edge of the paper. This tar keeps the next overlapping piece of paper secured flat and keeps rain from seeping back up to the joints. If it's a hot day and you're working directly in the sun, tarring can be messy. Tar runs and drips and sticks. On a cool day the paper is stiff and won't unroll or lay properly. It will, instead, crack and split and buckle. Ideal roofing weather is medium-hot and preferably cloudy.

After the first row of paper was on, nailed, and tarred, we carefully unrolled the next. It should overlap two to four inches and be pressed down firmly. If your roof will be exposed to a lot of strong winds, nail this edge down and put tar or Wet Patch over each nail head. Or you can leave a much greater overlap, a foot or more, and the weight of the paper will supposedly hold itself down. Our paper had an orange mark running where the overlap should end. If your paper is not marked or you want a wider overlap, you can measure up a certain distance and snap a chalk line where you want the bottom edge of your next layer to be. Positioning the paper carefully along the factory or homemade line reduces lumps riding up. Unrolling the whole piece before you nail it down also allows you to pull out any wrinkles that would later cause leaks. We learned to work in our socks or bare feet—boots and shoes can gash holes in the paper if it's being laid over old shingles or if it's cold. Boots will also rub the mineral granules off. When you have to join two pieces of paper to cover an area, allow a generous overlap of at least six inches and tar under the seam. Nail down the top piece and then tar over each nail head to seal it really well.

Another material that may be used over nail heads is Wet Patch. It holds its consistency, is waterproof and may even be put on a wet roof. Wet Patch works well over nail heads because it will not melt and run in the hot sun as tar does.

As we worked up this side of the roof, we devised a simple scaffolding of two by fours and used a long ladder to carry up the rolls of paper. Getting to the top of the roof, we left the last row undone—later the ridge piece would overlap both sides. The second side of the roof was more of a challenge: there was no stable to work from. The two men soon took charge of arranging that we women would go over the edge with heavy ropes secured around our waists. They would stand braced on the stable side with the other ends of the ropes tied around their waists—to keep us from falling off the roof if we slipped. Both amateur mountain climbers, they were using this opportunity to impress us with their techniques. And besides, we were lighter.

To our then unfeminist minds, this all seemed logical enough. We would take the risks and do the work and they would protect us. Working with rolls of tar paper, cans of hot tar, and nails and hammers when you are also tied by the waist to a man who cannot stand staying where he belongs and has to stand on the peak of the roof and supervise your work *should* be consciousness raising, besides being downright unwieldy. But it took me these many years later, retrospectively writing this book, to realize that they couldn't bear to let us do it alone.

But we did it. Sliding down the roof to the lower edge, trying not to look down, and finally doing it, and getting dizzy and high on the experience. Working on our hands and knees. We learned to work with the rolled up paper by putting ropes through the cores with a short length of two by four attached crosswise at the bottom. This way it

could be lowered and roughly guided into position.

When the roofing was on, we considered how to secure the edges. Edges of roofing may be held down in one of three ways. You can nail them down, running a vertical row of nails about an inch from the edge of each piece. A dab of tar or Wet Patch over each nail head will prevent leaks. Or you can do your whole roof and then run thin wooden battens along the edges. The third alternative is to use what we call flashing—in the hardware store it's called "roof nosing."

This is basically a long strip of galvanized metal, folded lengthwise. It comes in different widths (1″ ×1″, 1″×1½″, etc.). Choose a width that will adequately cover your edge. You can buy this flashing by the foot. It is not expensive. Before you take it up on the roof, though, punch all your nail holes. Holding the flashing against a piece of two by four while you pound works pretty well. Along the sides of your roof, nail the flashing so that it comes down over the edge of the roof and holds the paper securely.

On the lower edge of your roof, you may want to slide the flashing *under* the roofing paper rather than over it. The advantage of this positioning is that all water will be shed cleanly. If the flashing is placed over the paper, water may catch and be pulled back under the paper. This could eventually rot the edges of your roof sheathing. We placed our flashing over the paper because we were concerned with wind damage along this lower edge. A good gust of wind catching under even a section of roofing paper can rip a whole area loose. So far, this roof has been through many seasons of rain with no backup leakage problem. If your roof is less steep than ours, though, you might have a problem.

When our roofing project was completed, we climbed down in tar-spotted clothes, elated and exhausted. It was the first of many, many roll roofing projects, and of all of them, our most ambitious. We learned a lot about roofing, working together, and the courage to take things on. Our roof, many years later, is in fine shape!

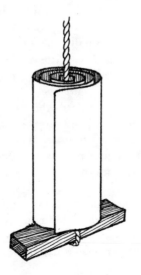

RECYCLING WOOD:
TEARING DOWN BUILDINGS

There are a lot of good reasons for recycling wood. Wood is expensive, and we have more time than we have money. We have more time than we have trees. As I watch the firs and redwoods roll down the highway on trucks, I think it senseless waste because enough trees have already been cut down to house us all. Lumber surrounds us in the form of old buildings no longer used but full of still usable wood. The wood that my family and I recycled from a house came years ago from trees in our very own hills. It seemed only just that the wood came to us and to others who are trying to restore logging-raped land.

Some research is necessary to find available houses, barns, garages, and then, of course, a fair amount of labor to tear down the building, but eventually you can get yourself a lumber stash.

Old houses and barns close to your home are probably your first choice. Be sure to find out who they belong to and make some arrangements before you start. Places where there is about to be new construction are good to check. At the local office of the Division of Highways, you can get on a mailing list that will inform you of the bids to be let on buildings to be torn down for highway construction. The auctions are held at the site of the building, so you have to go there to bid. Some of the minimum bids will be out of your price range, but you will find some that start at $1. Yes! But check this out; it is no longer available to "non-professionals" in some areas. It might be worth it to so some city footwork and find a building there for demolition. And then, rumor can be on your side . . . put out the word that you're looking for a house to tear down. Some people are glad to get old, unused buildings off their land. Especially if you sign a contract to clear and clean up the site.

Once you have located a building, you should do some preparation. A Demolition Permit is required from your county Building Department. If you deal with them, get one; they cost around $5.00. There are a variety of tools that you'll need for the job. If there are a number of people working together on the job, try to gather as many tools as you can so there will be tools for everyone (borrow them). Here's a list: crowbars, nail claws, light sledges, hammers, wrenches (for dismantling plumbing), screw drivers, wire cutters, flatbars.

Once you have all these tools, be sure to use the right tool for the right job. A flatbar is flatter than a crowbar and useful for getting into tight openings; with a nail claw and a hammer you can pull nails right up out of the wood, and then use a crowbar to pull them entirely out. Don't use a hammer for a job that a crowbar can do better; the crowbar gives you leverage. Consider obtaining a Sawzall or reciprocating saw. With the right blade, this power tool can cut right through nails. This makes it possible to remove wall or wood shingle sections to be trucked whole to your site to be reused as "new" prefab-style wall sections. Remember tools for your body, too. This is a demolition derby—dress appropriately: shoes, gloves, if you prefer, to avoid a million splinters, even a hard hat if there is a danger of things falling down on your head. If you're working on farm property, consider getting a tetanus shot ahead of time.

With all your tools in hand, you're ready to begin. You can learn a lot about construction by tearing down a building, so examine the house carefully before you start. Note beams that look rotted or weak, how rafters tie into the rest of the house, and generally get a picture in your mind of how the house is put together. Sometimes you learn what *not* to do in construction and carpentry, so as you go along, poke around the substructure, the framing, etc., and consider how well the construction has served. While you're looking at the house, take an inventory of the wood you'll be getting. Measure the house, individual rooms, the roof and note the type of wood in each place (i.e., 10′× 12′ room of 1″×8″ siding). Allow yourself 15–20 per cent breakage during demolition, depending on the condition of the house. Also, before you begin, be sure that utilities (if there were any) have been disconnected, and clean out the inside of the building. Broken chairs and old bedsprings can get in the way.

A note about safety. Demolition can be hazardous work. If there are several people working at once, be sure to arrange the work areas so that no one is in danger of rafters falling on them, tools being dropped on them, and so on. Keep track of who is working where so that you don't wander into someone's wrecking space.

Begin wrecking. As soon as possible, take out all

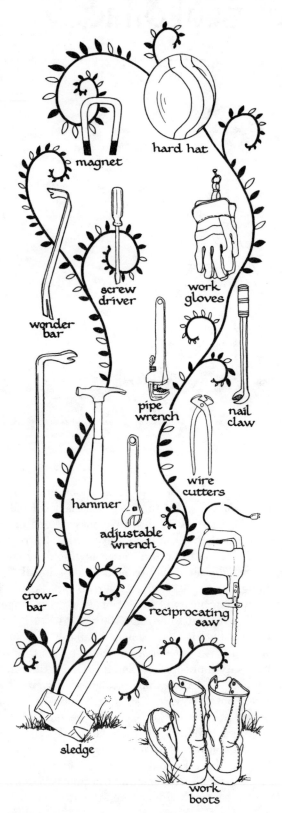

magnet

hard hat

wonder bar

screw driver

work gloves

pipe wrench

nail claw

hammer

wire cutters

adjustable wrench

crow-bar

reciprocating saw

sledge

work boots

the windows and put them in a safe place well away from the building. Sometimes some of the siding will need to be taken off before you can get to the window to remove it. Then, take it from the top. Whatever is covering the roof comes off first, starting at the peak. If the roof is particularly steep, you may need to rope yourself to the roof by securing a rope to the side opposite where you are working. It would be helpful to learn some mountain climbing knots that would give you mobility on a steep roof. Once the tar paper or shingles are off, begin again at the peak, with crowbars, nail claws and hammers, to take off the roof. Note which way nails are hammered in, and pull *with* that angle. After the roof is removed, the rafters will be exposed. Most of this work is a matter of pulling the nails and taking down each individual rafter carefully, as these are long pieces and could break. In most conventional houses, though not in sheds and barns, below the rafters are the ceiling joists to which the ceiling is probably nailed. If the ceiling is nailed into the joists, don't stand on it, as you will be putting pressure downward on the nails, and weak boards could give way underneath you. When you take these ceiling joists out depends on when the ceiling comes down, before or after the inside walls. Anyway, you can get down from the roof for a while and take the outside siding off, and then the inside walls.

At this point, the wall structure will be visible

and you should check the bracing and test the walls to see if they are wobbly. If so, put up temporary bracing to stabilize them. (In fact, if any time during the destruction, the ceilings, roof, or walls look shaky, for your own safety, a few braces or stud supports are in order.) At some point it will be easier to pound siding away from the nails, rather than pulling out nails with the nail claw. Use a block to pound on to avoid damaging the wood instead of hammering directly on the wood. Pound close to where the nails are rather than in the middle of the board. After all the siding is off, you will have the skeleton exposed. Then you can pretty much work where you want to, dismantling inside freestanding walls, other inside walls, and the ceiling.

There are a number of ways to take down the walls. You can pull them apart piece by piece, or carefully push the whole wall over so that it is lying on the ground or floor and more accessible to rip apart with crowbars. Lowering the wall (skeleton) to the ground with a rope pulley probably means less breakage if it's a large wall. In order then: you take down the ceiling, ceiling joists, top plates, wall studs, and braces downward to the flooring, subflooring, joists, and foundation.

The flooring is likely to be tongue and groove, and if you want to preserve it in that state, it's important to be careful removing the nails. If it's nailed into the tongue at an angle, be sure to take

Thursday morning, still in bed. Weaving co-operative day and I woke up feeling pleased. Five weeks ago only two of us could weave or spin. Last week we wove our first piece—all ten of us, each one taking a section so each of us could learn. Learning without fear, working together in co-operation and excitement is a whole new world for me. It makes me realize how much power and revolution may have been hidden in a quilting bee!

The first few weeks were hard for me. I couldn't learn to spin, was tense with trying, and quit in frustration. When I began to try again, there was Margaret beside me, letting fleece just flow through her fingers as though she'd been doing it for years. I was about to drop the spindle, give up forever, when I realized no one was watching me, measuring my progress. It was only with my own expectations that I was competing. I began to hear what Lea was saying near me, to rejoin the circle. By that afternoon I, too, was learning the feel of wool meshing together and becoming yarn.

the nail out that way. Gently does it. Use the hook end of the crowbar to pull the board toward you. Go along the length of the board at each joist and ease the nails out. If you pull hard on just one nail or section, you may damage the tongue, or break the whole board. If the tongue and groove is nailed to the top rather than through the tongue, then you take it out as you would any other board. If there is subflooring, it's probably nailed straight into the joists and comes up easily.

The foundation and floor joists require more muscle than some other parts of the house because they are bigger pieces of lumber and nailed with bigger, sturdier nails, but by this time you are working, hopefully, close to the ground and have most of the work behind you.

Probably the most monotonous part of the job is pulling the nails from the wood you are going to save, but you can use this work to take a rest from the more strenuous destruction. Space this task so that you won't have all of it to do at once. Make yourself a nail-pulling stand, or use a couple of sawhorses; your back will love not stooping to the ground to pull nails. Nails come out of wood much more easily if the wood is wet. We found our nail-pulling time cut at least in half after the first fall rains. Consider wetting down your wood!

This is the time to decide what wood you want

to save, and then stack it for trucking. You have to establish some standards about the quality of wood you want to save, or else you'll be standing there considering each individual piece and its possible uses. Rotten wood goes into the burning pile (or the dump), anything under two or three or four feet (depending on your needs) is junk, but good for kindling. Watch for termites! Separate termite wood from good wood, and consider the damage the critters have done. Either discard it completely, or if the piece is still basically good, give it a heavy dose of (ugh) creosote before you use it. This is especially important in structural lumber . . . a house built on termites does not stand.

When you've finished tearing everything down, clean up the area as much as possible. The earth under the former building will be pleased to breathe again. A big magnet will help to pick up stray nails. Remember fire seasons and burning permits. In some places you can burn, in some you cannot. Burn what you can (except for stuff that produces noxious fumes, like plastic, roofing paper, linoleum, rubber, etc.) and haul the rest to the dump. Plant some flowers.

Now you have all this wood. If you're planning to build with it immediately, stack it so that the wood you need first is easily accessible, i.e., structural wood like 4×4's, 2×6's, 2×8's, is needed

Peter has come home in the middle of the season. It's the first job he's ever lost; the captain hired his nephew. It's not surprising—fishing is family business and Peter has been a very lucky outsider. Yet he feels this failure intensely, personally. He is depressed and uncertain and frantically busy, trying to build the goat pen to build his ego. He gets angry at nails that bend and at me but won't recognize that the source of his anger is losing his job.

He says that he has come home to stay, and it's funny. He's doing what I begged him to do for years, but now I've gone somewhere else. It's almost as though we've changed places completely. He is unsure of himself, seeking companionship and communication and reassurance. I'm involved with my weaving, my job, the women's group, the garden. My life has demands of its own which take up my time and make me feel good. He gets up in the morning and wants me to pay attention to him; I want to get the garden watered before I go to work. He has a special dinner waiting when I get home and I'm not feeling hungry, only obligated. He needs me and I don't need him, though I love him. It's really funny—he's become my wife, and what he says to me is almost word for word what I might have said to him a year ago.

before siding or roofing. Anarchy in the lumber-yard is frustrating; stacking all the 2×4's separate from the 2×6's and separating tongue and groove from lap and regular wood will save you time searching when you are in the midst of building. If you are storing the lumber on the ground, be sure to put several pieces of 2×4 below the stack for the boards to rest on.

If you are planning to truck your lumber yourself, you might get some advice from a lumberyard, and some heavy chains and chain binders to hold the load on securely. Also, in most areas it is possible to rent, for a reasonable amount, a strapping machine. This makes it possible to pull the wood load tightly and securely together and hold it in place with flexible steel strapping, which is paid for, by the pound, when the strapping machine is returned. If you have lumber longer than the truck bed, put some long boards on the bottom to extend the truck bed. However, never attempt to load a truck with boards of one third more in length than the bed (minus the tailgate, if there is one) of the truck! Important! If a truck is overloaded, especially with boards that are too long for the bed, you will cause costly damage to the truck.

Truck it home and hear the voices of the swaying firs and redwoods thanking you.

choosing and handling wood

Wood is one of the most wonderful of all materials to work with. It is versatile and beautiful, pleasing to the touch and to the eye. A wooden shelter can merge gracefully with its environment, warm and delight its inhabitants, and in time return to the earth in an organically sensible cycle.

As a dweller on country land that is largely forested, I have developed a love for and respect for trees. As a farmer and animal breeder, I have many needs for lumber. Somehow the two have struck a balance in my becoming a little better carpenter. Building shelters, gates, feeders, and such that will last for years and years seems a good beginning in not wasting wood. Learning more of the nature of wood and lumber helps me to conserve, substitute, use minimum materials, be more creative. I would like to stress recycling lumber more than buying it. Our forests are limited, logging

practices are abominable—being really conscious of these facts comes to life in the simple task of building even a new goat feeder. It is well worth the effort to seek out an old structure that is no longer wanted and to carefully dismantle it for reuse. Reusing the lumber you already have is another possibility. In some areas, as in ours, there are small lumber mills being run by people who are willing to mill the wood from your land in exchange for a percentage of the finished lumber. You then have the choice of taking down a few trees from heavily wooded parts of your land rather than buying lumber that has been clear-cut in a devastating large-scale logging operation. There may be times, though, when you find that you have to buy lumber to build a new structure that will be a permanent part of your farm or homestead. In planning your building, consider the type of lumber you choose: are you able to use lumber from varieties of trees that grow quickly and abundantly, or are you unwittingly about to contribute to the decimation of varieties that take hundreds of years to mature? After this initial consideration there are many other things to think about and to watch for. In the paragraphs following, I'd like to share some of what I've learned about working with wood. None of it is fresh or original, really—it is material you could learn (much amplified) from any good carpentry book, a basic course in woodworking, or in your own building. My hope is to demystify the lumber store a little, and give you some ideas about evaluating and handling the wood available.

Grading and Types: The lumber available in your area probably differs greatly from that available in mine. Here on the Pacific Coast, the lumber most commonly used in construction is redwood, Douglas fir, and cedar. In other parts of the country, the lumber will probably reflect, as it does here, the ecology of the area. Wood imported from other areas will usually be prohibitively expensive. To decide among the varieties available, you will have to talk with some willing carpenters or lumberyard employees, or consult a carpentry book. Things to consider are the durability of the wood, the suitability for your project, price, sizes available, and, of course, the aesthetics.

How to Work with Tools and Wood, listed in Resources, is a book that will help you make your choices. It contains a table that classifies woods according to their characteristics (i.e., freedom from warping, ease of working, strength as a post) and lists their common uses. From this table, for instance, you can see that walnut is a choice wood for furniture making and Douglas fir a strong, popular construction wood. Another book, *Modern Carpentry* (also in Resources) has a more detailed chapter on building materials. Included is a section teaching "wood identification" which has color plates of some seventy-one different species of wood. Accompanying each plate is a brief description of the type of wood, its common uses and its important characteristics. Of course the best way to learn about wood is working with it, but these two books can help you from beginning projects right through more "advance" stages.

For most construction projects around the farm, you will be using woods classified as "soft." These are woods from coniferous (evergreen) trees. They are generally easier to work with and less expensive than the woods classified as "hard." Generally, hard woods are from deciduous trees (those bearing seasonal leaves). They are used for special purposes—i.e., flooring and furniture making. Commonly used soft woods include fir, pine, and cedar. Popular hard woods are ash, oak, maple and cherry.

You should also know the difference between sapwood and heartwood. Sapwood is the wood from the outer layers of the tree. It will not hold up quite as well as heartwood when exposed to out-

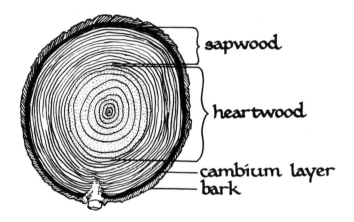

door weather, but is used extensively for interior sheathing, stud material and the like. Heartwood is the inner part of the tree, formed as the sapwood becomes inactive in the growth process. It is particularly valuable for areas of the building that will receive rough wear or weathering: window sills, thresholds, siding. If you are ever dismantling an old building, you will see a startling difference between heartwood and sapwood boards in terms of wear. Our old barn, for example, is sided in redwood on one-by-eight shiplap. For thirty years or more this building has been exposed to the coastal rains and fog, the salt air and sun. Some of the boards (the sapwood) are beginning to literally crumble. The heartwood boards surrounding them are still beautifully solid, though weathered to a dull gray old age. Heartwood is usually a good deal more expensive, and you must specify that you want it for a particular use.

In the lumberyard, wood is graded according to a system that considers its use, quality and size. You should consult a carpentry book for details on this grading before you go to buy lumber. Grades range from first or select (the best) to utility and economy (the lowest). Certain defects and weaknesses are allowed in certain grades. Most of your framing material should be "standard or better" if you are working on a code building or a building you want to last. There is structural lumber that is stress-graded for taking certain weight loads. This should be considered in choosing any members that will bear heavy weights. You can also find tables in carpentry books that will tell you what size piece you will need for a certain span—i.e., if you need a floor joist that spans a sixteen-foot space, you need 2×6, 2×8 and so forth. Plywood is also graded and stamped with a code that takes into account its thickness, the material used in its production, its relative freedom from defect, and the type of glue used. In general, you should try to choose the most economical grade of lumber for each job—i.e., the lowest grade that will still meet stress or weight or durability requirements.

You may also choose between rough and surfaced wood. Rough is, as the name implies, not finished smooth. It is a little more difficult to work with because of the splinters on its surface. It is cheaper, though, and may be used for exterior siding, gates, and so on. Surfaced wood is planed smooth. It is used for interior siding, rough framing, and so on. The planing makes the surfaced board slightly thinner than a rough board of the same apparent dimension.

Moisture and Milling—The Relationship of Wood and Water: the relationship between wood and water is an important one to understand when you work with wood. In years of learning the practical carpentry it takes to develop and maintain a small farm, I've been subtly and directly affected by this relationship. The siding that splits and draws loose from its nails after a period of apparent content, door that jams and sticks every winter, and stack of firewood that seems to lose half its weight after seasoning are all manifesting this relationship. The tree growing outside my window and wooden sill of that window both reach in their ways to receive the rain.

Up to half the weight of a newly felled tree is moisture. Once the tree is felled, a drying process begins as the tree cells give up the water they have held. The wood becomes lighter and also shrinks. Newly milled lumber is "green" (or unseasoned) and thus subject to a considerable shrinkage. If you use green wood to side a building, you will find your siding gradually shrinking, pulling and splitting where it is nailed, and leaving cracks and gaps. Knowing that green wood shrinks will allow you to compensate in advance: leave extra inches of coverage and place single nails in the center of the board rather than nailing edges firmly in place. Lumber rarely shrinks lengthwise, so your measurements for length will remain accurate.

Most wood you buy in lumberyards is seasoned; some is kiln-dried. Seasoning is accomplished by stacking the wood carefully with stickers (wood strips) between layers. The wood dries as the outside air circulates between the layers. Kiln-dried wood is put in large ovens and dried in a carefully moisture-and-temperature-controlled state. Kiln-dried wood is more thoroughly dried than air-dried wood, thus less likely to shrink or warp. It is used for door and window jambs and other precise work. In both of these processes, though, the wood undergoes most of its shrinkage and will hold the approximate size it is when you buy it.

Whether the wood you build with has come fresh from the mill or is recycled from an old building, it is subject to one more continuing relationship with moisture. Wood attracts moisture from or gives off moisture to the air around it until it achieves a balance: its moisture content approximately equals that of the air surrounding it. In achieving this balance, the wood swells and shrinks—sometimes noticeably, sometimes imperceptibly. The door that

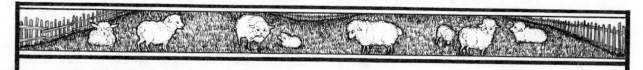

We are talking a lot about getting sheep lately. I don't know what it is about sheep that makes me love them so much. Maybe because this land is made for sheep, has had them for eighty years, and should have them now; maybe because I've learned to spin and weave. I know I should love goats who are so much more intelligent, but it's sheep I want. I sit watching my neighbor's sheep, loving their ungainly movements and placid faces. As I drive along the highway past fields of sheep, an image of our back meadow full of sheep flashes before my eyes. Maybe with Peter's help, I really will make this place a ranch.

sticks and squeaks through the rainy season is demonstrating this relationship. If you build things with wood that you want to swing open or slide, you have to take into account this swelling and shrinking that reflects the season and humidity.

Defects in Lumber: If you buy top-grade lumber and handle it carefully, you may never have to deal with defects. The best and most expensive grade of lumber is "clear." This lumber is supposed to be free of the defects described. If you buy clear or top-grade lumber and find pieces with knots or splits (described below), you have every right to return the piece(s) and demand replacements of the paid-for quality. Even the best grades of lumber, if improperly handled, can quickly develop twists and bows and crooks. Most likely, you will find yourself working with wood that is far from perfect—recycled lumber, lower grades, and so on. It helps to be able to recognize and compensate for some of the very common defects in your material.

The Wane: Bark is the protective outer covering of the tree. When the tree is milled, some boards may have pieces of bark along one edge, or have an uneven spot where the bark has fallen off. This is considered a defect in the board because it reduces the width of the board erratically, making it impossible to fit it tightly. This defect is known as a "wane." It is not very serious except where you need an absolutely perfect joint.

Knots: Knots occur wherever a limb has grown from the tree. They vary in size from a fraction of an inch to half a foot in diameter. They are considered a major or minor defect depending upon how large they are, how numerous, and the intended use of the wood they appear in. Knots may loosen and pop out, leaving holes in your siding. They may create a weak point in your roof rafter or floor joist. They are composed of and surrounded by very *hard* wood which is difficult to saw through and hard to drive a nail into. When you are measuring a board, try to plan so that knots fall in a space you won't have to nail into or saw through (between the studs, for instance, if you are measuring a piece of sheathing or siding). If you have to use a rafter or joist with a large knot in it, place the board so that the pressure holds the board together rather than pushes it apart at the knot.

Splits or Checks: These are sometimes found at the ends of boards that haven't been properly seasoned or have been left outdoors unprotected. The wood fibers literally separate along the grain. These splits will become more and more severe if the lumber is left unprotected and may totally ruin your board(s). If possible, the split part should be sawed off and discarded. If necessary, the board may be used if you are very careful when you saw, handle, and nail it.

Warping: Other common defects in lumber affect the plane surface of boards and are known as warps. Warps may occur in the wood naturally or be caused by improper drying or handling. They may be very minor or so severe as to ruin a piece of lumber for any use besides firewood. If you are buying lumber, watch for these defects and refuse any pieces that are badly warped. Even low-grade lumber should be reasonably straight and true. As you build, look carefully at each piece you work with, determine its warp, and use the board accordingly. Even the straightest-appearing board will often have a gentle crook or bow. The defects of plane surface are the crook, bow, cup, and twist.

The *Crook* runs this way:

If you are working with a 2×4 or 2×6 lift one end up to eye level and sight down along either the upper or lower edge. You should be able to see the crook of the board, even if it is quite subtle. This is something you can develop an eye for, and it helps at first to consult with a friend. Some boards are, indeed, straight—but most I have worked with have some crook to them. If you are using this piece for a roof rafter, floor joist, rough window framing, and so on, you should cut and position it according to its crook. You want the pressure on the piece to work to compensate for rather than accentuate the crook. For example, a roof rafter should be placed with the crook curving *up*, so that the pressure of the roof presses down and straightens the board.

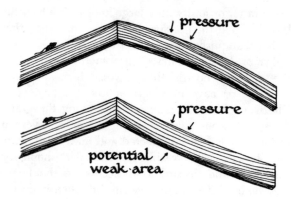

A *Floor Joist* would be positioned similarly, so that the pressure bears down upon the arc of the crook. If you are using the piece as a wall stud, it matters little which direction the crook goes. It will (hopefully) be forced into place when the siding is put on. Severely crooked pieces should not be used in your rough framing as they may weaken the entire structure by creating unbalanced pressure areas.

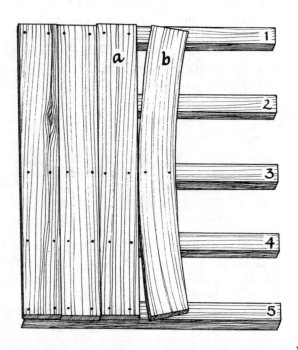

If you are working with flooring that has crooks, put some extra energy into forcing each piece as straight as possible before nailing it in place. Otherwise you will find yourself with gradually increasing arcs that may ruin your whole floor, or at least leave you with some problematic cracks. One method I've used to correct crooks in flooring employs leverage.

Board A is straight and nailed in place, but board B is crooked, leaving gaps at either end when it is laid in place. Board B may be nailed first to joist #3, where it fits tightly. Then go to joist #2 or #4 and use leverage to force the board back to join board A tightly. You can use an old chisel for this. Place the chisel just under the edge of board B as shown

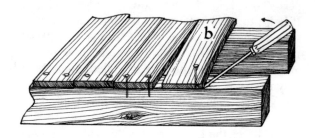

and use your hammer to drive it about a quarter inch into the floor joist. You should then set a nail lightly in B (not through to the joist, just enough to hold it upright). You can then push the chisel toward board A as hard as possible. This should force board B to straighten out and fit up against A fairly well. Nail it in place. Then go to joist #1 or #5 and repeat the whole procedure. Finish the other end similarly. If your floor is low enough to the ground, you may be able to use a crowbar pressed against the ground between the joists to achieve the same squeezing action. Sometimes you can even push with your foot to force a board into place. Roofing may be straightened with the chisel method, too.

The *Bow* is similar to the crook, but runs this way:

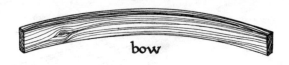

bow

124

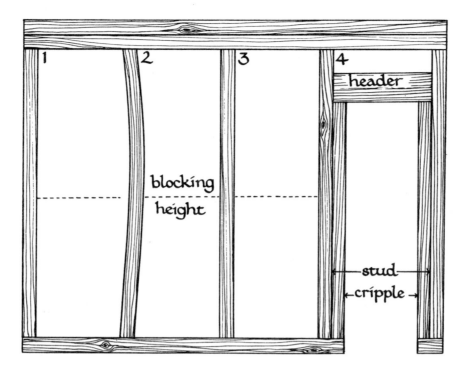

It can create problems in rough framing, siding, and so on. For example, in framing a wall, a bowed stud makes a problem.

Stud #2 has a bow that will make the distance between studs #1 and #2 larger, and the distance between #2 and #3 smaller. If you measure at blocking height to get the dimensions to cut your blocking pieces, you will find yourself with some odd-sized pieces. Make your measurement along your bottom plate where the studs are centered and nailed properly. This will give you the right measurement for your blocking lengths. When you go to nail the blocking in place, you will have to squeeze stud #2 over toward #1 in order to straighten it and fit the blocking snugly and properly in place. Once this piece is in place, the blocking between studs #2 and #3 will have enough room to fit properly. All of this seems simple to the point of absurdity, but if you've ever gone to nail blocking that's been measured to fit a bowed stud, you'll know how frustrating even a quarter-inch mistake can be!

Badly bowed two by fours can have a useful place in rough framing—though a minor one. Nailing them together for double headers over doors and windows is one possibility. You can also use them for cripples. In all of these cases, make your measurements very carefully, as the bow might throw your measurement off a half inch or so. Place the pieces with the bows this way

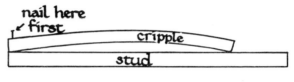

so that the joining will force them toward one another. Nail from one end and be careful to keep your edges aligned. Use a nailing pattern that zigzags to hold the wood together.

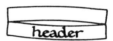

nailing pattern

Trying to nail a bowed piece of siding can also be frustrating—the board will often split when you nail it. Usually the board should be positioned with the bow out (curving away from the building) and nailed gently in place with blunted box nails.

The *Cupped* piece of wood looks like this:

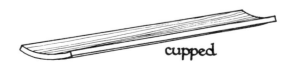

You are most likely to find a cupped piece of one-inch material, particularly in the greater widths (1×8, 1×12). Thicker pieces (2×8, 4×12) are rarely cupped. The cup creates a problem when you go to nail your board: a nail driven in the center will probably hold, but those driven near the edges will tend to split the board. Try placing the board with the cup arcing away from the building. Use blunted box nails and drive them very gently.

The *Twist* in a board is usually found just at one end:

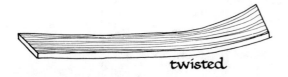

twisted

Again, this is most likely to be encountered in thinner material. The twisted end should be discarded or handled the way the cupped or bowed wood was handled.

Other Defects in Wood: Other defects in wood include decay, holes, and damage by fungi and insects. Wood that is decaying will be softer and sometimes spongy. It may be a different color in the decaying areas. As the decay advances, the wood begins to crumble. Old wood will sometimes have areas of decay interspersed with areas of good, solid wood. Most decaying wood should not be used, as you will find yourself having to redo all your work and replace the wood anyway. Some may last inside, though—old siding may be beautifully fashioned into cupboards and shelves, or used as interior siding.

Holes in new wood may have been caused by milling or handling equipment, or by insects. These holes lower the value of the wood without causing any further damage or weaknesses. Wood that has been exposed to rain or dampness or improperly stored may have a surface discoloration of blue or black mold. If this wood is dried out or used in an area with freely circulating air, it should be fine. The staining is unattractive but not really damaging. If you find exposed wood on your structure beginning to be discolored this way (spotty, irregular patches of mottled blue or black), you may want to treat the wood with a preservative. The growth of this fungus is an indication of exposure to too

much dampness—something you should deal with *before* your wood becomes badly damaged.

Buying Lumber: Lumber is sold in two ways: by the board foot, and by the linear foot. "Board foot" denotes a piece of wood that is one inch thick, twelve inches wide, and twelve inches long, or its equivalent. If you are buying 1×12 material, you are automatically buying by the board foot. Price on this lumber, quoted by the board foot, is easy to compute. If you are buying any other size lumber, though, you have to figure out the equivalent board feet. A foot of 2×6 material will be equal to one board foot. A foot long piece of 1×6, though, equals half a board foot. A piece of one foot long 2×12 material equals two board feet. Any material less than one inch in thickness is computed as one inch. Price may be given by the linear (or lineal) foot, which is simply the length of the material, regardless of size. A piece of 2×12 lumber, then, may be priced at thirty-two cents a board foot (a piece ten feet long will cost you $6.40) or sixty-seven cents a linear foot (the ten-foot piece then costs you $6.70). When lumber prices are quoted be sure you understand which system is being used.

Lumber size is also tricky. When you ask for a two by four, and go to work with your wood, you will find yourself with a piece that is actually one and a half inches by three and a half inches (or smaller). The "two by four" refers to the rough, unfinished measurement of the board. When it is seasoned and milled it comes out quite a bit smaller. This is important to know when you are calculating the amount of wood you will need or figuring out any dimensions. We recently bought 1×12 siding which is actually ¾"×11½"—it made quite a difference in the number of feet we needed to side and roof a sixteen foot by twenty-four foot building! Price, of course, is calculated by the rough, unfinished dimension—so you pay more and get less. If you are used to working with new wood and chance upon some old, recycled lumber, you are in for a pleasant surprise: two by fours that actually *are!* Some combination of abundance and integrity allowed early mills to turn out lumber that actually was what it was said to be. Now our lumber, like our resources, is shrinking by the year.

If you are buying lumber, you can economize greatly by planning your building's size according to the sizes lumber is milled in and cut to. There is an art to this, which I am only beginning to grasp. Plan your building for the least waste possible,

I am feeling crazy and pressured all the time now. Peter is supposedly helping build this place, but mostly just being here and needing me. I am earning all the money, working overtime to have enough for both of us. I resent him constantly: resent his not earning any money, am angry at his feeling free about spending money while I work and save and worry about the bills.

I have been spending most of my time with women these last few months, often going to meetings or going out. I have the women's group, the weaving co-operative, open women's meetings. With the women in the group I can just be myself. If my verbal fluency intimidates Lea, who is quieter, we struggle to hear each other because our sense of self isn't threatened by the other's independence. We've all been so well trained in sensitivity to others that working together flows easily. Rhythms mesh instead of clash. With Peter there is always tension now—he says that I give him nothing, that I am never here for him. I am frightened by his dependency, by his need for me to reassure him. And I feel angry that all I do is not enough. That even though I support us, he wants me to devote myself to him and our home, to give up all my time spent with women. I feel worn out with all the work and struggle, with always having to allay his fears.

knowing that lumber is cut in multiples of two (six-, eight-, ten-foot lengths). If you adapt your building to the size of the lumber, you can save a lot of waste and therefore a lot of money. Side and roof and floor your building according to a plan that takes the multiple system into account. This orientation is equally valuable when you are using recycled wood. By the same reasoning you can choose your lumber according to the size of your building. If you plan vertical siding on a shed whose walls are seven or eight feet high, try to buy lumber in eight- or sixteen-foot lengths. If you buy in ten-foot lengths, you will lose two or three feet on every board. This waste makes wonderful kindling or nice shelves . . . but at the price you probably pay per board foot, it's an expensive luxury. Similarly, when you figure out the pitch of your roof, consider the lengths of your rafter material. A slightly lower pitch might save you a lot if you can get by with fourteen-foot pieces instead of sixteens, or if you have half a foot of waste per rafter rather than three or four feet. In this same vein, it is sometimes thriftier to build close to building code dimensions, even though you are nestled deep in the woods and nurturing a healthy anarchy. This is true simply because code and companies go hand in hand. Codes require and companies mill their

wood according to certain set standards (who is to say in which directions the benefits flow?). Lumber in sizes that are standardized (wall studs, for example) are mass-produced and usually cheaper than equivalent pieces purchased by the foot. It might smart a little to build your dwelling with code-sized walls, but it might make the building considerably cheaper. Similarly, spacing your studs on more or less conventional centers will allow you to use pieces of siding, plywood, and so on with minimum waste and a lot of time saved in cutting. All of this takes some thought and investigating (and compromising), but it seems worthwhile in terms of resources both financial and natural.

Handling Your Lumber: Once you've chosen and ordered or picked up your lumber, you should prepare a storage area for it. Properly stacking and storing wood will help prevent a lot of loss and waste in twists, bows, and splits that commonly develop when wood is handled thoughtlessly. One lesson I've learned in ordering wood is to be there when it's delivered. You have a right to refuse any piece that is badly warped or is split when dropped from the delivery truck. If you aren't present when your lumber comes, you may end up paying for unusable boards. All lumber should be stacked in

the levelest area close to your building site. Use old boards or cement blocks or whatever you have to stack the wood on—it is important to keep it up off the ground. Make your stacks evenly and carefully. It will help when you actually begin building if you group the wood according to its various sizes and lengths and uses. Don't bury your studs under your roof rafters or stack all of your lumber on top of your piers! The lumber should be covered to protect it from dampness and the drying effect of the sun. After extensive use of black plastic in some big building projects, I've come to a sadder but wiser position concerning this material: it is not worth it, economically or ecologically. Though the initial expense seems fairly low, the plastic will not hold up. It tears and rips very easily, and once it begins to deteriorate it is one of the worst unnatural disasters you can inflict on the land. If you possibly can, invest in a few good canvas tarps. Properly cared for, they will last you for years and years—through many a building project. And in the end they will disintegrate and return to enrich rather than pollute the earth.

Any building material that must be perfectly straight (window and door jamb material, for instance) should not be bought until you are ready to use it. Even then, it should be stored indoors until actually used. If you are buying or salvaging lumber that you don't plan to use for a long time, stack it inside a barn or shed if at all possible. Careful handling of all of your lumber will make your work infinitely easier—repaying your efforts with a material that is a pleasure rather than a problem to work with.

FIGHTING FIRES

Where I live (northern California) the late summer, stretching into fall, is a time of no rain, when the golden grass starts shriveling and flattening in the heat. My first hot, dry summer here made me very conscious of the danger of fire, and fairly scared. The fear is real, and healthy if it helps you take precautions, but if a fire ever happens, you will need to move quickly and surely. Planning ahead, and having equipment on hand, is the best way to cope with your fear and the fire. I can tell you my own experiences and all the things I learned from and after them.

I lived in a commune in a rural community of about 150 people. The community stretched along a canyon—redwoods and eucalyptus in the bottom, changing to live oak, bay laurel, and madrone trees, then thinning out to dry manzanita brush and scrub pine across the hot ridge tops. One mid-October afternoon we were patching our roof when we heard yelling and car horns up on the ridge. We looked up and across our ravine to a huge column of glowing smoke rising behind the eucalyptus trees. There was no doubt that it was a large fire, uncontrolled. It was the first fire for most of us, and all we knew was the general principle—that *we* had to stop it. We grabbed the few tools around and started running, jumping on the first car or truck of neighbors that passed us. The whole community turned out as a matter of survival. The fire was in the manzanitas and scrub fir, about five to seven acres was already in flames over my head. Luckily there was already a dirt road up there, a fire break along one edge, so all we had to do was to keep it from jumping the road, and to stop the leading edge.

I started working heroically and frantically along the edge of the road, trying to keep the fire back, but after the first gust of air sent flames roaring over my head, followed by stinging, blinding smoke, I moved in closer to an experienced old-timer, asking for directions. After about twenty-five minutes, the nearest fire department arrived, but there was no water source so they were limited to their tanker truck and by the physical difficulty of access across rural terrain. Then a neighbor arrived with his bulldozer—it's not pleasant to order a swath cut through the land, but the only way to stop the fire was by creating a break where there was nothing to burn. Between us all, the fire was

contained by late afternoon, but we had to keep a vigil on the entire area for two more days to put out the flare-ups.

Since then I have been through another fire in similar scrub terrain, a grass fire, and several house fires—more than I hope you ever see. I've learned a lot, trying to prevent fires and to prepare myself.

Preparation is the Most Essential Thing. If you ever have to fight a fire, there will be *no time* to figure things out. You need to be prepared to move fast and efficiently to take care of everything, and forethought will really help you deal with panic.

First, look at your situation. Do you live in a fairly populated or isolated area? What type of terrain, and therefore, what type of fire is likely? Where is the nearest help—fire department, neighbors, Forest Ranger, etc.? What communications do you have—phone, electricity for a siren, a gong, or will you have to send for help? I lived in a fairly populated area with phone and electricity. Two households, one at each end of the community, were equipped with electric fire sirens. When there was a fire someone called the fire company and the siren houses, which were also the beginning of an

emergency phone tree for the community. If you have phones, organize a calling list that spreads across the whole community as fast as possible, and include special contingencies like calling the school, or notifying people without phones. Give your information concisely and clearly—exactly where the fire is, what type, how big. Our house had one of the sirens because, as a collective of ten people, there was almost always someone at home.

Next, know what you will need to do at your home before going off to fight the fire. The critical factor will be how close the fire is, so have several plans. Do you have small children or invalids who will have to be taken care of? If so, know all alternate escape routes and possible transportation—get them completely out of the way if there is any danger, so you won't be distracted. If the fire is at your house or has any chance of reaching it, turn off the electricity at the master switch box—a fire will sever wires, leaving you working with water and 120 volts . . . If the fire is coming toward you, do you have animals, and if so, should they be moved or turned loose? Horses panic in a fire and are apt to run straight into it, or refuse to leave unless you blindfold them and carefully move them to a safe location. Goats stay calm and can be used to lead other animals out of danger.

Next (all of this actually has to happen simultaneously . . .) get yourself ready. For rural terrain you need to wear sturdy boots, comfortable long pants, and a long-sleeved shirt. Even if it is hot, take a long-sleeved shirt, maybe worn over a T-shirt—it will make it easier going through brush and intense heat. Take a cotton scarf or kerchief—damp, if at all possible, to help with smoke inhalation. If you have one, wear a hard hat, especially in forest fires. Unless you are in imminent danger, take the extra three minutes to put on good boots and clothes.

Now, as you run out, grab the tool you need. I mean that literally—you will be running and you will need to have the right tools, in good condition, and always kept in a central, accessible place. The tools vary according to fire. Shovels are essential, and flashlights if it is night. For a scrub or low-woods fire, you also need axes, machetes, adzes, picks, or mattocks. If you want, mark your tools with tape so you can tell them from others' after the fire. For a grass fire, where sod is thick and hard to dig through, the best thing I've found is a wet blanket or gunny sack; otherwise, picks will be needed to break the ground. You should, of course, have several household fire extinguishers around the house; keep them in likely and handy places—the kitchen? . . . In addition to the usual small CO_2 extinguishers, we got two foam extinguishers (like they had in your elementary school . . .)

from a surplus store. If you can afford one (check surplus stores), get a water backpack—a five-gallon can that straps on your back and has a hose with a hand-pumped nozzle on it. Backpacks are really useful in areas a fire truck can't reach, but they are heavy, so practice carrying and using one. Salt tablets are also necessary when fighting large fires—they slow water loss and heat exhaustion. Think about whether your vehicle will block the road or get trapped by the fire, and then get there as fast as possible with your tools.

Once you are up against the fire, first check its direction so that you never get trapped. A FIRE BURNS UPHILL, almost never down—that is one of the most important factors to remember. The heat creates an updraft which pushes the fire up until it reaches a ridge; then it will usually run sideways (whichever way the wind blows) rather than going down the other side. So, if you are ever trapped above a fire, try to get over the ridge and down the other side, angling in the direction the fire came from once you are over the ridge, so that it will pass behind you. Be sure to approach a fire steadily and systematically—frantically rushing around will only tire you out.

There are three ways to put out a fire that I know of (besides chemicals)—water, suffocation, and a fire break. Your main effort will be to contain the fire, so that it burns itself out from lack of fuel. If you have a water source and hose or buckets—splash away. Concentrate your efforts on the edge because that's where the fire moves, and wet each area *thoroughly*. Most effective, though, is a fire break—clearing all growth in a wide swath across the fire's path. The swath should be slightly wider than the surrounding growth is high (i.e., if burning brush is ten feet tall, make fire break at least ten feet wide). Be sure you start the fire break far enough ahead of the fire to be able to get through safely—always sacrifice a little more area to save lives. Be generous in guessing the fire's speed, *especially* in hot dry weather. Don't try to put a fire break across a steep slope; try to take advantage of the terrain—build your line where fuel is thin, or where natural openings in brush or ridge tops will slow the fire down. As you chop and cut out brush, throw it back *into* the fire—that keeps you from building bridges for the fire, but doesn't significantly speed up the fire. Dig along the swath with shovels, throwing the dirt on the fire to suffocate it; again, watch where you throw, and concentrate on the leading edge. If the fire is on a steep slope, dig a trench along the bottom edge to catch burning brush that rolls downhill. Work in teams—first, brush cutters, with axes; second, shovelers; third, rakers and scrapers to completely clear the fire line. Shovelers should use the cutting edge of the shovel to be sure all roots are severed in the fire break.

For a grass fire, have two people holding opposite edges of a wet blanket walk along the fire edge laying down the blanket, picking it up, moving a few steps, laying it down, etc. Move slowly so that it suffocates the fire and *thoroughly* puts it out—otherwise you may just fan the flames with your flapping.

Always be aware of your situation—constantly check which way the fire is moving and what your escape routes are. Keep within seeing or shouting distance of each other. Don't underestimate heat and smoke—smoke especially can suddenly surround you, block your vision; so move out of it, keep your orientation. In addition, watch behind you. In our scrub fire, the fire traveled along the roots underground, and occasionally I'd find a small blaze starting behind me. If you are fighting a large fire, be sure to take rests periodically—it's exhausting work, and as you become tired you'll become a danger to yourself and others.

Know who to contact for major equipment: who to contact for a bulldozer; who to contact to get a borate bomber (this is a plane that flies in low spreading borate—a pink, decomposable salt—over the fire to suffocate it. A bomber may be absolutely essential in a forest or large grass fire, but it has to be authorized by a fire or forestry official).

Once a fire is contained or dying, remember that embers or burning roots will continue to flare up—so a long vigil must be kept to put out the spot fires.

The above information is about land fires. I have watched four houses burn down and know very little about how to stop them. The key is to catch the fire early and to have extinguishers and water available. If water is scarce, collect several fifty-gallon drums of rain water during the wet season, and save them for fire *only*. If you can't stop the fire, the first thing, absolutely, is to make sure everyone gets out. Sound the alarm, and turn off the electricity at the main fuse box. Shut all windows and doors to cut off air and slow down the fire. Beyond that, just try to keep it from spreading by building a fire break around the house. A good preventive measure is to cut back or clear the flammable grass and brush from around your structures (forestry laws suggest a thirty-foot perimeter).

It is hard for me to end here and pretend that I have shared enough information. There are other ways to fight fires—like starting a back fire—which I haven't explained because I don't know about them. Check with a local fire official to learn other practices. A fire is full of variables, so start thinking about what to do *now*. Your biggest opponents will probably be time and fear—so think about what to do and practice until it is automatic and efficient.

LADIES' AUXILIARY

After the first big fire in our area, we invited the nearest fire marshall to speak to our town meeting. He came and explained a lot to us about use of tools and fire-fighting techniques. As he finished, he beamed and said: "You ladies can be useful too! Those men will need coffee and sandwiches, and will really appreciate your help."

There was a short silence and then one of the women stood up slowly. She voiced our feelings: "I don't know where *you* were, but during that fire I was out there with my shovel and my hatchet stopping it. So don't tell me how useful my coffee is when my community is on fire!"

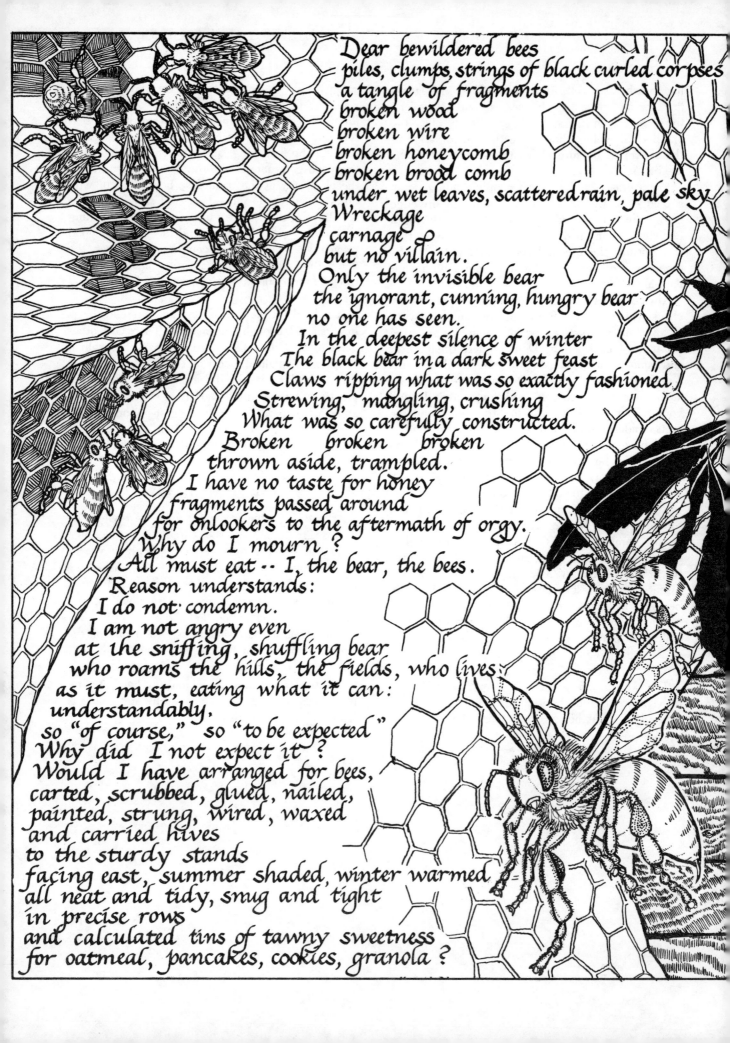

Dear bewildered bees
piles, clumps, strings of black curled corpses
a tangle of fragments
broken wood
broken wire
broken honeycomb
broken brood comb
under wet leaves, scattered rain, pale sky
Wreckage
carnage
but no villain.
  Only the invisible bear
  the ignorant, cunning, hungry bear
  no one has seen.
    In the deepest silence of winter
    The black bear in a dark sweet feast
    Claws ripping what was so exactly fashioned
    Strewing, mangling, crushing
    What was so carefully constructed.
      Broken   broken   broken
    thrown aside, trampled.
    I have no taste for honey
    fragments passed around
    for onlookers to the aftermath of orgy.
    Why do I mourn?
    All must eat -- I, the bear, the bees.
    Reason understands:
    I do not condemn.
    I am not angry even
    at the sniffing, shuffling bear
   who roams the hills, the fields, who lives
  as it must, eating what it can:
  understandably.
  so "of course," so "to be expected"
Why did I not expect it?
Would I have arranged for bees,
carted, scrubbed, glued, nailed,
painted, strung, wired, waxed
and carried hives
to the sturdy stands
facing east, summer shaded, winter warmed
all neat and tidy, snug and tight
in precise rows
and calculated tins of tawny sweetness
for oatmeal, pancakes, cookies, granola?

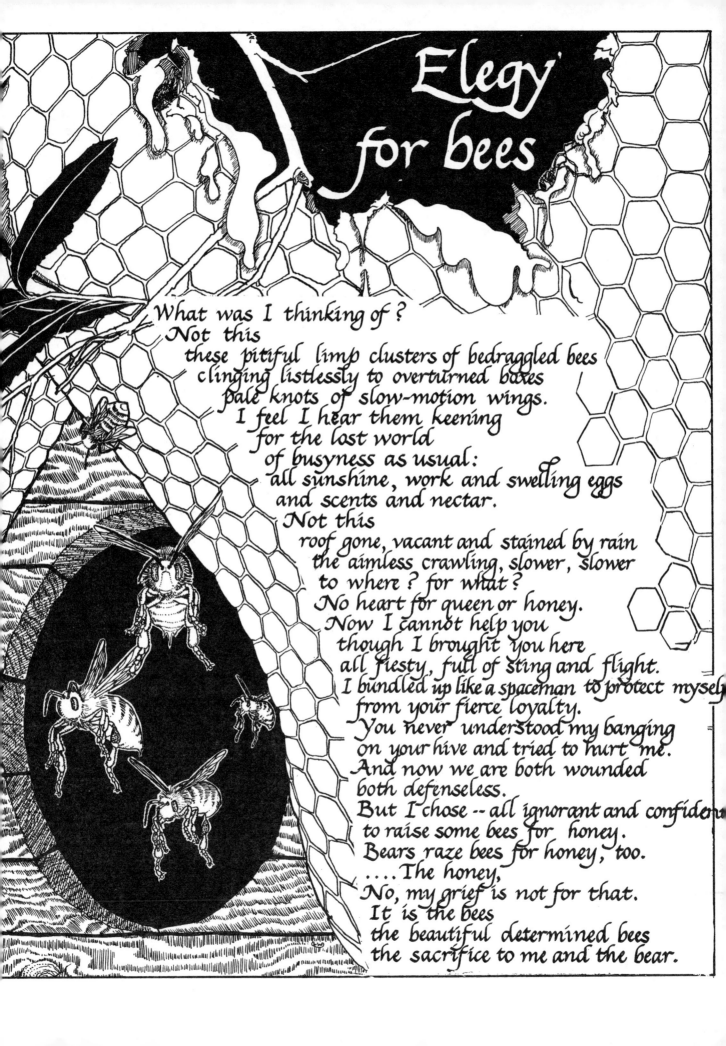

Elegy for bees

What was I thinking of?
Not this
these pitiful limp clusters of bedraggled bees
clinging listlessly to overturned boxes
pale knots of slow-motion wings.
I feel I hear them keening
for the lost world
of busyness as usual:
all sunshine, work and swelling eggs
and scents and nectar.
Not this
roof gone, vacant and stained by rain
the aimless crawling, slower, slower
to where? for what?
No heart for queen or honey.
Now I cannot help you
though I brought you here
all fiesty, full of sting and flight.
I bundled up like a spaceman to protect myself
from your fierce loyalty.
You never understood my banging
on your hive and tried to hurt me.
And now we are both wounded
both defenseless.
But I chose -- all ignorant and confident
to raise some bees for honey.
Bears raze bees for honey, too.
....The honey,
No, my grief is not for that.
It is the bees
the beautiful determined bees
the sacrifice to me and the bear.

Peter and I have separated this week, maybe for a while, maybe forever. I feel hurt, confused, torn. I don't know how we let things get this terrible. I remember our wedding and my certainty that if we ever separated it would be as friends. Now we can hardly stand the sight of each other—so many broken hopes, expectations, needs. And somewhere buried deep perhaps still some love and caring. It's hard to meet his eyes when they are gentle: there's too much experience there between us. I don't know what I want. More and more I am coming to believe that there are no men I've ever met who are really self-aware. There are a few I know who are open to growing and changing, but none who would be if they weren't afraid of losing their women. I know all that and know that there are no men I trust now, but somewhere lingers the expectation Peter will be back in my life. I can't conceive that it could be over; somewhere I feel that this is all temporary, but I don't know where that comes from. Maybe just the need to love and share with another, with others.

Yet, I've actually been enjoying the silence and slow pace of the last few days. Nothing to distract me from the moment—no focusing on tomorrow or dinner or what happened in bed this morning. Finding my own rhythm, flowing with it, I feel in tune with the earth. Moving slowly, I get so much done. Every moment is full. I think that I need someone to share my life with; I feel afraid of being alone. Yet this aloneness is feeling good.

speed composting

Speed composting is a method of rapidly converting organic materials into garden-ready humus. When we first began gardening we had hard-packed clay soil that we were in a hurry to improve. The rich, organically balanced soil produced by a compost heap could become our topsoil. When we read about speed composting in *Organic Gardening and Farming* magazine, we decided to try it. By following Rodale's simple method closely, we had perfect success!

To contain our compost heap and protect it from marauding chickens, we built a simple collapsible bin. A compost heap should be made above ground as a pit will tend to collect and hold water, making the heap too wet. The four sides of our bin were built as separate units and wired together. The bin could easily be taken apart when the heap needed turning, or it could be disassembled and moved to a new spot. Our bin was four feet wide, eight feet long and four feet high. Each side was a simple frame of 2″×4″'s covered with sturdy two-inch-mesh chicken wire. A sheet of half-inch plywood covered with tar paper acted as a top. A larger bin could be used, but ten to twelve feet wide and five feet high is considered maximum.

The first step was to turn up the ground where the bin was to stand. This activates soil microbes that will aid in the decomposition process. When this was done we set up three sides of our bin and wired them together. We then took four or five poles (at least two inches in diameter and four feet high), sharpened the ends and drove them in vertically at random points inside the bin. These poles, which will be pulled out when the heap is completed, make air shafts down through the heap. As an alternative method, use cylinders of chicken wire or other fine screen—these may be left in place.

We followed Rodale's Indore method of layering the different materials of our compost. First went a layer of straw and manure from our goat barn. This layer could also be plain straw or spoiled hay, garden rubbish, grass clippings, or leaves. Shredded material will decompose even more quickly than unshredded. However, we didn't shred ours. The layer should be about eight inches deep. Next goes a layer of fresh manure—an inch or so will do. This acts as an activator for bacteria and fungi which do

the work of decomposing the organic matter. We followed this with a layer of ground limestone and wood ash. Rodale suggests "four ounces of lime to each square yard of surface" (*The Complete Book of Composting*) or twice as much wood ash (the ash contains lime). The lime is necessary to keep the heap from becoming too acid. Soil bacteria do best under slightly acid to neutral conditions. As the fermentation process creates an overacid condition, lime is needed as a neutralizer. After this came a thin layer of earth. At this point we watered the heap thoroughly. The entire layering procedure, including watering, was repeated again, adding a few extras—old wool, household garbage, raw rock phosphate, etc. When the heap reached the top of our bin, we gave it a final watering, pulled out the air shaft poles, and put up the fourth side of the bin. We covered the bin. The end of the first stage.

Each of the next three days we watered the heap briefly. It is important to keep the heap damp and not let any part dry out. On the fourth day, we turned the heap. Our collapsible bin proved perfectly adapted to this. We took down the four sides and set up three close by. Again we turned up the ground with a hoe and set up our poles. We shoveled the heap into the relocated bin, watering it well as we went along. Turning the heap aerates it, enabling the bacteria to function at maximum. We found the lower third of our heap had already begun to decompose. Again, we removed the poles and covered the bin.

Four days later we again turned and watered the heap. Then we waited a week and repeated this. About four days later our compost heap was done. The straw, wool, manure, and garbage had been converted into dark, crumbly earth—topsoil for our new garden.

Speed composting will provide you with garden humus in about a month. If the weather is really cold, it may take longer—or you can make a more protected bin to avoid heat loss and slowing of microbial action. Normally, a compost heap left to itself will take about three months to decompose properly. One disadvantage of speed composting is the extra time it takes to water and turn it at first. And you must have all the materials on hand in the beginning. But for that special soil for early planting or problem gardens, try this method.

SOIL PREPARATION

I garden mostly by vibration and experimentation. My gardens have varied enormously, from my first "put the seeds in the ground—hope it rains—and fish the vegetables out of the weeds" effort to my last "seventeen truckloads of manure and hours of hard work and happy puttering." Along the way I've learned a lot, though I am only now beginning to label with theories what has come from actual experience. Most of all, I've learned that good soil preparation is what makes a garden. Not since the year I tended a meager garden in what looked like rich, dark soil have I taken the ground for granted. What comes out of it must be replaced, and even years of growing only grass take their toll.

Before you start to prepare your soil, you should take a good look at it. Soil is composed of mineral particles and humus—decayed organic matter. In the spaces between the mineral particles are water, air, and dissolved nutrients, all of which are essential to plant growth. Soils are classified in three types, though few soils are truly typical: clay, sand, and loam. Clay is made of minute particles that tend to bind tightly together, excluding air and water and keeping nutrients unavailable to plant roots. Sand is made of larger, irregular-shaped particles with large spaces between them that cause rapid water drainage and leaching of nutrients. Loam is a near perfect combination of different-sized particles able to retain humus (source of nutrients), water, and air. So, look at your soil. Pick it up and rub it through your fingers. Make a ball of it in your fist. Look closely at what it's made of. Clay soils are usually reddish brown or gray in color. They will stick together in lumps when you squeeze them and bake into hard lumps in the sun. Loam is dark brown to black, full of tiny bits of vegetable matter and grainy particles. It crumbles

Marriage doesn't end easily; it feels like being torn apart. Marriage is safety and security and insurance. It doesn't matter how bad the present is, there's a promise about the future. I can describe every detail of the farm in Idaho that Peter and I would have had fifteen years from now; I've dreamed and planned it all so many times. Now I am futureless, every dream and plan ruptured, torn in half. I can't just go on with my life; I have to begin all over again, to build a life that counts only on myself. No matter how much I know that this is what I have to and want to do—that being me, I have no other choice—I still wish that I was someone else, that I could believe there would be a "we" and a farm in Idaho fifteen years from now.

through your fingers and won't form a tight ball. Sand is pale brown, yellow, or red with little or no organic matter. It feels gritty, will dribble through your fingers, and not hold any shape.

Another thing you should be aware of is the acidity-alkalinity balance of your soil (soil pH). Most vegetables will grow well only in a neutral to slightly acid soil. Some (i.e., tomatoes) will grow well only in a really acid soil, while others (the cabbage family) only grow well in slightly alkaline soil. You can buy soil-testing kits to check for pH. Or you can get soil litmus tape, which will give a general idea of your acid-alkaline range. I've never used either but have relied on general local information. Soils that come from leaf (or pine needles) mold are acid in nature. (Leaves are acid in nature.) Acid soils also tend to attract other acid-loving plants, like huckleberries, rhododendrons, blackberries, and azaleas. Soils that derive from grasses tend to be neutral to alkaline. Find out what your soil is like by asking other gardeners and old-timers and by noticing what grows in and near it.

Now you know about your soil, what can you do for it? Clay soils can be improved by the addition of calcium which causes the clay particles to adhere to each other in small "crumbs," helping aerate the soil. In alkaline-clay soils, add gypsum, spread over the ground like a light snow and then dug in to obtain calcium. In acid-clay soils, lime will furnish calcium. It is mostly readily available as powdered dolomite. Sand is also a helpful addition to clay or clay-loam soil, especially where you plan to plant root crops like carrots and beets. It will loosen the soil and help these vegetables penetrate

easily. Carrots take the strangest shapes when stunted by hardpan clay! If you have the time, planting a cover crop of soybeans, caraway, buckwheat, or flax, which can be dug into the soil at the end of the growing season, will help break up a heavy clay soil too.

Sand can be improved by the addition of enormous quantities of organic matter, which will fill the air spaces and help the soil to retain water and replace leached out nutrients. Pure sand is difficult to improve to garden standards. But sandy loam is actually a desirable combination when generously fed with compost and manure. Lupine is an excellent cover crop, adding nitrogen and organic matter to sandy soils. It grows naturally in quarries and on dunes, helping convert sand to soil.

The acidity-alkalinity balance in any type of soil can be adjusted to suit the needs of your plants. Acid soils can be neutralized by adding dolomite or crushed limestone. Spread a thin coating over all areas—except where acid-loving plants will be planted—and dig this in when you turn your soil. Don't add your dolomite far in advance of planting as it will only leach out in the rain. Wood ashes also help neutralize the soil, and I add them, in addition to dolomite, to areas where I'm planting cabbages, broccoli, or cauliflower. Wood ashes on the surface of the soil also help keep away cabbage worms. Alkaline soils can be corrected by adding acid organic matter: pine needles, wood shavings, or leaf mold. If you add manure combined with wood shavings to your garden, remember that this will increase its acidity and compensate with extra dolomite if necessary.

The most essential addition to any soil is organic

matter, especially when it is high in nitrogen, the most essential plant nutrient. Clay and sandy soils need as much organic matter as you can gather and even good black loam responds to nourishment. The basic rule of soil preparation is to add as much organic matter as you can. Visitors who admired my twenty-two-pound cabbages and six-inch-diameter beets this summer didn't realize that the real magic was in seventeen loads of manure and not a green thumb. All your efforts before planting will be returned to you tenfold by midsummer. About the only way you can go wrong is by putting too much unrotted vegetable matter directly into your soil. It takes nitrogen to decompose anything; in this case, the nitrogen would come from the soil, robbing the plants. Green vegetable matter should either be composted first or worked into the soil several months before planting.

There are many fine sources of organic matter around you, wherever you live. In addition to compost, you can add well-rotted grass clippings and leaf mold, seaweed (an excellent source of trace minerals too), hulls and shells of peanuts, buckwheat and oats, stalks of any legume (very high in nitrogen), peat moss, and cottonseed meal. Many of these are available as by-products in agricultural areas. Any legume (alfalfa, red clover,

At the women's meeting last night, Judith said that she didn't believe that there were any relationships between men and women that were "honest and equal." I don't believe that there can be in this time: for a woman to be truly honest and equal challenges a man's deepest conditioning and there are no men that I know of who are able or willing to give up their dominance. The converse is true too—in relating to a man, all of my deepest and most basic conditioning comes into play and I always respond at some level as the Jennifer who was raised to be a "woman," to be acquiescent and pleasing and needing approval.

I find myself in the midst of changes I didn't choose or expect. I didn't set out to be where I am now. But here I am, really separated from Peter. My political reality has caught up with my personal life, and I find myself again a radical. I suppose all this was inevitable once I started to examine and admit to myself the dynamics of our relationship. Once I recognized how much I pay for those half truths—that the price of that relationship is my selfhood—then it was inevitable that I couldn't go on. Who wants to live out a perpetual suicide once she has started to feel the pain?

When I see Sarah tense because her time is filled with providing the meals, buying food, doing dishes, with "help" but no relief from the responsibility; when I see Kathy obediently molding her life to suit Danny, then I know why they are so afraid of Judith. Because if they let themselves see, then they will have to know what they are doing to themselves, and something in their life will have to change. Change is painful and taking responsibility for one's self is most frightening of all. How unprepared we women are to be whole people! I find myself respecting Margaret more because she's honest about the contradictions. She knows she wants to stay with Jim; she recognizes and chooses her compromises. So many of the others don't. It is so incredible—I have heard Sarah, Kathy, Lynda, Holly, all admit the pain of isolated incidents in their lives, but they are afraid to generalize, to see the whole pattern and what they're doing to themselves. I share all the fears of loneliness and alienation but I have no choice except to struggle for my own liberation. I have to find out who I am, and I can't do that if my focus is on someone else.

soybeans, vetch, as well as beans and peas) grown as a cover crop in the fall and dug into the soil in the spring will greatly enrich the soil. Legumes have the ability to take nitrogen from the air and deposit it in the soil. Nitrogen-fixing bacteria are available from seed supply houses and will double the amount of nitrogen deposited in the soil.

If you live near the coast, fish scraps and heads from the packing plants are wonderful fertilizer, buried treasure beneath your rows. Blood meal and bone meal are also excellent sources of nitrogen and other minerals, but they are fairly expensive to buy if you do not live in a meat packing area.

I myself am a manure fanatic. I think of my goats affectionately as speed composters. The corn stalks from the garden fed to them on Monday return to the garden as manure on Friday, a perfect symbiotic arrangement. Manure is high in nitrogen and is a good soil conditioner, but my love of it comes mostly from its ready availability. It takes a lot of collecting and effort to produce enough compost for a large garden. It takes time and energy to haul and dig in manure, but then you're done: rich, productive soil and "all" you have to do is water and fend off predators. If you don't have animals of your own, you can often get manure from farms and stables in return for cleaning it out of their barns. Last year, three women friends and I cleaned a sheep rancher's main barn, hauling over thirty pickup truck loads to our two gardens. It was hard, sweaty, satisfying work.

Manures are classified as either hot or cold. Hot manures literally give off heat as they decompose. They are also generally higher in nitrogen than cold manures. Hot manure should not be put straight into the garden while still fresh as the heat can actually burn the roots of plants. Hot manure can be stacked in compost-like heaps with a hole down the center to let air in, or it can be worked into the soil several weeks before planting. Hot manures are (in order): chicken, horse, rabbit, sheep (this is borderline). Cold manures are goat and cow and can go straight into your planting areas. Once you are aware of the rule, you can always hedge a bit. I dug huge quantities of sheep manure into my garden and planted ten days later with no problems. I have also added horse manure that was pretty fresh (heated up in the pickup truck bed) two weeks before planting and had the plants do fine.

Manure almost always comes mixed with straw or wood chips. This is nothing to worry about. In fact, they will enhance the soil and help break up clay particles. At first I worried that the decomposing straw would rob the soil of nitrogen, but the manure seems to provide plenty for straw and plants too. I do try to make sure I use mostly manure with some straw mixed through it. The parts of the barn that are mostly straw with manure mixed through, I save and use as mulch between the rows. That way, it has time to decompose slowly and will fertilize next year's plants.

The other principal nutrient your garden needs is phosphorus. I was unaware of this and wondered why my vegetables grew so slowly in virgin soil one year. Finally, bulging tumor-tipped zucchinis pointed to a phosphorus deficiency. Rock phosphate (the natural form) can be bought crushed in twenty-five- or hundred-pound sacks. A twenty-five-pound sack will cover one hundred square feet of garden. Rock phosphate decomposes slowly over several years and is best added to the soil several months before you plant. Bone meal is another good source of phosphorus. It costs more than rock phosphate but contains seven times as much available phosphorus. Bone meal, gypsum, rock phosphate, and dolomite are all available at nurseries and garden supply stores.

Once I have gathered together all the organic matter I can find for the garden, I still have to get it into the soil. With a new garden this means breaking the sod first. A Rototiller, if you can get one, will break the sod into tiny pieces and turn the soil at the same time. Without one, clearing sod is a long, backbreaking job, but I have learned to do it at my own slow rhythm, listening to wind and birds.

First I clear the sod using a mattock or an adz (a lighter weight, sharper tool usually used on wood). It's easiest to do this after a moist or rainy spell, but not directly after a rain for the the ground is heavy and sticks to itself. I lift the grass clods up, beat them to return the soil to the garden and then pile them in a wheel barrow to carry them to the compost heap. They will return to the garden next spring as soil. Then I sprinkle dolomite (if needed) and phosphate over the soil. On top of this goes a good thick layer of manure. If I have lots of compost, I add it too. If not, I save the compost to put on top of my planting rows. I usually prepare a small (8′×6′ or so) patch of garden at a time, and then turn it all under using a spading fork. The dirt falls through a spading fork so it is lighter to lift than a shovel. I turn the soil and dig the manure in a foot deep if I can. By the time I've hauled and spread the manure and turned the soil

> I am alone now after months of focusing on Peter, on visitors: interacting, re-acting, acting—always a little off balance, my energy given to and altered by those around me. Only now, alone, I am rediscovering my center, feeling how-ever so slightly crazy I become when I am relating to someone else—always a lit-tle off center, a little out of touch. Yet just a little, so little, that I never really know. I give my center away so easily; no one even has to try and take it from me; I'm already there, hands out: here it is. I abdicate. You can direct my energy as well as yours. Here, now feeling the incredible serenity and peace of aloneness, of being self-directed, I want to learn how never to give that up. I hope someday to feel that harmony so closely that I never lose it, even when I am with people. But it seems I will need much aloneness to get there.

in a patch, I am ready for a rest. So I stop for a few minutes, feel the sun and breathe the air; then I do another patch.

Always before, I have done this process in the spring, just before planting time and in some haste. This year, I turned the soil and dug in manure in the late fall. Then I covered the whole garden with a layer of newspaper on top of which is a straw-and-manure mulch. When spring comes I will push the mulch aside, to put between the rows, and have a garden ready to plant. The newspaper keeps weeds from growing and will eventually decom-pose and join the soil. I love this new cycle because it lets me putter and poke in the garden without haste or pressure during seasons that are normally times of dying and inactivity.

When I actually begin to plant, I use the *Basic Book of Organic Gardening* ($1.25, Ballantine Books) as a reference and tool. It lists every com-mon garden vegetable and describes the soil condi-tions, the amount of water, the kind of weather each one needs. So I consult The Book and then adjust my soil to suit the particular tastes of each plant. For nitrogen lovers, I dig in extra manure. For root crops, a little sand. For alkaline fanciers, an extra dusting of my wood-stove gleanings. For deep-rooting tomatoes, a little encouragement: post-holes two feet deep with fish heads and manure in the bottom. This book has proved invaluable to me as more and more I've tried to provide carefully for good plant growth.

Perennials (plants that continue from year to year) need extra attention paid to their soil prepa-ration as they will be feeding from it for a long time. You can dig in extra manure after a year or two, but you will have to be careful not to damage the roots. For most perennials, I just work a lot of well-rotted manure as deeply into the soil as I can and then mulch around the plants each year with very manurey straw (a cold manure).

Whenever possible I try to follow a simple plant rotation in my garden to make best use of the soil. Areas that have received a heavy application of manure and compost that year get the heavy feed-ing vegetables (cabbage, chard, celeriac, cucum-bers, onions, spinach, squash). The next year, these areas grow beans and peas, which help restore ni-trogen to the soil. The following year (with ma-nure or compost added each season, of course) come the light feeders (beets, carrots, parsnips, radishes, rutabagas, salsify, turnips). This way, the soil has time to rest and replenish itself. Looking at soil that I've been planting in for five years now, and seeing it richer and more fertile than ever be-fore is testimony to the effectiveness of these methods.

Your plants will let you know if the soil is lack-ing anything they need. Slow-growing or stunted plants in good warm weather may reflect an im-proper pH balance. I usually add a little leaf mold or wood ash, depending on the plant's taste, and see if that helps. But I am careful: overcompensat-ing can be as bad as under. Nitrogen deficiencies show up in slow growth with leaves or stems fad-ing to yellow. Often plants blossom but don't fruit. Squash and cucumbers will be pointy on the blos-som end. This can be cured by manure, compost, and watering with "manure tea." Manure tea is ma-

141

nure soaked in water overnight, then diluted to a pale brown color. Phosphorus deficiency is the other common lack. Its signs are a reddish or purple tinge to leaves and stems. Cucumbers and squash are narrow at the stem end and bulging at the tip. Ears of corn will be pointed and missing kernels at the tip. Add rock phosphate. Any time your plants look sad or hungry, feed them manure tea or fish emulsion. Fish emulsion is a concentrate made of ground up fish by-products and is rich in minerals. Dilute it one tablespoon to a gallon of water and never use it straight or it will burn the roots of your plants. Pay attention to your plants and let them teach you about the soil.

Before I became a gardening fanatic, I wondered if days of shoveling and hauling tons of mucky, sloppy, wet manure and hours of spading and turning heavy spring soil was worth the effort. Then I reaped an overwhelming harvest of every vegetable I had ever eaten and several more. Now I love the whole process from start to finish, this constant cyclical involvement with birth and death. Perhaps it is the cycle that I love most of all. As the summer's plants are dying, I am digging in manure for the spring—promise of a harvest yet to come.

Naked Summer

I look in the spin
and shake of the baler
and see your body
naked, lovely,
like cool rain
while you sit on the tractor
waiting for the baler
to get fixed
so we can get on
with the haying.
Under the hot sun
I see your body
like cool rain.

pLaNting and traNspLaNting

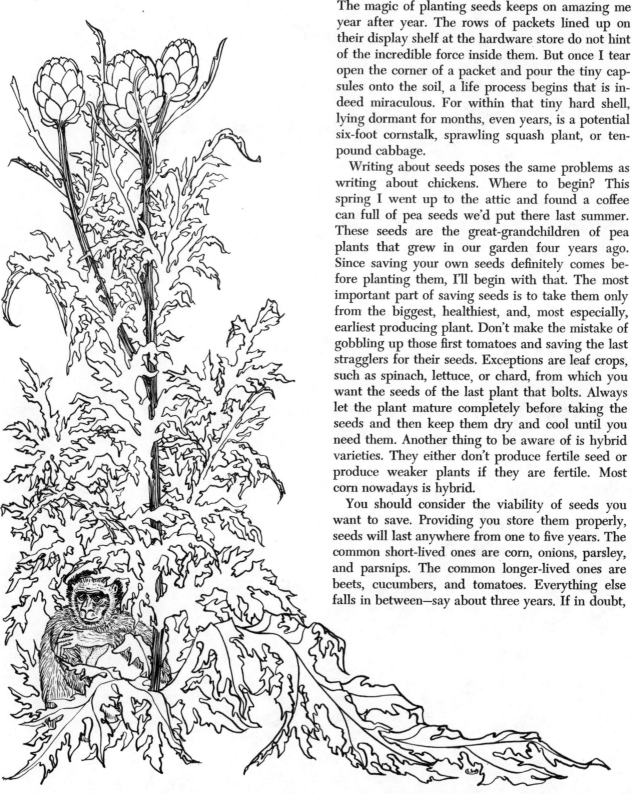

The magic of planting seeds keeps on amazing me year after year. The rows of packets lined up on their display shelf at the hardware store do not hint of the incredible force inside them. But once I tear open the corner of a packet and pour the tiny capsules onto the soil, a life process begins that is indeed miraculous. For within that tiny hard shell, lying dormant for months, even years, is a potential six-foot cornstalk, sprawling squash plant, or ten-pound cabbage.

Writing about seeds poses the same problems as writing about chickens. Where to begin? This spring I went up to the attic and found a coffee can full of pea seeds we'd put there last summer. These seeds are the great-grandchildren of pea plants that grew in our garden four years ago. Since saving your own seeds definitely comes before planting them, I'll begin with that. The most important part of saving seeds is to take them only from the biggest, healthiest, and, most especially, earliest producing plant. Don't make the mistake of gobbling up those first tomatoes and saving the last stragglers for their seeds. Exceptions are leaf crops, such as spinach, lettuce, or chard, from which you want the seeds of the last plant that bolts. Always let the plant mature completely before taking the seeds and then keep them dry and cool until you need them. Another thing to be aware of is hybrid varieties. They either don't produce fertile seed or produce weaker plants if they are fertile. Most corn nowadays is hybrid.

You should consider the viability of seeds you want to save. Providing you store them properly, seeds will last anywhere from one to five years. The common short-lived ones are corn, onions, parsley, and parsnips. The common longer-lived ones are beets, cucumbers, and tomatoes. Everything else falls in between—say about three years. If in doubt, you can germinate a few seeds in the kitchen to test their viability before you plant your whole crop in the garden and have nothing come up. Essentially, to do a germination test, you must keep the seeds moist—but not wet—warm, and dark. If they sprout, they are viable.

The moment of magic comes when I kneel over a newly prepared seedbed, warmed by the spring sun, and carefully lay those tiny specks of dormant life all in a row, cover them with the finest dirt, water lightly and wait. But the long-awaited sprouts may never appear if you don't perform the ritual properly.

First of all, the seedbed will need some attention. Large seeds, corn, beans and peas, can be planted in coarsely worked soil and still manage to push their way through to the sunlight. The small seeds —carrots, onions, and most herbs—can only sprout in the most finely worked soil. Another essential element for this life to begin is the sun's warmth. Cool weather crops can germinate with less heat, but all seeds will germinate faster when the soil is thoroughly warm. Spacing the seeds as carefully as you can will save waste and make thinning easier later. I usually plant three seeds for every one plant I'll keep. Of course, with those minuscule carrot seeds you just have to shake the package lightly and later thin *a lot*. My general rule for covering seeds is: large seeds (i.e., corn, peas) get one inch of soil on top; medium (such as beets, chard, cabbage) get half an inch; and all small ones (i.e., carrots) get under a quarter of an inch. In hot summer soil, seeds should be planted a little deeper than usual to stay moist. With the larger seeds, instead of making a trench and filling it in, I just press them into the ground as deep as the first joint of my index finger and pat some soil on top. I have also discovered that large seeds will germinate faster if they are soaked in a jar of water the night before planting. Firming the soil on top of the covered seeds brings them in contact with the life-giving moisture of the earth and speeds up ger-

145

pLANTING AND

VEGETABLE	WHAT AND WHEN TO PLANT (unless otherwise indicated, plant in mid to late spring)	TYPE OF SOIL
Beans	starts from seeds	needs little fertilizer, any good soil will do
Beets	starts from seeds	needs alkaline soil, slightly sandy, well drained
Broccoli	starts from hotbed or nursery; plant early or late (Feb.-March or July-Aug.); move location yearly	needs lots of calcium (from lime)
Cabbage	starts from hotbed or nursery; plant Feb.-March or July-Aug.; likes cool weather; don't grow in same place twice in a row	needs lots of calcium and nitrogen
Cantaloupes	starts from seeds	sand-humus mix, alkaline to lightly acid
Carrots	starts from seeds	rich sandy soil, free from stones
Cauliflower	starts from hotbed or nursery; plant Feb.-March or July-Aug.; likes cool weather; move location yearly	needs lots of calcium and nitrogen
Chinese Cabbage	starts from seeds; plant very early or in August	lots of lime and nitrogen
Corn	starts from seeds	needs little fertilizer except calcium (lime); can take alkaline or acid soil; good for breaking new ground
Cucumbers	starts from seeds; cool, shady location	very heavy feeders; need lots of nitrogen and phosphate; moderately acid soil
Dill	starts from seeds	alkaline to moderately acid
Eggplant	starts from hotbed or nursery	very heavy feeders; need lots of nitrogen and phosphate; slightly acid
Kale	starts from seeds; southern exposure; plant in July-Oct.; move location yearly	lots of nitrogen and lime
Kohlrabi	starts from seeds; plant in March or August; move location yearly	lots of lime

hARVESTING ChART

WATER NEEDS	HARVESTING
moderate	keep picked to stimulate new growth
average, deep watering	for winter storage, gather just before first frosts; in warm climates, can winter in the ground
abundant water	once main head is cut, small edible side shoots will form
abundant water	harvest when heads are full-sized before they begin to split; store with roots on in root cellar or in a trench 3' deep lined with straw
abundant water	ready when stem parts easily from fruit
moderate	harvest after first fall frosts; in warm climates can remain in ground, harvest when new leaf growth starts
abundant water as seedlings and when forming heads	when flower head forms, bend surrounding leaves over it and tie them in place; harvest about one week later
abundant water	may harvest outer leaves only until plants mature
can stand drought in early stages; needs water just before, during, and just after tasseling time	ready to harvest when silk turns brown
huge quantities; become bitter with drought	gather at any stage until they turn yellow-green (bitter then)
moderate	use green leaves in salad; store dried leaves and seeds for seasonings
abundant	gather when skin is glossy purple
moderate	harvest outer leaves only for long growing season
abundant	

VEGETABLE	WHAT AND WHEN TO PLANT (unless otherwise indicated, plant in mid to late spring)	TYPE OF SOIL
Lettuce	starts from seeds; plant early or in partial shade	almost any well-drained soil; has very shallow roots
Onions:	plant early to get spring rains	need lots of nitrogen, moderately acid soil
yellow	starts from sets; plant Dec.-Feb. where there is no snow	
red	starts from seeds or from hotbed or nursery	
green	starts from seeds—Evergreen Bunching type lasts two years	
Parsley	starts from seeds	lots of nitrogen, neutral to slightly acid soil
Parsnips	starts from seeds	any good soil, not too acid; prefers clay subsoil
Peas and Snow Peas	starts from seeds starts from seeds—Dwarf Gray Sugar fruit sooner, Mamouth Melting fruit longer; move location all types yearly	not rich in nitrogen; alkaline to lightly acid
Peppers	starts from hotbed or nursery	alkaline, few nutrients added
Potatoes	starts from seed potatoes with eyes, cut in quarters or halves; Irish Treasure produces large quantities; use same bed several years	rich, slightly acid soil; don't use lime or cow manure
Radishes	starts from seeds	need very little food; do well in most soils
Rutabaga	starts from seeds; plant July-Aug.	rich in nutrients, alkaline soil
Spinach	starts from seeds; plant March or Aug.-Sept.	needs large amounts of nutrients, especially nitrogen; can take alkaline or acid soil
Squash	starts from seeds; plant summer squash in March-May; plant winter squash in May-June	needs abundance of nitrogen and phosphate; slightly acid soil; add humus and sand to clay soil
Swiss Chard	starts from seeds	thrives in any soil; needs lime
Tomatoes	starts from hotbed or nursery; strip bottom leaves and set stalk deeply in the ground; likes to remain in same place year after year	lots of nutrients, *very* deeply dug; alkaline—lightly acid soil
Turnips	starts from seeds; very early (Feb.-March) or mid July-Aug.	thrive in poor soils, so can follow spring crops with no extra feeding; add sand to clay soils

WATER NEEDS	HARVESTING
abundant; tastes bitter with drought	thin and eat entire young plants; then harvest outer leaves only till heads start to bolt
abundant; need to be kept constantly moist	yellow and red onions: harvest when tops fall over and are dry
	pull onions and let dry several days in the sun; store in a warm, dry place
	green: Evergreen Bunching type break off at the roots and send up new shoots; other types are pulled out, roots and all
moderate	harvest outside leaves only; will reseed itself and spread
moderate when small, very little after that	may be left in ground all winter; harvest after first frosts for best flavor
moderate	harvest daily to encourage new growth; for Snow Peas harvest pods before peas develop
lots during growth; little to none during fruiting	leaving peppers on vine until they turn red gives them a sweeter flavor and higher Vitamin C content
moderate	harvest when vines die back, before frosts; let dry in sun one day; store in cool dark place; sunlight causes poisonous enzymes (greenish color)
abundant	reach maturity very early
needs early watering	harvest late fall, before frosts
moderate	pick largest leaves; pick off seed heads to discourage bolting
moderate	summer squash: keep picked for new growth
	winter squash: harvest before first frosts
moderate to low	pick outer leaves; pick off seed heads to discourage bolting
needs no water except when seedlings first start (will root 6' down)	pick until frost; green tomatoes picked before frost will ripen on window sill; after first frost, use green tomatoes for chutney and pickles
moderate, deep watering	can survive hard frosts; pick before ground freezes solid

mination. Now, water with a light spray until you're sure the soil all around the seed is wet. This one watering is sometimes all the seeds need until they sprout, unless the weather is particularly hot and dry. If the soil begins to appear dry, spray gently again.

Now begins the wait. No matter what the average time it takes for any particular seed to germinate (usually a week to ten days), I'm always out there on the third day, nose pressed to the ground, looking for the first shoots. After they're all up, watering is very important. Use a light spray so as not to disturb the tiny root system and water often enough to keep the top few inches of soil moist. This is especially important in summer planting.

Soon there is a crowded row of small plants that obviously need a lot more space to grow to full size. Even in the face of this, I sometimes can't bear to thin them. After tending the seedlings so well, now I must ruthlessly pull them out of the ground. But it's absolutely necessary to thin. Otherwise, you'll have a garden with masses of stunted, unproductive plants. One thing that makes the job easier is to let the plants get to edible size before thinning. Tiny carrots, tender lettuce leaves and baby onions make a good salad. With other plants, especially those sown farther apart, transplanting is the answer to the thinning problem. Plant short rows of seeds in the beginning and you can save all the seedlings in the end by filling in the rest of the row with transplants.

Transplanting is not so tricky when you're doing it with seedlings grown in your own garden. Plants from the nursery undergo much greater shock. It should be done on a cloudy, cool day if you want them to survive. Have the new hole worked and waiting. Dig up as much soil and root system as you can for each plant. As you work, try not to touch the roots themselves at all. If the main root is damaged chances are slim for survival. Set the seedling in its new hole so that the roots aren't too crowded and then firm the soil well around it. This way the roots will be in contact with the moisture of the soil. Now water well. Soak each transplant until the soil around it puddles. If hot weather follows, cover the transplants with boxes, pots, or caps for a few days until they've adapted to their change. Be sure these plants, tiny as they are, are spaced as far apart as they need to be for full growth. I usually give roots and bulbs one more inch than their final size (always hoping, of course, for the biggest). Leaf crops need about six inches. The cabbage family about eighteen, and tomatoes and peppers about two feet. Squash and cucumbers I usually plant in hills, thinning to four to six plants per hill. These do not transplant well.

And now the garden is transformed from an expanse of dark, barren earth to a maze of green plants. Beds and circles and rows of exotic plants: ferns (asparagus) and thistles (artichokes); flowers (sunflowers) and grasses (corn); bulbs (onion) and roots (carrots). All of these are wild plants carefully bred for thousands of years to yield bigger and more nutritious offerings to those who plant their seeds and tend their growth.

hotbeds

what　A hotbed is a miniature greenhouse for starting your summer crops (tomatoes, etcetera) early. It will save you money that you would otherwise spend on seedlings at the nursery and will allow you to begin to save your own seeds from year to year and experiment on varieties best suited to your garden. A manure-heated hotbed is simple to make and use.

where　Well-drained soil and a good southern exposure to catch the low winter sun are necessary. Also locate it near your house. Keeping the hotbed properly ventilated is essential, and you will probably neglect this if it's far away.

when　Generally, start seeds six weeks before you want to put them into the ground. If the seedlings are properly spaced, they can continue to grow in the hotbed (and will, like a jungle) until you're ready to transplant them. We start ours in mid-February on the coast. By then the severe frosts are over.

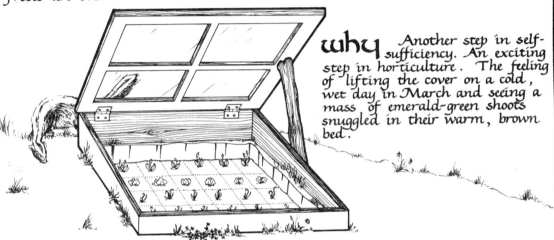

why　Another step in self-sufficiency. An exciting step in horticulture. The feeling of lifting the cover on a cold, wet day in March and seeing a mass of emerald-green shoots snuggled in their warm, brown bed.

how　Dig a pit two and a half feet deep, whatever size you'll need (when thinned each plant needs 3" x 3" so plan accordingly). Ours is 2' x 6' and it easy to reach the seedlings in the middle this way. Fill the pit one and a half feet up with manure and straw. The manure must be hot (fresh horse or chicken manure is best) and mixed with about one-third straw. Pack this in firmly and water thoroughly. Then put on a six-inch layer of planting soil, a rich compost if you have it. The manure will begin heating and kill any weed seeds in the soil. The cover for the hotbed can be a frame of saplings covered with plastic or a piece of glass. It must have a slope for the rain. A more ecologically sound frame would be an old wooden or aluminum frame window. Again, you must build a box sloped, as illustrated, for shedding rain. Hinge the window to this box. When the temperature inside has dropped to 85° (on a cool day with the cover on) it is time to plant. I mark out a grid of 3" x 3" squares with dolomite and scatter the different seeds in three large sections (later to be thinned). We have started tomatoes, sweet peppers, and cucumbers with success. If you can grow them, eggplants and melons are good too. The bed temperature (65° to 75°) depends on the weather. When it's cold keep the top down and the composting manure will keep the bed warm. On sunny days the top should be raised slightly. Keeping the moisture consistent is important too. In order to prevent fungus damage, it is best to water on days when the top will be up.

COMPANION PLANTING

Companion planting appeals to my strong sense of plants as living things; letting them clump together with friends rather than be separated in sterile, suburban-like rows. Companion planting is actually a somewhat scientific practice—the grouping together of plants for their mutual benefit and protection. Some of the reasons that plants become companions are purely mechanical. Plants with the same nutrient needs (potash, extra nitrogen, potassium, or phosphate) may make good companions because you can plan to meet those needs, adding extra minerals to the soil in that area. Deep-rooting plants are often partners with shallow-rooting ones (Swiss chard and beans, kohlrabi and onions). The deep-rooting plants break up the subsoil, releasing nutrients for the shallow-rooted ones, and neither competes with the other in the same strata for nourishment. Plants with similar water needs should also be planted together. Summer flowers in the garden attract insects which help pollinate the vegetables.

There are other less-obvious reasons behind companion planting systems. All plants excrete organic compounds from their roots and leaves; healthy plants excrete greater quantities of these compounds than plants that are under stress. In some plants, these compounds are actually antibiotic or fungicidal—not only repulsive but also poisonous to plant predators. Companion planting allies weak and more susceptible plants with natural insect and disease repellers. It also tries to combine plants for the most vigorous growth—so that all are strong and healthy. Thorough soil preparation and companion planting provide a good measure of plant protection—they probably won't save you from a horde of locusts, but they will discourage aphids or a menacing black beetle. Much of the studying about companion planting has been compiled in a wonderful little book: *Companion Plants,* which is fine reading even if you never follow the system (see Resources.)

The first year after I read it, I followed it religiously. I drew charts of friends and foes and planned the whole garden on paper, double and triple checking to be sure everyone would be happy. Following the advice of the book, I surrounded the whole garden with onions and garlic, which are an anathema to rabbits. Little did I know

I am overwhelmed by all the hostility and rage I feel for men right now. I know all the real and good parts of that anger: the part of me that feels that Peter and every other man has no right to come through all this scot free, no right not to confront and experience his part in making this culture what it is. I will no longer be a willing victim. It is more than that too—it is my rage at my own complicity making me angry at men; angry at them not just for how they act, but also for what I do to myself in order to be liked by them. I don't like being in this extreme position, but being in it feels inevitable and overwhelming.

I feel that my individual relationships with my women friends are really good and strong. I trust them and am beginning to realize that they are not just a stop-gap until I find another man. When Tory asked if she could come live on my land it was like a blinding revelation. Perhaps I could live and work with my *friends?* Create a life in which I have space and time to be myself. I find myself more and more relating only to women, rather than being subjected to the daily pain of always being treated as less than equal, a "female." Never have I so clearly felt the oppression of it, never before have I so clearly wanted to be free.

COMPANION PLANCING CHARC

A BASIC GUIDE TO PLANT COMPANIONS

PLANT	FRIENDS	FOES
asparagus	tomatoes, parsley	—
beans	beets (bush beans), carrots, cauliflower, cucumbers, potatoes, savory, petunias, rosemary	onions, garlic
beets	kohlrabi, onions, bush beans	pole beans
cabbage family (broccoli, Brussels sprouts, cabbage, cauliflower, kale, kohlrabi)	beets, celery, onions, early potatoes, most aromatic herbs (esp. dill, nasturtium, peppermint, rosemary, and sage)	tomatoes, beans strawberries
carrots	lettuce, onions, peas, radishes, tomatoes, rosemary, and sage	dill
celeriac	leeks or onions (alt. rows)	—
corn	beans, cucumbers, melons, peas, potatoes, pumpkins, squash	—
cucumbers	beans, corn (alt. rows), peas, sunflowers	potatoes, strong herbs
eggplant	beans	—
lettuce	carrots, cucumbers, radishes, strawberries	—
onion, garlic	beets, carrots, celery or celeriac peas, beans (alt. rows), lettuce, tomatoes, summer savory	peas, beans
peas	most vegetables, esp. beans, carrots, cucumbers, corn	onions, potatoes
potatoes	beans, corn, cabbage, peas, horseradish at corners of patch	cucumbers, squash, sunflowers, tomatoes
pumpkin	corn	potatoes
radishes	lettuce	—
raspberry	most vegetables	blackberries, potatoes
spinach	strawberries	—
squash	corn, nasturtiums	—
strawberries	borage (few), lettuce, onions, spinach	cabbage
tomatoes	carrots, onion, parsley, marigold, nasturtiums, basil, mint, borage	cabbage, kohlrabi, potatoes
turnips	peas	mustard

scattered throughout the garden: marigolds, calendulas, yarrow

153

that they are gourmet delicacies to California gophers, who came charging into the garden at my express invitation. All I can hope is that the one who ate forty feet of garlic in one afternoon was ostracized by her friends for a long time to come!

Now, I don't plan my garden quite so carefully, but after several years the basic rules have become familiar to me. Medieval gardens were a jumbled mixture of herbs, vegetables, and flowers, an untechnical but equally reliable practice of companion planting. Companions work best when they are *together*. So I often plant in beds, rows of mixed plants—two rows tight together with space on either side, or with rows nestled together zigzag fashion.

The chart included here will give you more detailed companion information, but there are some simple guidelines. It's not a good idea to plant members of the same family together (i.e., squash-cucumbers-melons or cabbage-broccoli-cauliflower-kale-kohlrabi) as they can cross-pollinate and will reinforce each others' vulnerabilities. I once got a quite interesting zucchini-colored, pumpkin-shaped something out of such a relationship—but it didn't make good pie or tempura! The cabbage family are among the most insect prone of all plants and should be combined with strong, odiferous herbs and flowers.

Herbs can actually be the salvation of a garden, as many of them have fungicidal properties. Fennel

onions

cauliflower

cabbage

marigold

is the only one which is completely unwelcome in the garden as it inhibits plant growth. Yarrow, on the other hand, is almost universally loved, increasing plants' resistance to adverse conditions and insects. Both stinging nettle and yarrow have the ability to increase the pungency of other herbs which are planted with them. Stinging nettle is said to help vegetables store better (carrots, beets, tomatoes, etc.) when they are grown as companions. Yarrow, lemon balm, and valerian make good borders around vegetable beds. Soft hemp (Cannabis sativa) is beneficial to most plants and helps repel insects; it also makes a good border. Dill works wonderfully with cabbage, improving its growth and vigor. Rosemary, sage, and peppermint work to repel cabbage moths from that whole family.

Flowers should be liberally integrated with your vegetables. Marigolds, calendulas (pot marigolds), foxglove, and nasturtiums are most particularly desirable as companion plants. All are known as insect repellers. Last year, my squash blossoms were attacked by a particularly gross black beetle. When the nearby calendulas got to be a good size, the beetles vanished completely. Not the least reason for planting flowers in the garden, though, is the sheer beauty of the blossoms against the green vegetable leaves.

Companion planting makes for physically and symbolically beautiful gardens: a little private dabbling in herbal magic, a return to the medieval woman's lore. If you're interested, study this chart a bit and read more; then plant your garden with some instinctive attention to plant relationships and tastes. Watch your plants and try to learn what makes them happy. If the borage wants to live with the tomato, who am I to separate them?

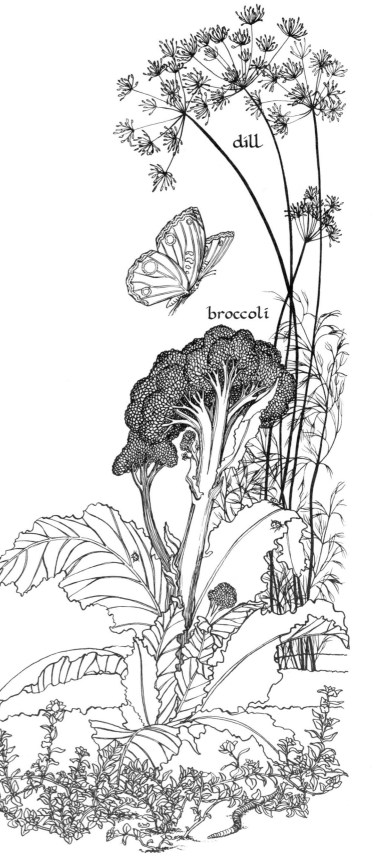

dill

broccoli

lemon thyme

CARING FOR YOUR GARDEN

Once the seeds have sprouted and the plants are growing, all the garden needs is mulching and possibly watering. I have been converted to mulch as the gardener's very best tool. Essentially, mulch is just some sort of ground cover set between the rows of plants to prevent weeds from growing and to conserve moisture. No more weeds! That's what convinced me. My garden is neat, orderly, beautiful, and all it takes is a few hours effort once the plants are up. All the time I would (or maybe would not) have spent weeding is now spent tending vegetables and enjoying my garden. There are few activities that I dislike more than weeding; it's one of those things that no sooner done, needs doing again—providing very little satisfaction. Weeds are the most successful and pushy plants on this piece of ground. So I mulch and return those plants to their rightful homes; they're perfectly lovely on the other side of the garden fence.

Mulch can be made from almost anything, organic or not: boards, black plastic, old carpets, wood chips or sawdust, straw, grass clippings, etc. I like to use the organic varieties since they also double as fertilizer next spring. Mulching is definitely the easy way to garden. Boards, carpets, or plastic will stop weeds from growing by preventing light from getting through. Not so the other materials. What I do is to set down a layer of newspaper between the rows, four pages thick, and then put my mulch on top of that. The newspaper stops weeds from coming up or the straw in the mulch from growing down (it is possible to get a crop of oat hay between your rows), and it will decompose into the soil by the following spring. Just remember that paper is acid and add extra lime where you need it. On top of the newspaper, I like to put something that will really enrich the garden later. So I usually use the leftovers from my manure loads: the parts that have more straw or wood chips than manure, but do have some of both. Using them as mulch gives the straw or wood chips time to decompose without robbing growing vegetables of needed nitrogen. Wood chips are acid, so I try to use them as mulch between acid-loving plants and use the straw everywhere else.

Besides mulching between the rows during growing season, I also mulch all the planting beds before winter, after the last plants are gone. I use newspaper again, covering the ground with it, but this time put a very manurey mulch on top, since this will be my basic fertilizer in the spring. When warm weather comes again, I have weed-free, easy-to-work soil, which makes spring planting a joy. I dig all the mulch (newspaper and all) into the soil and then make my rows. It's important to dig in or remove your mulch in the early spring as it can be a haven for snails and slugs. Once the plants are up, I mulch again.

Living in an area of summer drought with a low well has been a major determining factor in the size of my garden and what and when I plant. I remember my first garden in Pennsylvania nostalgically—where, like Camelot, the rain came promptly every evening. If you live in an area of summer rains, this watering information won't be so essential to you. Nevertheless, you probably will be faced with dry spells and wilting plants, and you will need to know what to do.

Gardens need a lot of water. I have read that they need an average of $\frac{3}{5}$ of a gallon per square foot of garden. This means a 30'$\times$40' garden requires 720 gallons of continuously flowing water; one 100'$\times$100' garden needs 6,000 gallons.

If you have a shortage of water, you will have to compensate in various ways. I plant the vegetables that need large amounts of water as soon as the soil is warm, so that they will get as much natural rainfall as possible and need little help from me. I am careful to plant my rows parallel with the ground slope so there is little erosion or run off. I plant in flat or slightly depressed rows so the water will puddle around the base of the plants; raised rows require a lot of water.

Whether you have a water shortage or not, very thorough, infrequent waterings are the best. The perimeter and depth of a plant's root system are determined by the size of the moist area. Deep root penetration gives plants a greater resistance to drought or freezing and produces larger vegetables.

What this means specifically in your garden is subject to a lot of variables—soil type, watering method, weather conditions, and age and type of plant. Young seedlings of all types need a lot of water. After that, needs vary with the type of vegetable. For shallow-rooting plants (lettuce, spinach,

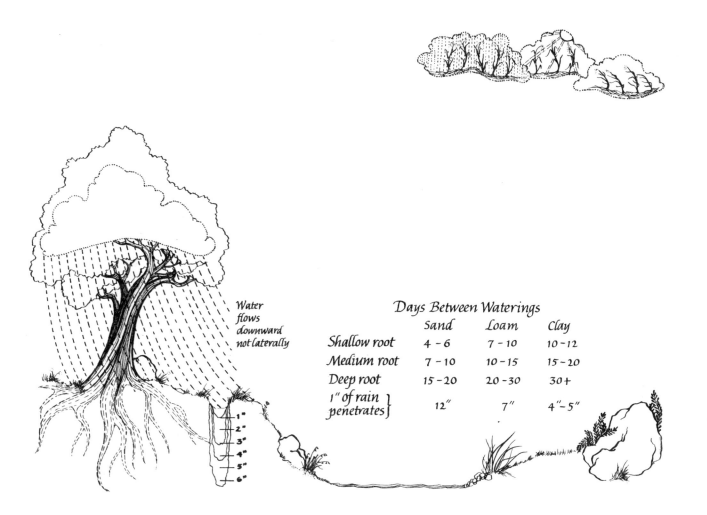

	Days Between Waterings		
	Sand	Loam	Clay
Shallow root	4 – 6	7 – 10	10 – 12
Medium root	7 – 10	10 – 15	15 – 20
Deep root	15 – 20	20 – 30	30+
1" of rain penetrates	12"	7"	4"–5"

Water flows downward not laterally

onions, squash, cucumbers), I water when the top inch of soil is really dry. For deep-rooting plants (cabbage family, carrots, beets, turnips, etc.), I push my index finger all the way into the soil and water when it is dry to that depth. Overwatering can be just as bad as drought. It floods air pockets in the soil, robbing plants of oxygen, and can cause fungus diseases, mildew, and rot.

My soil is a sandy loam that drains very quickly (it seems to gulp down gallons). I have found that watering very slowly with a drip method or shutting the sprinkler off at thirty minute intervals to allow time for penetration works best. Non-porous clay soil requires more water less often since it retains moisture. With clay, though, there is often a problem getting the water to penetrate at first. Very slow, steady watering is best.

I used to water my garden—using a hose—at the base of plants, but this was a really inefficient system in terms of time and water. The hose water (even turned low) was of too great a volume and tended to run off along the surface and wash the

soil away from plant roots. There are three better methods:

Flooding or soaking in furrows is good for shrubs, trees, or when you have an abundant water supply. For fruit trees, dig a shallow basin a few inches past the drip line (the outer perimeter of the branches) and water thoroughly. For vegetables, make wide, flat furrows between the rows and flood there.

Perforated plastic pipe or porous canvas hose between the rows allows for slow, deep soaking. It's very good for clay soils. I now use this method in my garden, since it conserves water by putting it exactly where it's needed. But the fifty-foot hose must be turned off or moved every few hours and so takes regular attention.

Sprinklers are the best way to distribute water over a large area if you have a good water supply. Inexpensive bar sprinklers that can cover a 40'×50' area in one sweep are available. The disadvantage is the potential for plant diseases such as mildew caused by the water on the leaves. This is especially a problem with squash and cucumbers. Watering on cloudy days or in the early morning helps, so leaves have a chance to dry before the sun scorches them. A fine spray won't damage big-leafed or large-headed plants.

Working this summer with two first-time gardeners, I have realized that there is more to growing things than the bare facts of what type soil or how much water. But this "more" that I feel and experience is the hardest of all to teach or to explain. It's what happens when you really care for your garden, when you take the time to watch each plant closely, get to know it, meet its needs. It came as no surprise to me when I first read the research being done on "the secret life of plants." I have literally seen squash burst out in blossoms or asparagus shoot up tall overnight after concentrated attention from me.

On the more practical and mundane plane, good garden care means paying close attention. It means noticing the too-slow-growing plants in a healthy row and correcting the soil around them; it means noticing what needs water that day and what can wait; it means noticing what needs pruning or picking now so it will go on bearing, what should be let grow or go to seed. My relationship with the garden is a very special, mystical one. But it is not one based on special skill, knowledge, or expertise. It comes from simple caring and time spent. I have learned a lot about gardening techniques over the years, but I count loving attention as at least as important.

I try to spend an hour each day in the garden—freely given time when my energy is concentrated right there and I'm not distracted by other people or obligations. For me, early morning is the best, and I gain from the garden as much as I give—a sense of peace with which to start the day. Whatever time you choose, try to be really *there*. Watch the plants, poke around, start the sprinkler or set the hose, eat a few berries—just feel the garden and let it feel you. Your plants will know the difference between care and a perfunctory duty; they will show it.

Bare feet in wet soil
connecting electric
earth and air –
so much green from one seed.
In my garden
I am a continual witness
of life, death,
beauty in a scarlet runner bean.

GARDEN PESTS

That first garden was a wonder: we planted seeds, plants grew, we ate them. By the end of that first summer, I naïvely wondered what all this fuss about fertilizers and pesticides was for. I didn't even know what compost was, and I hadn't seen a bug in months. The second garden was not so simple: the worn-out soil on the new place needed fertilizer; the rainless California summer meant watering; aphids had discovered my tender young cabbages. To make things worse, gophers were sucking onions and carrots beneath the ground before my very eyes! I had thought that only happened in cartoons. So I began to formulate a conscious policy toward garden predators; it was a softhearted socialistic version of "from each according to her means, to each according to her needs." We would all share. After all, how many cabbages could we really eat? This policy reigned (unsuccessfully) for a year. The problem was, my visitors interpreted this share-and-share-alike situation differently than I: they quickly invited all their

friends, cousins, sisters, to join the feast. I was left with the Swiss chard, some half-gnawed carrots and beets, and a few lovely rows of corn for my share. Since then I have been learning strategy and a certain ruthlessness instead. I have tried most nonchemical techniques known to humans to rid my garden of little beasts.

But through the years, I have also discovered that my initial naïveté touched on a very basic truth. Healthy plants growing in rich, fertile organic soil are not prone to disease or insect infestation; larger animals are another matter altogether. Most of my garden protection happens now, before the plants are even above ground. Thorough soil preparation and companion planting are the basic techniques. A soil that contains a balance of minerals, instead of the heavy nitrogen concentration in commercial fertilizers, produces much stronger and more resistant plants. It is also possible to buy varieties of seeds that have been hybridized to make them resistant to common plant diseases

(leaf mold, fungi, etc.). You may not need this protection; it depends mostly on local climatic conditions (dampness and lack of sunlight, for instance). I rotate the location of the annual plants in my garden each year, partly to prevent soil depletion and partly to discourage predators like cutworms and cabbage moths. Each spring I am also careful to clean the garden of old rotting mulch (this can be dug into the soil or added to the compost heap), and to cut down weeds and grasses around the fence. All of these provide hiding places for slugs and snails, and breeding places for insects; their removal lessens the likelihood of problems later in the summer. These initial efforts have repaid me well and I have had very, very few problems with insects over the years and had enormous, beautiful vegetables. But, as I said, larger animals are another story.

Birds are probably my greatest enemies. This hostility has been growing for several years as every bird in two counties has come to make its home in the trees around my garden, just waiting till my back is turned to grab a tasty snack. It's their deviousness, their gentle looks of innocence, that infuriate me the most. From vultures I would have expected such behavior, but vultures are the most ecological of creatures, recycling what is dead and no longer needed, whereas sparrows, starlings, titmice, and jays prey upon the newborn and very much alive. Connoisseurs, they prefer baby peas and early (two-day-old) lettuce! I have read in gardening books that putting out seed will attract birds to eat your insects. My experience has been that they follow a main course of seed with a dessert of my strawberries! At the heights of my passion, I have crouched motionless in the garden until a bird settled in, then jumped up screaming like a banshee—a very live "scarecrow." I have spent hours planning how to build a bird-proof garden: a fine screen cage covered on all sides and on top. But, in general, I have settled for less extreme measures. Scarecrows have worked for my friends, but they haven't done much good against the audacious sparrows in *my* garden.

I have discovered, though, that the principle of fear does work. I have hung tin can lids, tin foil, strips of cloth, and Christmas tree balls on strings above new seedlings; their movement in the wind keeps away the birds. My fanciful screened cage works well on a smaller scale. Fine chicken wire, wire screen, or nylon netting (Ballerina type) can be stretched over plants to protect them in the early days. I protect my strawberry plants all season long with green netting stretched at ground level and staked every few feet. For special young plants (like new broccoli or cabbage) I have sometimes had to build small cages—boards laid on edges, with wire over the tops works well. I have also learned the virtues of giving in now and then. If there is something the birds particularly like, I let them have it, hoping to save the rest of the garden. I learned this the hard way one year when I pulled up their favorite Chinese cabbage to make room for the winter garden; they promptly ate the spinach and lettuce in its place.

Gophers are my next greatest enemies, though my rage is tempered by a humorous respect for their gall. They have even eaten the very grass that I was sitting on, right out from under me. That they manage to zero in on my garden with a hundred other acres of tender, vulnerable roots to choose from, never ceases to amaze me. When I first came to California, I heard wild tales of the madness that gophers have driven gardeners to: of an old man who moved his chair and his shotgun to the garden and waited all day long for a gopher to show its face; of a neighbor who got her truck stuck in the mud driving it close enough to pump carbon monoxide down the hole; of another who dug a two-foot trench around the garden and filled it with a year's worth of broken beer bottles. I found all these tales both funny and unbelievable until one day I watched the twenty-ninth consecutive onion disappear from sight. I grabbed my pickax and began hacking madly at the ground.

If one does not want to be driven to such crazies, some preventive efforts help. The trench method actually does work. It can be filled with broken glass, tangled barbed wire and rusty tin can lids. Or you can just bury *fine* chicken wire two feet deep, all the way around your garden. I did that one year, using 1½″ mesh and when I finished, realized it was just the right size for a hungry gopher to slither through. For plants that you really care about (perennials, for instance), you can dig out your beds about two feet deep and line them with hardware cloth (¼″ mesh). This is hard work but will last forever. A small-scale version of the same principle also works. A friend's grandmother taught me to set the plastic cartons in which strawberries are sold into the ground. I fill these with dirt and set special seedlings, like flowers and parsley (gopher favorites), inside them. This at least insures that part of their root system will remain intact.

Traps are really the most consistently reliable way to rid your garden of gophers. They come in several types (I have not developed any preference) and cost about $1.25 each. I am not good at trapping. I check my traps reluctantly, hoping I have *not* caught a gopher. But gradually I am learning to set traps as soon as I see any signs of digging and before any damage is done. I try to check the garden each morning for fresh diggings (moist earth kicked up from a tunnel) and set the traps right away. I usually catch the gopher by noon. I always use two traps. The surest method is to dig down until you find a tunnel going in two directions, then set your traps facing in both directions. The surface diggings you first see are usually just a dead-end exploratory passageway, so keep digging until you find the main thoroughfare. Gophers are partial to artichokes, parsley, and peanut butter, so I usually bait my traps with one of those. There are two schools of thought as to whether or not you should cover the tunnel over again once the traps are set. The theory is that the gopher interprets light from an open tunnel as a cave-in and hurries along to fix it. I am not sure about this, so I sort of compromise; I cover most of the tunnel over (without filling it in; a board weighted down by a rock is good for this) but leave a few holes for light to trickle in. The one consolation in all this is that gophers are smart; once one of their number has been killed, they spread the word and move on to less treacherous pastures. And why not? Every plant is a potential victim. So, I remind myself as I set the traps that the first poor dead gopher has saved many of its fellows.

Moles are harder to catch than gophers, but they also seem to do less damage. At least with them, the plants don't vanish from sight completely. Moles make long trails of fresh-plowed earth right on the surface. Frequently when you dig, you won't find a tunnel since it often caves in right behind the mole. They knock plants over rather than eat them, and most plants can be set back in place again. Sometimes, however, the mole tunnel creates an air pocket around the plant's roots, which doesn't kill it but thoroughly stunts it. If you have only occasional mole trouble, you can probably just pack the earth firmly around the roots again and ignore the moles. If they are more persistent, you can set gopher traps for them, too (where you can find a tunnel)—though they often push their way around the traps. I have also heard that you can get rid of moles by mixing one ounce of detergent and two ounces of oil together and beating them until they are the consistency of shaving cream. Then add an equal amount of water and store. When you need it, put two tablespoons of this concoction in a sprinkling can and fill it with warm water. Sprinkle this over the areas of heaviest infestation.

There are also deer fanatics in the world. Our local newspaper once contained a proposal to fence the deer out of the five-hundred-square-mile stretch of coast. I have heard of deer that have learned to jump eight-foot fences to get at something they wanted. But I don't share this prejudice. Deer are reasonable beings and will respond to your feelings and vibrations as well as fences. I have never lost a garden to deer, unless I didn't care about it any more. One year, I left a sagging gate in my garden fence hang open for several months without the deer ever venturing in. The night after we moved from that land to this, the deer jumped the fence (the gate was closed) and

I am apprehensive about telling my parents about Peter and me separating, breaking all the myths I have made about myself and him. I feel like I am admitting inadequacy and defeat. I need their support and approval so badly now; I am terrified of how much power I give to them, to him—my father—living my life as Electra. I cannot bear not to be perfect. I feel this need to keep everything together, to do everything on a superhuman scale. I set impossible standards for myself and never quite measure up. I always somehow feel I've failed. I want to be normal and to be great, and I am neither.

The women's group is ending. At least, I am not going any more. For weeks now the meetings have been painful, if only in their emptiness and repetition. The women who helped me grow the most, Judith, Tory, Lea, have already quit. We sit here now saying the same things, or nothing. I am tired of helping release the pain of Sarah, Carol and Kathy's relationships with their men so that they can go on living with them. I'm angry that they only seem to remember me when *they* have a problem. Their lives are so bound up in their lovers that they have little time for their friends.

Perhaps this ending is right, something to accept as a part of the process. We all have grown and learned and questioned and changed together for more than a year. Now we have to take all that awareness into our separate lives, to make of it what we will. For me, it's not that I feel that I'm finished with discovering myself or that I know all there is to being a woman. But I have talked enough. I'm immersed now in living this new consciousness—in the land, my weaving, alone and with women who are committed to helping each other grow.

ate the entire garden! So, I start by getting to know the deer around me and making them aware that I *do* care about the garden. Then I reinforce this first impression with a six-foot-high fence, and a gate which is usually closed.

I used chicken wire for my first garden fence, but have now been converted to medium-weight poultry and garden fencing, a type of field fencing that is graduated from one-inch spacing at the bottom to six inches at the top. This stretches easily, lasts longer, and costs about the same as chicken wire. If you have deer that are persistent jumpers, slanting the top of the fence outwards will stop them. Do this by nailing pickets to the top of your fence posts pointing both upwards and outwards; then stretch several strands of plain wire across them. There is one thing I must admit about my vibration method of deer control: it only works for gardens. I've had every flower I've ever planted (including the rose bush by the front step) gobbled to the ground and the young fruit trees are, at this point, a tossup. But the deer do stay clear of the garden.

Rabbits can be as big a pest as birds (bigger, since their stomachs are), but they are far more easily controlled. A good, narrowly spaced fence is needed—either chicken wire, poultry and garden fencing, or wooden pickets driven into the ground. Boards can be squeezed under. If you have persistent problems with rabbits, try digging a shallow trench at the base of your wire. Lay saplings in this, nail them to your fence posts, and nail your wire to the saplings. Then pack the dirt tightly all around. This will discourage rabbits from digging. And remember to keep the gate shut, too! Rabbits do *not* respond to vibration.

With insects I have a more ambivalent relationship than with the larger pests. I can't identify them all well enough to decide who should live and who should die. So, except for the ones I can identify, I try the optimistic method of carrying them thirty feet from the garden, pointing them in the opposite direction and wishing them well.

I've never had a serious insect problem, and perhaps that's why this lighthearted effort has been enough. Insects can be terribly destructive, however, decimating whole rows in a few days, and it is important to recognize the worst garden enemies and to know how to deal with them. Sucking insects attack tender buds, leaves, and stems. Signs of their presence are small black mold specks which grow on droplets of honeydew excreted by the bugs. Suckers cause stunted leaf and flower growth. They include ants, aphids, leafhoppers, mealybugs, mites, red spiders, scale, spittlebugs, thrips, and whitefly. Leaf chewers are the most obvious. They eat holes in growing things. They include caterpillars, diabrotica, earwigs, grasshoppers, oak moths, snails, and slugs. Soil pests are not so visible, but can be detected by your plants' failure to grow despite much water and fertilizer. Common soil pests are cutworms, lawn moths, and wireworms. If you watch your plants closely, you can see many of these insects so it is a good idea to get to know what the specific bugs look like. For many there are specific things you can do to get rid of them; for the rest, there are organic sprays and natural predators that are effective.

Cutworms will attack almost any garden plant, chewing more than they eat or wrapping themselves around the stem of the plant and strangling it. The cages that I mentioned for keeping out birds, made of boards laid on edge, will also work to keep cutworms away from the base of plants. If you have no bird problem, a simpler deterrent to cutworms is to cut both ends off a tin can and set it around the plant.

Earwigs like to crawl inside dark places and will collect in crumpled newspaper or the cardboard tubes from toilet paper. Even more effective is a square of plastic two or three feet square, folded twice and placed under artichokes, tomatoes, or other good cover. The earwigs will collect inside it and can be shaken off each morning into a bucket of water and kerosene or soap. If you can stand to wait two or three days between each harvest, you'll find many newcomers have been attracted by those already present.

Diabrotica look like lovely chartreuse lady bugs. At one time, I thought they were as friendly as lady bugs and paid the neighborhood children a penny apiece to import them into my garden. Now I pay the same rate to have them removed. Diabrotica are great chewers and will go for anything with tender leaves. Though they fly, they are relatively slow moving and I've found the easiest way to get rid of them is to carefully check the gnawed-on plants each morning and pick the diabrotica off.

Slugs and snails are perhaps the most voracious of the insects and can infest a garden by the hundreds. They can be identified by the trail of silvery slime they leave behind them on the plants. Snails and slugs love dark, damp places and tend to be attracted to gardens surrounded by tall weeds, full of old mulch, or placed up against a woods without a lot of direct sun. Their sensitive bodies cannot cross a rough or gritty area, so they can be stopped from entering a garden by a border of sand, lime, ashes, or rock salt. Mulches with tan oak bark or oak leaves will also repel slugs and snails. If your visitors manage to get past these barriers, you can turn their natural inclinations against them. Wet boards or wet cardboard laid out in the garden will provide irresistible hiding places for slugs and snails when they have finished their nightly prowling and settled down for the day (these boards can double as mulch). In the morning you can go out and scrape the congregation into the same bucket you used for earwigs. This water can remain standing until you have plenty of corpses to bury before starting over. One friend told me she collected two or three hundred slugs each morning

163

This afternoon the Anderson brothers began teaching me how to graft fruit trees—the careful joining of life with life. Even more than I loved gaining a new skill, I loved learning from two old men who have so very much to teach me. I admire the audacity of eighty-three-year-old men setting grafts that will not bear fruit for years: the total involvement in a process they love. Those trees will stand and live; I doubt whether Jake and Fred even stop to wonder if they'll pick the fruit. I want to live my life with that kind of harmony and purpose. I want to be planting seeds on the day I die.

for about two months last winter. "The slugs were so thick they covered the ground like a spring rain." Now, when she goes out and turns over the boards, five is a big bounty. She found that snails love to crawl into cardboard boxes too, and can be easily caught inside them each morning. Another way to insure plenty of victims when you go out on your slug patrol is to put shallow dishes (jar lids work well) of beer out in the garden. Drunken slugs make easy victims. Nighttime's the right time for collecting them by this method.

Whitefly and aphids, both sucking insects, are repelled by nasturtiums, so companion planting in areas of heavy infestation may solve the problem. I have used a homemade spray to stop aphids from eating my cabbage, and I have read that such sprays will stop almost all of the leaf suckers and chewers. I grind up garlic bulbs in the blender with a little oil, some water, and some cayenne. Some people also add chili peppers. Adding a little oil-based soap (such as Palmolive) to this mixture will make it stick to the plants' leaves. I let the mixture stand at least twenty-four hours in a glass jar and then strain it before spraying on the plants. Some pests are killed by this solution and others are just confused by the odor. One application is usually enough and will last until the next heavy rains, but if one doesn't stop your guests, repeat the process. Another common recipe is to mix one pound of tobacco with five garlic cloves and one hot pepper (liquefied in a blender) in a large pot. Fill the pot two thirds full of water and let it sit in the sun for a week. Strain this and add a half cup of dishwashing soap before spraying it full strength on plant leaves. The advantages of both these sprays is that neither is the least bit harmful to humans or to higher animals.

You can also follow the companion planting principles by spraying herbal teas onto plants to kill or repel insect predators. Teas should be made by covering the herb leaves with water; bring this to a boil, remove from heat, let stand about ten minutes, then dilute with four parts water and use. The following herbs are said to repel these bugs: stinging nettle, nasturtium, and spearmint for aphids; mint and pennyroyal for ants; sage, rosemary, hyssop and thyme for cabbage moths; horsetail for slugs and fungus diseases; nasturtiums for squash bugs; garlic for weevils.

Rotenone is a pesticide derived from a tropical plant. It is effective against insects but is not toxic to humans or other animals. It is available in its pure state from veterinary supply houses, veterinarians, and pet shops. When sold commercially, it is often mixed with more toxic compounds, so read the label carefully before you buy it.

Diatomaceous earth is another natural compound that works well against almost any insect. It is a fine, powdery substance that consists of the fossil remains of microscopic, one-celled marine animals, marine algae called diatoms. Milling cracks the particles open, exposing razor sharp silica needles which scratch the bodies of insects coming into contact with it. Without their protective shell, these insects dehydrate and die. Diatomaceous earth is non-toxic and inert, and kills insects mechanically, not with poison. It is commonly used on fruit trees, but in gardens it will kill a number of pests, including the ever-present earwig. It can be bought at pool and building supply stores, ceramic supply stores, or from Perma-Guard Corp., 1701 East Elwood Street, Phoenix, Arizona 85050.

Sprinkling wood ashes around the base of cabbages will stop root maggots and cutworms, as well

Yesterday I went out to feed the goats and found the plastic from the goat house roof torn loose and whipping about in the wind. I spent the next hour fighting to pin it down before it ripped to shreds. I hoped, at least, to save it until I could get some roll roofing to replace it, all the while feeling a nagging tension about how to juggle next week's money to include roll roofing. I cursed myself for ever using plastic in the first place, for not doing it "right" once and for all. There is always that decision: whether to build well or cheaply? Too often, I have opted for cheaply, still seeing myself as an eternal transient, only to find myself still here three years later, frustrated from patching old mistakes.

I worry about this and my sheep fencing: if I use saplings, they will last only five to fifteen years; if I buy heartwood, it will cost an extra $300 that I don't have. Fifteen years feels like a lifetime to me now. Yet I also know that posts that only stand for fifteen years will not make a farm for anyone else. And to not believe in the future of this land as a farm is to accept the inevitability of encroaching vacation homes and environmental destruction.

There is always that question: how to treat the land well? Shall I build like our ancestors, to last forever? Or like the Indians, to leave no trace? Where to get the money—and the time after working for the money and the energy to live up to my loving—to not betray this land with my carelessness or expediency?

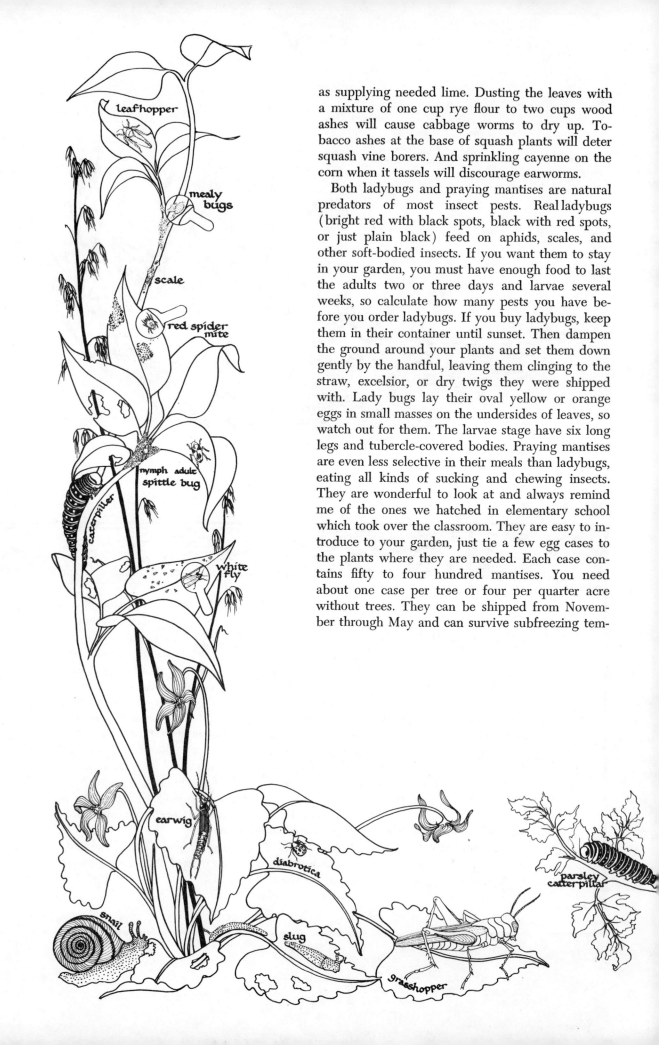

as supplying needed lime. Dusting the leaves with a mixture of one cup rye flour to two cups wood ashes will cause cabbage worms to dry up. Tobacco ashes at the base of squash plants will deter squash vine borers. And sprinkling cayenne on the corn when it tassels will discourage earworms.

Both ladybugs and praying mantises are natural predators of most insect pests. Real ladybugs (bright red with black spots, black with red spots, or just plain black) feed on aphids, scales, and other soft-bodied insects. If you want them to stay in your garden, you must have enough food to last the adults two or three days and larvae several weeks, so calculate how many pests you have before you order ladybugs. If you buy ladybugs, keep them in their container until sunset. Then dampen the ground around your plants and set them down gently by the handful, leaving them clinging to the straw, excelsior, or dry twigs they were shipped with. Lady bugs lay their oval yellow or orange eggs in small masses on the undersides of leaves, so watch out for them. The larvae stage have six long legs and tubercle-covered bodies. Praying mantises are even less selective in their meals than ladybugs, eating all kinds of sucking and chewing insects. They are wonderful to look at and always remind me of the ones we hatched in elementary school which took over the classroom. They are easy to introduce to your garden, just tie a few egg cases to the plants where they are needed. Each case contains fifty to four hundred mantises. You need about one case per tree or four per quarter acre without trees. They can be shipped from November through May and can survive subfreezing temperatures. Once introduced to an area, the mantises will stay, laying their eggs in the fall and hatching in the spring. These insect predators can be ordered from: Eastern Biological Control Co., 104 Hackensack Street, Woodridge, New Jersey; Gothard Inc. Box 367, Canutillo, Texas; or the California Bug Co., Route 2, Auburn, California.

Toads are another friendly predator, eating insects as well as cutworms, slugs, yellow jackets, wasps, spiders, ants, moths, flies, caterpillars, and more. They can be collected after spring rains in swampy, marshy areas or around shallow pools. You should keep them penned up the first few days they come into your garden so they can get used to it and resist their strong homing instinct. Toad houses can be made by setting clay pots upside down and burying them several inches in the ground. Break holes in the sides for doors and put them under shrubs or tall plants. Keep a shallow pan of water in the garden, too, as toads must sit in the water to drink through their skins. I have not been able to use toads myself, because of my cats which sit in the garden trying to catch gophers. But given a safe haven, toads can make a substantial dent in your pest population.

Lest all this detail sound discouraging, let me emphasize that you will probably have to deal with only one or two of these predators in your garden during any one season. And given good fencing, good soil preparation, and a little luck, you may not have to deal with any of them at all. Organic gardening by itself is the best protection there is against plant predators and disease.

pLANCTING pERENNIALS

Perennials—plants that last year after year—were the very last thing that I planted in my garden. After growing up to rented homes and serial jobs, it was hard and even frightening to feel settled down. Even having bought the land and talked a lot about living here for "a lifetime," the actual growing of a plant that wouldn't produce for three more years was a very alien act. It was hard to dream in three-year terms, much less to plan or act. So it was the third year on this land that I began the asparagus bed which might by then have been feeding us. And it is now, in the fifth year, that we have an abundance. Gradually in those years, the perennials have filled up the whole original garden. I have come to love them best of all—their fruitfulness and faithfulness and persistence. And their taste, for they are among the most special of garden delicacies: artichokes, rhubarb, strawberries, asparagus, raspberries, and blueberries. There's something so satisfying about plants that keep growing and multiplying for years and years, almost literally for a lifetime. It gives me pleasure to know that whether or not I stay, these plants will be here and will transcend my presence—a fine gift to the land.

The only real secret to growing perennials is the same as that of all gardening, good soil preparation, only more so since these plants will be growing in the same soil for ten, twenty, maybe thirty years. For whatever perennial you plant, as much cold or well-rotted manure and compost should be dug as deeply into the ground as you can. It should be the richest, most fertile soil you can make. I also mulch around the plants with a thick manure-straw mixture each year. This mulch later decomposes, becoming fertilizer for the succeeding year. As with my other plants, the ailing or slow-growing perennial is revived with doses of manure tea. Fill a bucket half full of manure and the rest with water; let it steep overnight and then feed it to plants. Because perennials are just that, they'll be there year after year, you should plant them all in one section of the garden or along fences so that you can easily cultivate the rest of the garden. Be sure to leave room around your perennials for expansion—they will keep increasing. I try to water the perennials that need it with a long, slow, steady trickle for deep saturation which encourages deep rooting and stronger, more productive plants.

All perennials need to be divided and thinned periodically, so flourishing, long-time gardens are often a good source of new plants for your garden. There's a limit to how much of even a good thing any one family needs. When I do have to buy new stock, I prefer to do it from one of the large reputable seed companies that guarantee their stock and will replace roots that don't grow.

Strawberries were the first perennial in my garden. Following the advice of a long-time gardener, I valiantly let them grow a whole year without bearing fruit (I pinched off the blossoms). Later, I tried the other method—letting the young plants fruit. This way I verified from experience that it is worth the wait. The second-year plants are large, hardy, and abundantly productive. The ones I let bear early have never in succeeding years reached the same size or production as the first batch. If you can't bear to wait, I'd advise doing some each way, so that at least part of your patch is a good perennial one.

Strawberries will grow and expand for years if they are well tended, so leave room beside your first plants for future growth. Strawberries may be gotten from friends who are thinning their patches or bought as young plants, but they are most often bought as roots. The roots arrive looking dry and brown—very dormant, if not dead. But a few days after planting, bright green leaves begin miraculously appearing from those dry crowns. Strawberries, like all other berries, are acid lovers, so acid soils should not be corrected and base soils should have leaf mold or sawdust added (add extra manure to compensate for the nitrogen taken by the

decomposing matter). Roots or young plants should be planted in wide deep holes, with the roots extending vertically and slightly spread. Only the crowns should protrude above ground. Once the plants are growing well, I mulch heavily with a manure-wood chip combination to provide nitrogen and acid. Strawberries are very heavy feeders and need a twice yearly manuring to meet their needs. In addition, I give any plants with yellowing leaves regular waterings with manure tea (the mulch makes its own tea too). If the top layer of mulch is mostly chips or sawdust, the growing berries will have a clean dry place to rest and will be less likely to rot.

Strawberries may be planted in hills (one foot apart) or in rows (two to three feet apart). In the fall of each year, old plants will send out "runners," new plants at the end of a long lead root. With the row system, these runners are allowed to fill in the area between the rows, forming a solid patch. With hills, the runners are removed and transplanted to new hills each fall. I find the hill system somewhat easier to mulch, water, and pick, but there's no particular virtue to either system.

Strawberries come in several varieties—everbearing ones and ones with specific fruiting dates. The usual plan is to plant all everbearing plants or a succession of the specific types (a few June, midsummer, and late bearers). I have chosen the everbearing type and they have done well for me. I get two heavy crops (late May–June and September) and a respectable harvest the rest of the summer. I began with twenty-five plants which is enough for two normal consumers or one strawberry junkie; now I have fifty plants plus their runners, being myself a very confirmed strawberry addict.

The second perennial that I put in the garden was artichokes—a wonderful vegetable for West Coast (mild winters, mild summers) inhabitants. The plants grew very quickly and soon became my very favorites in the garden—large, exotic, junglelike bushes. The first summer they produced artichokes off of every stalk.

Artichokes are easily acquired from friends as they must be divided regularly. They can also be bought as plants. If planted in rich, not acid soil,

they will bear the first summer and every spring and fall after that. The first artichokes of each season are the largest; after that they gradually diminish to the "twenty for a dollar" size you see in California roadside vegetable stands. An artichoke is ready to pick when the leaves first begin to open slightly—like a bud unfolding (which, in fact, it is). After that, they grow tougher and many of the leaves are wasted. Watch this on the last small ones of each season as it is easy to wait hopefully for them to grow larger, while they remain small and grow tougher. It's important to give the plants long, deep waterings during dry spells too, to prevent the artichokes from tasting bitter. And just once, after you have eaten more than you ever thought you could, let one artichoke open all the way into a lovely purple flower.

Artichokes are favorites of gophers as well as humans. If you have the energy, digging a two-foot-deep by two-and-a-half-foot-wide trench and lining it with fine mesh screen will protect your plants forever more. I didn't do that and have learned the hard way (a beautiful three-foot-high bush van-

I went to the movies last night to see *Lady Sings the Blues*. Rode home in a car full of talking, laughing women holding myself tightly. And then I began to cry. I cried and cried.

I cried. So simple to say that now. I have not cried in two years. Everything just dried up. Peter with his anger and his silence taught me not to cry, not to care, not to let pain touch me.

I am feeling lonely and alone now. I feel all the distances, the not touching, the not seeing, between us now. With these women who are my friends there is not one with whom I share trustingly, intimately; there is not one with whom I don't retain some hidden measure of reserve, whose sociability I don't match with my own. What we had was real, but now there are more spaces than there is closeness. There are the forms of intimacy without the content. There is no blame—we are each of us what we are. I can respect and love the intensity of how we grew together—we who met almost by chance. In a city, given choices, many would never even have known each other. We have been drawn together in our isolation, and we have made much of it, but not enough.

It's a lie that there is no blame—it's only true that there shouldn't be. Sometimes I am so consumed by dislike of Sarah, Katie, and Kathy that I can't even stay in the same room with them. Yet I also know that my resentment and my anger at them is a reflection of my need. I dislike them for not being what I want them to be for me. We made too many myths within the group; we promised ourselves everything.

ished overnight) to check constantly for gophers and keep traps set.

Because artichokes are so bushy, they should be planted two to three feet apart in rows three feet apart. Six plants will feed four people well. They are by nature communal, and new plants should be set in clumps of two or three depending on size. As the plants grow, they send up new plants from their roots and the clumps become more and more dense. Every three years, these groups should be divided and the smallest plants removed to new rows. Try to thin the groups back to the three strongest plants at that time.

Don't worry if your plants seem to die back in the summer or winter. If the leaf color is still good (gray-green but not yellow) and they are getting water and nutrients, this is a natural process. New artichoke-bearing stalks will shoot up amazingly quickly at the start of the next season.

Asparagus came to the garden with berry bushes the third spring. It takes more dedication than strawberries or artichokes, as it is much slower to produce. Asparagus may either be planted by seed or roots, either of which will probably have to be bought. With seeds, you may take a light cutting the third year and steady cuttings each year after that. The roots you buy are already two years old and will produce a light cutting the second spring. Asparagus beds last a long time once they are started and grow denser and more productive. I still remember picking asparagus as a child from the corner of my uncle's corn field. That patch was bearing well when he and my aunt bought the farm, and they had been there seventeen years.

I chose to buy roots rather than seed and have been glad I did. I followed a rather elaborate planting procedure recommended by *Organic Gardening* and have had wonderful results. My second-year stalks were large and tender and plentiful enough for many dinners. I have seen other gardens where third- and fourth-year stalks are thin and spindly. First, I dug trenches two feet deep and a foot or so wide, two and a half feet apart from each other. The first few inches were the richest manure I could scrape from the goat shed. (You may use a hot manure in this first layer; after that, use a cold or well-rotted hot one.) Next came an inch of dolomite lime (or crushed limestone). Then several inches of topsoil (or compost humus) mixed with rock phosphate or bone meal. Then I put in six more inches of manure, another inch of lime, and a thick layer of compost humus and phosphate. Repeat the layering one more time if you

need to, until you are six inches from the top. Then, I planted the asparagus roots in the trenches, twelve inches apart. I put in twenty-five plants, which have fed two of us well. But it's hard to have too much asparagus and I'm planning to get twenty-five more. As the plants began to grow, I gradually filled in the trenches until they were even with the ground. That first year, the shoots were very thin with fine feathery leaves. I just let them grow and watered occasionally, very thoroughly. Asparagus will root four feet deep, so after the first year, it is unnecessary to water the plants during the dry season.

The second year, the shoots began poking through the mulch right after the last frost, thick, full-sized shoots. The shoots should be picked when they are six to eight inches above ground, but before the tips begin to open and spread for leaves. Asparagus should be picked as soon as they are ready, every two to three days, or your will lose the special garden tenderness and find yourself with tasteless, stringy stalks. Commercial asparagus is harvested below ground level, but since I always cut off those tough, white ends anyhow, I don't bother. In areas that have rain all summer, asparagus can be harvested until mid-July and then let go to seed. In areas that have dry summers, you have to be careful not to pick more than two weeks after the last reliable rains, so that one last batch of shoots will have a chance to grow to maturity. As the years go by, you will get more and more shoots in your rows for long, steady cuttings.

Asparagus requires lime to grow. The first two years are amply provided for by the lime in the trenches. After that, lime and manure should be dug in every spring between the rows. Use a good deal of lime, more than you would for any other plant. During the winter, the rows should be well covered with a manure-straw mulch, which can be pushed aside in the spring and used to mulch between the rows.

Raspberries and blackberries are among the easiest of perennials to acquire. If their fruits weren't so good, I'm sure the plants would be readily classed as pests. They spread really rapidly, aided by birds who inadvertently cast the seeds, and the berries will take over a whole garden if not ruthlessly controlled. They also need thinning every year—so almost any gardener who has some will gladly pass on a bunch (one should be warned by such ready generosity). The life-sustaining properties of berries are amazing. I've seen "dead" brown canes pulled out of the ground in October

I am going to stay in this retreating place for a while, going nowhere, seeing no one. I don't want to close myself to help or support, but I must close myself to tension and to alienation. I need to take care of myself now, and that means much silence and much peace. I weeded the garden for hours today and then cleaned the house. I tiptoe through it and drink tea for dinner so as not to have to deal with dirty dishes. Reasserting physical order calms my soul. I need order so badly now, cannot handle chaos or confusion. I want security and trust and love, and there is no one to get it from but myself and this land.

spring back to life when replanted in April. So take whatever you can get. I've had no such luck with blueberries, which grow on neat, orderly bushes and are most easily acquired with cash.

All berries love acid soil, so I work manure and wood chips (sawdust, peat moss, leaf mold, or any other acid producer will also work) as deeply into the beds as I can before planting. The best source of this combination I've found is horse stables which often use the chips for bedding. But if worse comes to worst, I've been known to scramble around on my knees scraping up chain saw gleanings to mix with my own purer goat manure! Raspberry and blackberry canes should be planted in beds with two and a half to three feet between each plant. This spacing is somewhat arbitrary, though, as they will rapidly fill in that and any other available space. Blueberry bushes should be planted in rows with three feet between each bush. They come in many varieties, with different fruiting dates, so it's a good idea to get several kinds for early summer to fall crops.

There were about a dozen raspberry plants growing here when we first came and that first spring we just sat back and waited for the crop to appear. It never did. Next winter, I mulched heavily between the plants, transplanted another dozen runners, and the plants did produce that spring (actually, July is our berry month). They begin ripening almost on the first of the month and stop almost exactly thirty days later. I don't water the plants at all, ever, and doing so might prolong their season. Anyway, that second year we had berries, but they were small and not too tasty. Also, I hadn't really grasped the concept of a biennial plant and didn't understand why my transplants hadn't produced. This last winter, I happened upon

the really crucial aspect of berry culture—thinning between the plants so that they are really two feet apart, at the very least. Also the crop doubled as the last year canes of the transplants were filled with berries. The younger plants seem to produce more heavily, so every year I will start a few more. But my ruthless thinning (even had to take out whole three-year-old plants that had grown up too close) paid off in larger and tastier berries.

Every spring, any old dead canes should be pulled out, and the plants should be pruned back to three or four feet high at that time. These second-year (or older) canes are the ones that will bear fruit. Most of the new shoots should be dug up and transplanted or given away to friends. Once the bed is thinned, I spread as much manure and wood-chip mulch as I can.

Blueberries require much less effort than the other berries. They need to be pruned every few years to keep them a reasonable height and to keep their energy into berry producing. In the spring I dig manure in between the rows, being careful not to chop off roots. And I keep a year-round mulch down of wood chips or sawdust and manure.

Rhubarb is the latest perennial addition I've made to the garden, having discovered it on a visit to Iowa last summer. As an intolerant child, I refused to even try it. "Ooo, rhubarb" I would groan; now I sigh instead with pleasure, and eat it as often as I can as sauce or custard or by itself in pie. The rhubarb plants have a special meaning to me too—they were brought to this ridge in 1906 by my neighbor's mother, transplanted to Oregon after her death in 1947, and returned to this ridge (and my garden) by her eighty-four-year-old son in 1974. "It's nice to have Mother's rhubarb back," he said.

From my limited experience, rhubarb grows easily and well. The plants are a lovely mixture of red and green with floppy elephant-ear leaves. It is a heavy feeder and needs a lot of manure; in the soil, as mulch and as tea. Rhubarb also likes a moderately acid soil. If you buy roots, they are planted much like strawberries, in deep holes with only the crowns above ground. The plants and rows should be at least two feet apart as they grow very large leaves, very low to the ground.

Rhubarb needs lots of water for abundant stalks and should be watered well and often. It grows something like artichokes, with new plants shooting off from the base, forming clusters. The stalks are ready to eat when they are large and red (sometimes the base is deep red, shading to green at the leaf). The books say the stalks should be pulled, not cut from the plant. If only full-grown stalks are taken, the plants will stay strong and keep on producing for many months. If the plant starts to flower, those stalks should be broken off and discarded. (Rhubarb leaves are poisonous to animals —and people as well—so watch where you put

them.) During the winter, the roots should be protected from freezing by a thick straw mulch.

I have been told by my neighbors, though I have not tried it yet, that the plants should be divided every five years. Then, the large clumps should be dug up and separated in the fall. Reset the groups of several stalks each—one bunch in each of the original holes and the rest in new rows.

The only other perennials I have grown are flowers and herbs, and I love them almost as much as the vegetables. They are useful as companion plants and make beautiful borders or alternating stripes among the vegetables. Though the vegetable garden seemed an unlikely place for flowers when I first came, I've come to love the splashes of color among the green. The herbs in the garden have now grown large and shrublike with delicate flowers and richly green leaves.

What more can I say about perennials except, "Grow them!"? They become dependable old friends, always to be counted on for winter color, early spring growth, a little something alive in an otherwise barren garden.

My father is here this week. Wearing his competence like armor, he marches through all my Waterloos: fixing the drain that never really worked, installing electricity in the bedroom—ending my extension-cord life—working on the truck whose engine has been alien to me. Simple matters to one who has knowledge, and the energy and confidence that knowledge brings. I am learning now what I could have learned at 7, 11, and 15. Beneath my truck, side by side, lie his seven-year-old son and his twenty-five-year-old daughter, both of us learning for the first time how bearings fit together, how to remove pistons. And here beneath this truck the patriarchy stops: he has passed his knowledge to his daughter, and from me it will pass to sisters, from sister to sister to sister.

173

CANNING AND FREEZING

CANNING

Canning is one of the most satisfying activities for me—I suppose it's because it is one of the few things that leaves lasting, tangible results from the hours of food-producing labor. The vegetables that go straight from garden to stomach satisfy all kinds of needs, but not the need to gaze proudly at well-filled shelves that remind me of all I've done. In fact, I'm not precisely sure whether I can in order to have food to look at or to eat each winter!

It's fairly easy to keep surplus fruits and vegetables from going to waste. There are three main types of canning: foods that are poured from cooking pot to canning jars with no other processing; foods that are put in jars and processed with boiling water; and foods that are placed in jars and processed in a pressure cooker. Which type canning you do depends on the susceptibility of a particular food to spoilage. Pressure canning protects against all types of bacteria, including botulism; unprocessed jars are protected against simple spoilage only. Any food that has a high acid, salt, or sugar content, though, is not susceptible to botulism or ptomaine bacteria. This means that jams, jellies, and pickles are the easiest foods to can, requiring no special equipment. Fruits, because of their natural sugar content, are also safe from the really dangerous bacteria but must be processed in a boiling water bath (the second type of canning). Vegetables and meat come under the third type of canning (pressure cooking) to protect them from botulism. Tomatoes, because of their high acidity, are the one exception to this rule and may be processed like fruits.

The equipment needed for canning is not expensive, except for the pressure canner. An ordinary canning kettle (for fruits and tomatoes) is simply a large enamel pot with a wire rack to lift jars in and out. They come in two sizes, one that holds five quart jars, and one that holds eight. If you think you will be doing a lot of canning, the larger kettle is well worth the initial expense. A pressure canner is a very large pressure cooker with a rack to hold the jars; they are much more expensive. I have chosen to pickle or freeze all my vegetables rather than buy a pressure canner, so I have no experience with this type of canning.

Your jars need not be special "canning" ones (sold at grocery and hardware stores). Any glass jar that will fit the standard lids (mayonnaise jars, for example) can be used. Ordinary jars are weaker than canning jars and occasionally one will crack during processing, but this rare loss does not justify the greater expense of bought jars. Once you get into canning you will find that jars become precious possessions. I collect and hoard them all year round. Before the start of canning season each year, go through all your jars and discard any that are cracked or have nicks along the rim—they won't seal properly.

In addition to a canner and jars, you also need lids and rings. The lids are metal disks with a rubber seal which fits on top of the jars. Rings thread onto the jars over the lids, holding them in place. Rings must be used until a jar has sealed and cooled, but then they can be removed, leaving only the lid in place. This means you need many fewer rings than lids. Lids can be bought separately from rings or in combination; they are much cheaper alone. Check to make sure you are buying the right size rings and lids for your type jars; jars come with either wide or narrow mouths. If you reuse old lids, always make sure they are not bent, punctured, or dented and that the rubber is uncracked. It's hard to reuse lids more than a few times, but rings will last almost forever, unless they become badly rusted.

There are several basic primers on canning fruits, vegetables, jams, and pickles. They are all very inexpensive. The Ball and Kerr canning jar companies each publish their own as does the U. S. Government Printing Office (see Resources). I have all three and find it is useful to cross-check them when I can. Since I usually want to use little or no sugar and the shortest processing time possible, I use the books to decide on safe limits.

Canning jams, jellies, preserves, and pickles is the easiest process of all. All you need is a pan for sterilizing jars and a large pot for making the jam or boiling the pickling brine. The first step is to make your recipe and have it boiling hot. Then sterilize your clean jars and lids by placing them upside down in an inch or so of boiling water for twenty minutes. Sometimes the jars will suck all the water out of the pan; just lift the jars and you can break the vacuum. Remove the jars one at a time from the boiling water and fill them with the recipe. Standing the jar in hot water or placing a knife in it while you are filling will prevent its cracking. Fill the jar to a half inch from the top. Then quickly wipe the rim clean, cover with a sterilized lid and tighten down the ring. Invert the jar to cool. Then fill the next jar. When the jars are thoroughly cooled, check to see if each one sealed. The lid should stay depressed when you push on it; you should not be able to "pop" it up and down.

Pickles don't really have to seal (remember ye olde pickle barrel?) because vinegar is a preservative, but I feel superstitiously safer if they do and try to eat the unsealed ones first. Jams and other sugary goodies will get moldy and ferment if they don't seal. I once got some nice "brandied" peaches off a bad seal, but in general, sealing failures look, taste, and act weird. They should be refrigerated right after canning or thrown away.

JAMS AND PICKLES

Making good thick jam, jelly, or preserves is hard because it takes a lot of sugar, more than I am willing to use. Most recipes call for equal parts sugar and fruit. Two things cause jams to thicken—sugar and pectin, acting in conjunction with each other. Pectin can be supplied by adding green apples or lemon juice to your fruit or by buying commercial pectin derived from fruit sources. There are recipes for making your own pectin from apples in the canning books, but I've never done it. I use the commercial powdered pectin after making very runny jams without it for several years. Unfortunately, even the commercial stuff only really thickens with a lot of sugar, so I've learned to live with partly runny jams. To make jam or preserves, you basically cook the cut-up fruit with as much sugar or honey as you want until it is rapidly boiling. Then add pectin (if you want). The mixture will thicken as much as it's going to pretty quickly. I've found further cooking usually thins it. The jam should be poured into the jars while it is rapidly boiling. Apricots are the only fruit I've found that makes good thick jam with little or no sugar and no pectin. Just cook very ripe apricots to boiling.

I once was reduced to adding cornstarch and arrowroot to jam, in an attempt to make a thick, lightly sweet concoction. Instead, I got a very thick, strange blackberry "pudding," and I haven't tried those additions again. There's an art to keeping runny jam on bread—peanut butter helps! The *really* runny stuff, I call sauces and use on custard, pound cake, and ice cream. Substituted for honey in homemade ice cream, it gives good fruit flavors.

Fruit butters are also easy to make and very tasty—they are thicker and somewhat less sweet than jams. To make apple butter, you quarter whole apples, skin, core, and all, and steam them in a small amount of cider vinegar until they are soft. Purée the cooked apples in a blender or strain them through a sieve. Then put the saucelike pulp back in a kettle and cook it, adding sugar or honey, grated lemon peel, lemon juice, cinnamon, and other spices to taste. Cook the butter until it is quite thick. Stir to prevent burning or scorching of the butter. While boiling, pour it into sterilized jars and seal. Pear, peach, and apricot butters are made the same way, except the fruit is steamed in water or fruit juice. Peaches and apricots must be pitted but need not be skinned. These less-common butters are really delicious!

There are two ways to make pickles: by cooking the vegetables in a vinegar solution or by packing raw vegetables into jars and adding boiling liquid. Except for chutney and one zucchini recipe, I prefer the raw vegetable method. This makes much crisper pickles and they have as much flavor as cooked ones if allowed to cure for several weeks. Cucumber pickle recipes abound in every cookbook, and it's fun to try a variety, carefully labeling each jar so you can make your favorites again next year.

My favorite pickles no longer come from cucumbers. During our first year of overwhelming garden surpluses (before we learned about successive plantings), we invented dilled beans. I've been dilling almost every garden vegetable ever since, and though beans remain my absolute favorite, there have been many other successes. Almost anything can be pickled: beans, broccoli, carrots, cauliflower, small onions and onion pieces, green tomatoes, all kinds of squash. Dilled vegetables are crunchy and wonderful to munch on and delicious in salads too. To make them you:

1. Heat to boiling a mixture of:
   2 parts vinegar
   1 part water
   1 tbsp. salt to each 3 cups liquid
2. Fill sterilized jars with the vegetables. Beans, small onions, and green cherry tomatoes can be used whole. All other vegetables should be cut into thin pieces.
3. Add two cloves garlic, a generous pinch of both dill weed and dill seed, 8 peppercorns, and a pinch of mustard seeds to each jar.
4. Place a knife in each jar to prevent cracking. Fill each jar with the boiling mixture, remove the knife, wipe the rim, seal, and invert. Let pickles stand at least two weeks before eating.

There is a zucchini pickle recipe that is another favorite of mine, both because it is delicious and because it solves a chronic zucchini problem—overpopulation. To make it, you combine the following ingredients and boil them for twenty-five minutes:

3 medium squash	2 tsp. turmeric
2 onions	1 tsp. celery seed
2 tsp. mustard seed	2 pinches alum
2 tsp. salt	2 cups or more cider
2 cups sugar or honey	vinegar

Pour the boiling mixture into sterilized jars and seal immediately.

My other favorite pickle recipe is homemade chutney, a treasure after years of buying meager and expensive half pints in fancy food stores. Chutneys can be made from many fruits, but my favorite is this combination of pears and (or) apples and green tomatoes. I cook up gallons in my big canning kettle and pack it all in jars at one time. This recipe should be adapted and altered at will. I use it only as a rough guideline and often make five times this quantity. Where the stars (*) appear, the spices are often doubled:

3 lbs. firm pears and/or green apples	½ tsp. cayenne
1 lb. green tomatoes	2 tsp. salt*
1 lb. brown sugar	½ tsp. cinnamon*
1 pint cider vinegar	½ tsp. ground cloves*
2 onions	2 tsp. mustard seed
1 cup raisins or currants	¾ tsp. curry powder
¼ cup preserved ginger, chopped (I sometimes substitute 1 tbsp. chopped fresh ginger	1 clove garlic*
	Allspice, coriander, cardamon, if you wish.

Core the fruit. Dice the unpeeled fruit and the vegetables. Combine the sugar and the vinegar and bring them to a boil. Add all the ingredients and cook slowly, stirring from time to time until thickened, about one hour. If you double the recipe, use only one and a half times the vinegar. While boiling, pour into sterilized jars and seal. This recipe makes about four quarts.

FRUITS AND TOMATOES

To can fruits or tomatoes, you fill clean jars and then place them in a canning kettle full of boiling water for processing. There are two different methods for filling the jars: hot pack and cold pack. In hot pack, you cook the fruit in juice or syrup until it is boiling and then place it in the jars. Cold packing is simply placing raw fruit in the jars and then filling them with boiling water, juice or syrup. I prefer cold packing, since the fruit is minimally cooked.

Fruits may be canned whole, halved, or quartered, depending on their size and your taste. Apricots and peaches are best halved, so the pits may be easily removed (and this is when you will learn the meaning of freestone and cling peaches!). The core may be removed from a pear once it is cut in half by scooping it out with a small spoon. Peaches and tomatoes can be blanched in boiling water to remove the skins if you wish (I don't bother). Only unbruised, ripe fruit should be canned. Damaged fruit can be used in jams or sauces.

Damaged fruit can be made into sauces by simply cutting it up in small pieces and cooking it in a little water until it is mushy and thick. You can add sugar and spices to taste, but none are necessary. When the sauce is thick and boiling, place it in clean jars, seal and process in the canner as you would whole fruit. Plum, peach, pear, and apricot sauce is as delicious as apple. They make wonderful quick breads and cakes, too, when substituted in recipes calling for applesauce.

Most canning books will tell you to pack your whole fruit in a sweet sugar syrup. Boiling water works just as well, forming a light syrup with the natural fruit juices. More sweetening can be added when you eat the fruit, if you want it. I find fruits canned this way have a more natural flavor and texture than the too sweet store-bought varieties. Honey can be substituted for sugar in any recipe, of course.

Once the clean jars are filled with either hot or cold pack fruit, the rims should be wiped clean. Then cover them with lids. Screw the metal rings on tightly and place the jars in a canning kettle full of boiling water. While one batch of jars is being processed, fill the next. This is when the advantages of an eight-quart canner will become apparent. The jars in the canner should be covered with water. When the water returns to a full boil, begin to count the processing time. Each fruit requires a different amount of time. Your books will tell you this. When the time is up, remove the jars from the kettle and let them cool. When you hear the lids "pop" (forming a vacuum seal), you can remove the metal rings, leaving only the lids in place. A jar is sealed if you cannot press the lid up and down. Jars that do not seal should be refrigerated when cool and used fairly soon. That's all there is to canning. It is often an all night orgy (or ordeal) for me when all the fruit ripens at once, but that one night's forty quarts of pears is a year's worth of food.

Pressure canning of vegetables (except tomatoes) is something I know about second hand. The reference books will give you all the necessary information.

FREEZING

I am now a great believer in freezing, which is often quicker and easier than canning vegetables and many fruits and produces more palatable results. The initial cost need not be very high either if you hunt around for a secondhand freezer (my chest type cost $25 and has run well for five years now—left outdoors under a big sheet of plastic). If you can find one at a reasonable price, an upright freezer is a better choice. Its freezing coils and insulation have a much higher survival rate than the chest type has. Containers for freezing can be ex-

pensive (the plastic or glass freezing jars), but you only need those for soups and sauces, if at all. Freezing paper works fine for meat. Clean milk cartons filled with water are wonderful for fish (the water keeps out air and protects the flavor). Plastic freezing bags (especially the quilted, "insulated" kind) do well for everything else. I even use them for liquids like soup or tomato sauce, placing filled bags inside the cardboard sections of a canning jar box. Then I put the whole box in the freezer. When the bags are frozen solid, they can be removed from the box, neat solid squares easy to stack and store. This method prevents strain on the bags which could cause them to break, or the oozing of still unfrozen liquid between the coils in freezer shelves or baskets.

Most freezing instructions tell you to cook vegetables, then cool them, place in containers, and freeze. The cooking destroys enzymes in the vegetables that would damage color and texture (even while in the freezer) if left alone. But the well-cooked vegetables end up mushy and tasteless, anyway. After some experimenting, I've discovered that the enzymes can be stopped without overcooking the vegetables. First I wash and slice the vegetables if they need it (like broccoli or cauliflower). Then I prepare a pot of rapidly boiling water and a bowl of ice water. Next I get my freezing containers ready for packing. Thin vegetables (like peas or snow peas) need only a swift (ten seconds) blanching in the boiling water. Other vegetables need slightly longer (about thirty seconds). Green vegetables will turn bright green when ready. As soon as they are ready, remove the vegetables from the boiling water, plunge them into the ice water and then pack in containers. Keep adding ice to the cold water as it heats up. Place the filled containers in the freezer as soon as they are ready. To speed the freezing process, don't stack newly filled containers together in the freezer. Spread them out on shelves or cookie sheets until they are solidly frozen. Vegetables treated this way will still be crisp enough to stir-fry or steam when you go to eat them. They are far more like fresh vegetables than traditionally canned or frozen ones.

I have also discovered that tomatoes for cooking freeze really successfully. I never can them any more. Just blanch them in boiling water, plunge them in cold water, pull off the skin if you want, and seal in plastic bags. They make wonderful perfect red balls in the freezer. Instead of freezing tomatoes whole, I also sometimes cook up large pots of fresh tomato sauce (including spices) and then freeze that, but I'm not sure it's worth the extra effort.

Freezing books will also tell you to cook fruit in sugar syrup and then freeze it. Instead, I simply place the fruit in the freezer on cookie sheets and then pack it in plastic bags when it has frozen solid. This method uses no sugar and freezes each piece quickly and individually for the best texture and flavor. You can also remove as much as you need from a package and then reseal it, since each piece is separate. To prepare fruit for freezing, you just wash it and cut out any bruises or rotten spots. Berries can be frozen whole. Larger fruits should be cut into pieces and pitted or skinned if you wish. Spread the fruit in a thin layer on the cookie sheet, preferably with no piece touching any other one. When the fruit is frozen (in about six hours),

place it in plastic bags. It will get frostbitten if left uncovered too long, losing flavor and texture.

Always keep your freezer temperature at below 0° F. Keeping foods longer than a year in a freezer isn't very functional—except to satisfy a full-larder complex. Vegetables get icy, bread soggy, and meat tough.

Though I certainly would never want to spend my life in the kitchen, like a good farmer's wife, I get a lot of pleasure from canning and freezing. It provides me with food all winter, delicacies I could never afford to buy, and gifts for city-bound friends: good use of summer surpluses. Once you become comfortable with the process, canning and freezing can fit into spare hours between other farm chores.

Drying Foods

Many of us think last, if at all, about drying as a means of preserving foods. I was certainly among the "us," being city-bred and "educated." My de-education and re-education in country living, however, has included learning this oldest of methods for storing foods, and I'm regretting the many years I spent without this knowledge. The foods are delicious, nutritious, and convenient. It's often easier, cheaper, and less time-consuming to dry foods than to either can or freeze them.

With a relatively small garden, yielding a steady flow of small surpluses, and a limited budget, drying foods is, in every sense, *the* economical method for preserving them. I still do some canning—fish, pickles, etc.—and some freezing—meats, fish, baked goods, etc.—but I'm not faced with a need to buy sugar or honey, pectin, and jars at current prices. And I do not find whole days taken up with processing.

These savings are important to me, especially in view of the often added nutritional value of dried foods. There is some loss of vitamins A and C in drying, and one must plan menus with this in mind, but other vitamins and nutrients are preserved at considerably higher levels than canning provides. Freezing is supposed to capture more food values for longer periods of time than either canning or drying, but my own experience was negative on texture and flavor with freezing of many vegetables, most especially after several months had passed. So far, experience with dried fruits and vegetables indicates fine texture and excellent flavor, even *two years* after drying.

Fruits, vegetables, herbs, and teas are all on my list of drying subjects. There are several ways to handle the process—strings, poles, bunches, racks, cookie sheets—in the sun, in the oven, or indoors in out-of-the-way spaces. I string, pole, bunch, rack, and cookie sheet—all indoors, because here on the California coast we are blessed (or cursed, according to one's inclination) with considerable fog, especially during the summer months when fruits and vegetables are plentiful. I seldom use the oven since I've found the slower air-drying to be the best insurance against mold, the only hazard I know of associated with the process. This can only occur if foods aren't thoroughly dry before storage.

Strings: Vegetables suitable for stringing include snap beans ("leather breeches"), snow peas, and almost all the leafy greens—cabbage, chard, spinach, mustard, radishes, etc. Snap beans and snow peas to be strung need to be washed only if they are dirty. To preserve maximum inner integrity, I leave about ¼″ of stem on each—nature's perfect package is thus utilized fully. Greens are, of course, separated and washed carefully. I also dry each side, so the evaporation of the plant's own water can proceed forthwith.

After preparation, vegetables are strung on a strong thread with a knot at the end, using a large needle which is run through the center of the vegetable. Beans and peas so strung may be hung from a nail on the wall. Strung greens are suspended between two nails on opposite walls. In every case it is important to allow space between *each* vegetable or leaf for air circulation while drying.

Poles: Any fruit or vegetable that makes circles with holes in them when sliced can, I believe, be poled, with the obvious exception of avocados (which don't dry well due to their high oil content). I use a broom handle for pumpkin and winter squash circles (though both of these can be stored whole), thinner dowels for onion and apple rings. Ends, or pieces that are not circles, are dried on racks.

Slice pumpkin or squash ½″ to ¾″ thick after removing seeds (which may be dried on racks or cookie sheets). Remove outer onion skin and slice into rings ⅛″ to ¼″ thick. Wash apples, core, and slice into rings ¼″ thick. Space circles on poles to allow air circulation and suspend between rafters or curtain-rod holders.

Bunches: Most herbs and teas can just be bunched together at their stem ends, tied with a rubber band or string, and hung, tips down, from a nail to dry. Herbs and teas thus bunched need to be examined carefully and any cobwebs or foreign matter removed before washing to remove dust. I use the spray nozzle at my sink, which feels right to me, its sprays penetrating to the inner reaches of fine-leafed plants. (Speaking of fine-leafed plants, carrot tops, fresh or dried, are excellent "herbs" for flavoring soups, stews, gravies, etc. They taste just like carrots.) Shake bunches well and pat dry with paper or cloth toweling before bunching and hanging.

Fine-leafed and flowered plants should be hung with large bags (paper or cloth; *not* cheesecloth, though) covering the foliage, or much will be on the floor and lost. It is important that covers be large enough to permit air circulation while drying. (Do not use plastic, which causes moisture to collect and will insure a moldy product!) Larger-leafed plants (eucalyptus, bay, Labrador tea, etc.) do not tend to fall off as they dry, so need not be covered during the drying process—just bunch and suspend from nails.

Racks: My racks are in constant use—there is no season that doesn't present abundance and seasonal flavors desirable all year round, and seldom does a week pass without the laying out of some delicacy on the screens. Almost all fruits and vegetables (including those strung, poled, and bunched) can be successfully dried on racks. Mine hang over my kitchen stove, where heat from the wood stove that heats my kitchen and on which I do most of my cooking, rises to provide a fairly even and constant drying source. Racks hung from almost any ceiling in living areas will obviously suffice, though the drying process will be slower if far from heat source.

Blanching: Mushrooms, onions for seasoning, and most fruits should never be blanched. But short blanchings of many vegetables and some fruits seems to help them keep their color better and dry faster. It also stops the ripening process. They are as flavorful as if unblanched and, for some, the additional beauty and faster drying may be worth the effort. Whether or not I blanch is inevitably determined by time available. If the surplus appears at a time when I'm in the middle of a construction or gardening project, I don't blanch! And I *never* blanch as long as recommended below.

Steaming is preferable to boiling for the blanching process—fewer nutrients are lost. The U. S. Department of Agriculture recommends the following blanching times.

For vegetables:
  Asparagus—10 minutes
  Beans (snap)—20 minutes
  Beans (lima and soy)—15–20 minutes
  Beets (whole)—45 minutes
  Broccoli—10 minutes
  Brussels sprouts—10 minutes
  Cabbage—5–10 minutes
  Carrots—8–12 minutes
  Celery—10 minutes
  Chard—5 minutes
  Corn—10 minutes
  Kale—20 minutes
  Onions (for eating)—5–10 minutes
  Peas (shelled)—15 minutes (6 minutes if boiled)
  Peppers—10 minutes
  Rhubarb—3 minutes
  Spinach—5 minutes
  Summer squash—7 minutes
  Tomatoes—5 minutes
For fruits:
  Grapes—½ minute
  Plums—2 minutes
  Prunes—2 minutes

All except tomatoes would be rack-dried after blanching. (Tomato pulp is dried on cookie sheets or glass or plastic sheets.) Some (especially old-timers) peel everything before drying. I don't and find textures and flavors fine. I'm also confident that food value is considerably higher.

Whether blanched or unblanched, both fruits and vegetables should be unblemished, washed clean, and sliced thin (or, in the case of berries and grapes, left whole) for rack-drying. It is important to space slices carefully to allow full circulation of air around each piece, so do not allow them to touch each other. This insures more even drying and prevents mold from developing. Pieces should be turned over several times during the drying process to hasten even drying.

Cookie Sheets: For drying fruit leathers (fruit purée that is dried as explained below) or vegetable pulps, cookie sheets are ideal or, as mentioned above, glass or plastic sheets may be used. For fruit leathers, fruit is steamed over low heat, with fruit juice or water added if necessary, until the fruit is soft. Fruit is then strained and pulp run through a food mill or sieve to further break it down. Pulp may be sweetened to taste with sugar or honey, and flavoring extracts may be added. It is then spread, about ¼″ thick, on oiled cookie sheets, covered with a single layer of cheesecloth, and placed in a warm, dry place to dry. The drying process takes from one to two weeks. When the leather is dry enough to lift from the cookie sheets, it's done and can be dusted with cornstarch or arrowroot powder and stacked, with wax paper be-

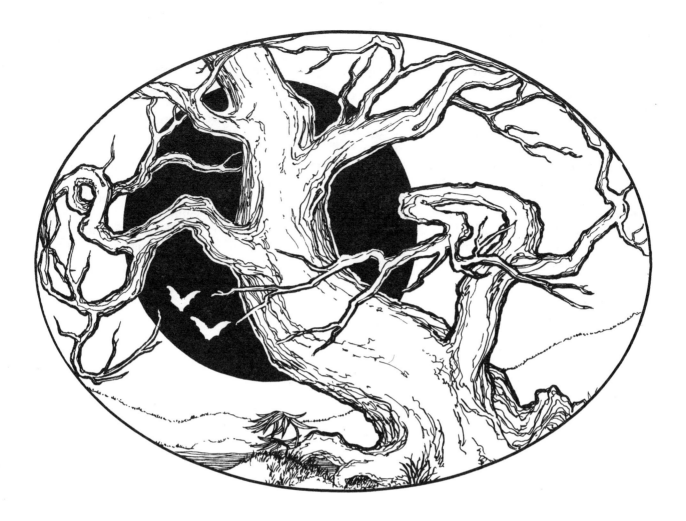

tween each leather, or can be rolled up, like a jellyroll, and wrapped in wax paper for storage.

For tomato (or other vegetable) pulps, wash, cut into pieces, and steam for about five minutes, then run through a food mill, removing seeds if desired. Strain through cloth, removing as much juice as possible, and spread on cookie sheets. As it dries, turn pulp often until it flakes.

Storage: When drying is properly done, there is almost no limit to ways of storing the end products. They take up much less space than with other preservation methods. For me, herbs and spices go into colored, air-tight bottles; beans and peas into paper bags; teas into jars with tight lids; vegetables and fruits into coffee cans. I do not use plastic, having been told that mold is more likely to develop with it.

Rehydrating: Replacing the water lost in drying

is necessary for all vegetables and sometimes is desirable for fruits, if they are to be used in compotes, for instance. Some references suggest "1½ cups of boiling water to each cup of vegetable or fruit," but overnight soaking in the same amount of lukewarm water achieves the same end. They should be soaked until they will hold no more water. If there's water left over after soaking, use it!

Using: Dried fruits and vegetables are excellent for everyday use in the same ways as fresh fruits and vegetables. They adapt well to favorite recipes. Experimenting is fun, with texture differences often enhancing otherwise smooth or fine dishes. They are indispensable for backpacking or camping. I use dried fruits as they are for snacks, for baking, in puddings, on cereals, etc. However, many references suggest rehydrating before use in baking, and this step would be essential for sauces.

For vegetables "as vegetables," they should always be rehydrated before use. After soaking until they will hold no more water, cook them for about *five minutes* in a minimal amount of water *if they have been blanched* before drying. If they were not blanched at the time of drying, cook them somewhat longer.

Dried vegetables can be really delicious. I vary the cooking of each vegetable to enhance its own flavor. With rehydrated beet slices, for instance, I add a little lemon juice and salt to the cooking water, and some honey too just before serving. When using spinach, I crumble the dry leaves to make instant chopped spinach. Then I soak and cook the leaves, adding green onions and sour cream to the final product. You will be surprised to find that dried vegetables are second best only to the original fresh state.

To use pumpkin as a vegetable or for pies or other baking, and winter squash too, soak overnight, bring to a boil, simmer for three to four hours. Mash and serve or use as usual in baking recipes.

When using dried vegetables in soups, stews, etc., where there is plenty of liquid, there is no need to rehydrate them ahead of time. Just toss them in. However, *do* keep an eye on the liquid level, as your veggies will be absorbing, rather than imparting liquids to the dish—you'll no doubt need to use more broth or water or juice to compensate for this fact.

SECTION III

ANIMALS
AND POULTRY

RAISING BABY CHICKS

No puzzle at our farm: The chickens came before the eggs. One of our first orders from our new Sears Farm Catalog was for twenty-five Plymouth Rock chicks ("steady layers of good size brown eggs"). They arrived lively and healthy in a cardboard carton at our post office—tiny, fluffy yellow Easter card specials. They were the beginnings of our chicken flock. Raising those chicks was another learning-by-doing experience. Five years and hundreds of chicks later, I'm still learning about chickens and chick raising, but some of the basics are quite familiar.

Buying Chicks: We fancied our farm incomplete without a few solid, sensible hens scratching around the barnyard and a rooster to crow us awake. Most people won't want to part with good laying hens, and we didn't want to buy commercially raised pullets, so we decided on day-old chicks. These can be ordered from Sears or Montgomery Ward, bought at a local feed store, or ordered from hatcheries. You have your choice of buying chicks that are "straight run," or buying "ninety-five percent pullets." Straight-run chicks are unsexed and will usually be about half hens (pullets) and half roosters (cockerels). If you eat meat, this is a good way to simultaneously start a laying flock and raise some healthy meat. If you just want a flock of laying hens, buy the 95 per cent pullets. These will be more expensive per chick, but you won't have a surplus of unwanted roosters at the end of six months. You will probably end up with a few roosters out of every twenty-five chicks, but if you want fertile eggs you'll have to keep one rooster for every twenty or so hens.

Some places sell only debeaked chicks; others will debeak chicks upon request. The end of each chick's beak is clipped short, supposedly to keep it from pecking at and harming its companions. We never bought debeaked chicks, and never had any problems with pecking; the practice seems to me to be unnecessary and ugly. Debeaking also makes it difficult for chickens to peck and pick up grains.

There are many breeds of chickens. Each breed has certain characteristics; coloring, body type, egg type and production, and so forth. There are dual-purpose breeds such as Rhode Island Reds and Barred Plymouth Rocks. These are heavy enough birds to be good meat sources and are also good layers (their eggs are large and brown). White Leghorns are a lighter body type and more suitable as strictly layers (their eggs are large and white and they are prolific layers). Bantams are small, hardy birds that lay small eggs but are very resourceful at foraging for their food, and excellent mothers for their own or foster chicks. There are innumerable "fancy" breeds of chickens with distinctive feathering, mops and flounces, and beautiful coloring. You would probably do well to start with a basic laying flock of Rhode Islands or Leghorns before yielding to the aesthetics and interesting possibilities of the fancier breeds.

How many chicks to start with? In her first laying season, a hen will usually lay an egg a day in good weather and drop to one every other day in winter. In winter she will also take a month or two off to moult. As a hen gets older, her egg production will drop off some and may become more erratic. But unlike the commercially kept "egg machine" chicken who is kept at top production with medicated feed and total confinement (and burns out in about a year as a result), the healthy farm hen will lay fairly well for three years or more. You can either buy enough chicks for your own needs or buy enough to produce surplus eggs to sell. If you keep roughly twice as many hens as you need, you can usually sell half your eggs for a fair price and make enough money to pay for all of your chicken feed. Don't buy more chicks than you can adequately house, though. Overcrowding is certain to cause more problems than it is worth.

If you're buying chicks at a hatchery or feed store, make sure they look alert and healthy. Pick up each chick gently and check its rear end—it should be clean and fluffy. If the rear around the vent is wet, sticky, or soiled, don't buy that chick. It is sick, either from being chilled or improperly fed, or it has a disease. You should think twice about buying any chicks from that batch. Diseases in baby chicks (and even older poultry) are usually highly contagious and often fatal. If you get "mail order" chicks from a hatchery or through Sears or Montgomery Ward, be sure to open the chicks' packing crate while still at the post office. Chicks that are dead on arrival should be replaced by the hatchery. Usually a few extra chicks will be added to each order to cover this possibility of fatalities, but now and then the chicks will come through in very bad shape. You will need certification from your post office people to insist upon replacement chicks if the hatchery you are dealing with is difficult about this. We had one very bad experience with a hatchery which taught me this caution. In general, I think the chicks shipped out by Sears have been the healthiest and arrived in the best condition.

Housing for Brooding: A mother hen is the ablest baby chick raiser. She will keep them warm under her feathers, protect them from predators, and teach them to forage for food and drink water. When they're old enough, she'll teach them to fly or hop up to roost. As they mature, the hen will cease protecting and herding her youngsters, gradually forcing them to fend for themselves. If your chicks were incubator-hatched, though, you'll have to fill in on all these matters.

One common brooder is a metal canopy with curtained edges and a heating element inside. The chicks can go under it whenever they get cold and should probably be confined under it for their first few days. If you have electricity connected to your outbuildings, a brooder can go in the corner of your chicken house or in any protected room.

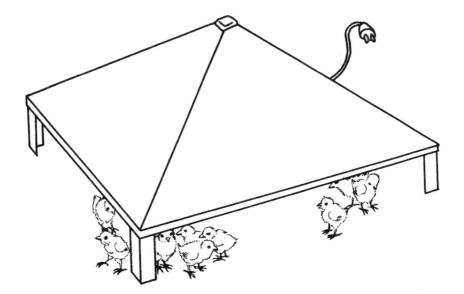

Brooders sell for from $30 up, depending upon size and elaborateness. You may also be able to design and build one yourself quite easily. Kerosene-run brooders are also available, though a little harder to find. You can also buy brooder bulbs of various types (price ranges from $5.00 to $15) and make your own brooder using one of these. If you want your house alive with chirps and feathers and want to really see your chicks grow—raise them inside! A simple cardboard box with a wire (screen) top is perfect for little chicks. The wire top will keep cats out and chicks in. The box can be easily replaced as the chicks grow and a new box used for each batch (this helps prevent disease). Line the bottom of the box with newspaper and change it daily. Wood shavings or shredded straw can also be put in as litter. For warmth you can use a small reflector lamp over the box. These are inexpensive or can be handmade, using a five-pound honey can with a bulb inside.

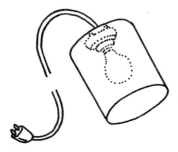

Choose the wattage of the bulb according to the room temperature, size of the reflector, size of the box, and number and age of the chicks. Very young chicks kept in a fairly small box usually need a 75–100-watt bulb. As they grow their feathers and can keep warmer, you can cut them down to a 50- or 25-watt bulb or move them to a larger box. Chicks that are too cold will huddle together in a corner and crush or suffocate one another. This piling up is noticeably different from their normal sleeping close to one another.

If you don't have electricity or a kerosene-run brooder, you can keep the chicks warm by placing them near the wood stove, or by putting a large jar full of very hot water and wrapped in a towel in the center of the box. This will have to be replaced often, though—particularly during cold nights. A double-walled box insulated with straw (stuffed between the walls) will also help to keep the chicks warm.

As the chicks grow, their downy fluff is replaced with feathers and they become hardier and more chill resistant. When they are about a month old (wait longer if the weather is cold), you can begin hardening them off. This means that you gradually reduce the temperature of their brooder or wattage of their brooder bulb, and then begin introducing them to outdoor temperatures. At first, leave their lamp off during the day and put it back on at night. They may be put outside on sunny, warm days if they have an area sheltered from too much wind or sun.

After a few weeks of this treatment, the chicks will no longer need their lamp at night and may soon be moved to a permanent house outdoors, or to a small pen with a shelter attached (lock them in at night until they are accustomed to sleeping in the shelter).

Tory said to me today that loving is knowing that you don't always have to be strong. And here I am, separating myself from others because I fear the vulnerability of my weaknesses. I am frightened of the dependency I once felt. I have to learn to be a whole person with others—to learn that doesn't mean caving in, doesn't mean becoming a leech. I want to find a way to share my uncertain parts, my feelings, doubts, fears, needs. Like a man, I wear only my strong face saying, "See, I really am independent," yet I never test the depth of that independence by being vulnerable.

Feed and Water: Newly hatched chicks usually must be taught to drink water. A flat jar lid or dish will do for a waterer at first. Take one or two of the chicks and gently dip their beaks into the water. Once they learn to drink, others will learn by imitating them. Chicks that have just been shipped in the mail or driven home from the feed store should be given water and taught to drink immediately, as they become severely dehydrated during transport. You can provide your chicks with a simple plastic waterer or a mason jar with a plastic waterer base for a dollar or less. This type of waterer provides a flow of fresh water as the chicks drink, and does not get as quickly and easily soiled as a dish would. It must, though, be cleaned and refilled at least once daily. Trough-type or automatic dome waterers are available for large numbers of chicks. These range in cost from five to fifteen dollars.

A ready-to-hatch chick must peck its way out of the shell. This pecking motion is like the food-finding peck once the chick is hatched, mobile, and hungry. Scratching uncovers insects, grains, and such, and though the mother carefully teaches her chicks to scratch and search, unmothered chicks seem to do this instinctively. Scattering their feed on the floor of their box is a good way to encourage the chicks' natural instincts. You can provide your chicks with special feeders (circular or rectangular tin, hopper, or trough types), but they will usually scratch the feed out if they possibly can—and seem to enjoy this most natural of activities.

Commercially made baby chick mash is a fully balanced diet containing all necessary ingredients. It may, however, be made with unhealthy meat scraps (diseased parts of animals and birds considered "unfit for human consumption"). It is also commonly medicated with penicillin and/or contains preservatives. Chick mash made from soybean meal as a protein source and containing neither penicillin nor preservatives is available and is probably a healthier feed. If you feed a soybean-based mash, you should give your chicks a regular supply of milk in liquid or powder form, as they need some animal-based protein. "Baby chick scratch" is finely ground grains (corn, wheat, milo, etc.). It is not a complete feed but can be given with the mash or with other feeds, or used as a base for your own feed recipe.

You can mix your own baby chick feed by finely grinding grains and adding wheat germ, bran, sunflower seeds, powdered seaweed, alfalfa leaves, and so on. The chicks must have adequate protein—15–20 per cent is desirable—in the form of powdered milk, fish meal, linseed meal, and so on. Chicks must also have fine grit or sand to be able to digest their food. Ground oyster shell will provide calcium. Green food—comfrey, lettuce, chard, clover, and so on—should also be fed them. Feed the chicks a little at a time as often as possible or give them free-choice with automatic or trough feeders. They won't overeat, but may waste a lot of food when free-choice fed.

As the chicks grow, they can handle coarser grains and are gradually switched over to laying mash and hen scratch. Compare ingredients of various commercial mashes, looking at content analysis versus price. Different grains will vary in price from area to area and season to season. Usually it's cheaper to buy grains separately and mix them yourself than to buy already mixed scratch. According to a table I found in an old poultry book, it takes about twenty-eight pounds of feed to raise a

Rhode Island Red chick to age six months. A lighter breed chick (White Leghorn) will consume about twenty-four pounds.

Health: A healthy baby chick is bright, energetic, and lively. If a chick doesn't rush up with the others to feed, or if it holds its feathers fluffed out or its wings away from its body and slightly down, it is either sick or suffering from a nutritional deficiency. Most of these signs also indicate a chill, and sometimes all the chick needs is more warmth. An insufficient amount of protein in the diet will make the chicks droopy or slow to grow and feather. Correcting the diet should have a noticeable affect fairly quickly. A single baby chick that is ailing should be separated from the others or it may be trampled by its stronger companions. It may also have a communicable disease. Keep it extra warm. A few drops of honey dissolved in warm water and given by eyedropper has a startling restorative effect on ailing chicks. Bloody droppings and diarrhea are signs of a common disease that affects young chickens—coccidiosis (see "Keeping a Healthy Flock"). Chicks may also fail if they become infested with worms (again, see "Keeping a Healthy Flock") or external parasites. Buttermilk is said to be an excellent addition to the chicks' diet—discouraging parasite infestation and promoting rapid growth and feathering. The *Herbal Handbook* (see Resources) gives good, sensible advice about raising and keeping healthy poultry without medicated feed. If you keep your chicks in clean quarters, give them fresh, clean water daily, and feed them a diet both balanced and varied (with particular attention to protein level), you should have very few health problems.

I finally buried the septic tank and the drainage lines today, almost a year since I finished the plumbing, a year of tripping over the ditch without even noticing it. There is so much I get used to rather than fix. During breaks in the digging, I kept wandering out to the back meadow, looking at the land, measuring it into fields with my eye. Back and forth all day between the work of the septic tank and the dream of the sheep ranch. For days I have been measuring and calculating. Off and on each day I struggle with the decision, play with the fantasies. I have to decide soon, before the ground gets too hard whether or not to dig postholes, whether and how much land to fence. Partly, I am afraid of the cost. Ten acres of fencing will cost me $280, five months of saving at my current rate, and still I will have only bare grass and no sheep. So funny how little $280 really is; how monumental it feels to me now.

Even beyond the problem of the money, I still hesitate. I think of stripping, soaking, and setting 220 posts, think of hauling, stretching, and nailing eight rolls of fencing, wondering whether I will day after day sustain the dream through the reality of the work. Will I be able to find friends to help with the work I cannot do alone? I have for two years now thought of these sheep—the crimp and color of their fleeces, the yarn I have spun and will spin, the dye colors latent in my field of daffodils. For so long I've dreamed about doing this, while slowly other bits of the vision have come about—goat pen and barn built, gardens grown, orchard pruned and picked, fence posts cut, stacked, and waiting, a silent reminder. Yet still I hesitate, worry, plan, calculate. I cannot afford to have animals just for the pleasure; I must do this big enough and well enough that they will pay for themselves and more. Do I have the courage or the craziness to do this alone? Soon I have to decide, stop hesitating at the edge of the orchard and commit myself.

housing and equipment

Chickens need only a simple structure, but one constructed for their needs and protection. The location of the house is the first consideration. If possible, it should face the south and have openings on this side so that maximum sunlight can be admitted. Sunlight acts as a disinfectant, provides for vitamin production and assimilation, and stimulates egg production. Hens enjoy sitting or stretching out in the sun and basking in the warmth. A chicken house set in a low or damp place or too exposed (as on an open hilltop) will make its inhabitants uncomfortable, if not outright unhealthy.

The house must be well ventilated. I had always followed this principle without really thinking about why it is so. Rereading our poultry books before writing this, I found the answer: a hen has a high respiratory rate and a high body temperature but does not excrete sweat or urine. She therefore breathes out a (relatively) tremendous amount of moisture and uses proportionately more oxygen than other creatures. Thus, the chicken house must provide a lot of fresh air and good air circulation. Large screened areas on the upper front walls of the house will provide for these needs. Cool air will enter these windows, sink to the floor, then rise and flow out as it is warmed by the hens' body temper-

atures. In cold or damp, windy weather, these openings should be partially closed with canvas curtains, wooden shutters or the like. The rest of the house, especially roosts and nests, should not be drafty. An even flow of air is healthful; drafts make for chills and subsequent disease.

The floor of the house can be concrete, wooden, or dirt. A concrete floor is permanent, rodent and predator proof, easily cleaned, washed and disinfected. It is probably the best—though most expensive and difficult to build initially. The one disadvantage is that the concrete tends to hold the cold. We just built a chicken house with a raised wooden floor. It seems to be working well—it is easy to clean and is rodent/predator proof. We sealed the wood with a preservative and allowed for ventilation underneath to keep it from rotting. Dirt floors are the simplest and cheapest but have many problems. They become pitted by the scratching hens and are difficult to clean. They tend to get and stay damp, and cannot be thoroughly disinfected. Weasels and rats can tunnel in to steal hens, feed, and eggs. If you have a dirt floor, scrape or dig it down once a year and and put in about six inches of clean sand, dirt, or gravel. Check regularly for predators' holes! It might help to dig down a foot

I walked all over the back sixty acres today looking for fence posts and saw such indescribable beauty. Some unknown trees by a spring were bursting with fuzzy pale yellow blossoms and new gray-green leaves. Wondrous muted shades which made me dream of dye plants and yarn. Weaving is becoming the core of my life. I shut my eyes and see ancient Aztec patterns. At the loom, archetypal images pour out of fingers, the loom and yarn drawing out from me long hidden secrets. I'm awed by this last hanging—its delicate subtle colors form a mirror to the earth. Afraid that I would find within me sterility and emptiness, I have found instead the joy of making a hanging which truly pleases me. I have learned the magic of real earth colors from natural dyes, which I could not see and did not care about when I started. And together in the co-op, we have woven more than objects; we have woven ourselves, our creative energies into something greater than any of us could have made alone.

or so around the perimeter of the house and bury a continuous protective line of hardware cloth or small mesh chicken wire. You might also line the dirt floor with special poultry house floor wire.

Whichever type of floor you have, a litter of some sort should be placed in your chicken house. Straw or pine shavings are ideal. Start with six inches to a foot and build it up as much as another foot. The hens will scratch about in this and keep it fairly dry by turning it. Occasionally you may have to get in there with a pitchfork and give it a good thorough turning. We've found that shavings stay drier and are much easier to work with—straw tends to mat when it gets wet and form difficult-to-break heavy layers. You should clean this litter out every six weeks or so—depending upon how dry and clean it stays. Use lime on the floor to disinfect it and make a fresher smell.

One way to keep the floor of your chicken house cleaner is to build dropping boards or screens under the roosts. These will catch a good deal of manure produced at night. Dropping boards may be hung about eight inches below the roosts, and should extend at least eight inches beyond the roosts in all directions. They can be hinged and suspended with wires so that you can drop them for cleaning. An alternative is a screened "pit" below the roosts. This can simply be a box with wooden sides to keep the chickens out and a chicken wire top which keeps the chickens out but

allows manure to drop through. The pit should be cleaned out periodically (the very best method I've seen was a pit that opened to the outside of the house and could be scraped right into a wheelbarrow).

The size of the house you need will depend upon the size of your flock, the type/breed of chickens you have, and whether they are to be allowed out or not. Large flocks (fifty or more) need proportionately less floor space than smaller flocks. (Larger breeds need more space per bird, though.) A small flock needs six to eight square feet per bird. A larger flock needs about four square feet per bird. A 14′×16′ building will comfortably house fifty heavy breed hens—or sixty Leghorns. Overcrowding is one of the worst conditions you can expose your flock to.

A good addition to your chicken house is a wire run. These are outdoor yards or pens that can be made with poles and poultry netting or poultry/garden fencing. The area fenced will depend upon your flock size, your energy, and your resources. Most heavy breed chickens will not fly out of a six-foot-high enclosure, but if you have lighter or gamey types, you may have to make a netting roof to keep them in! A roof will also be necessary if you have hawks in your area. A small door opening from the chicken house into the run should be kept closed at night to keep predators out of the house. If the chickens have been first given a period of a

Just finished three days of hauling manure from Koski's barn, working with Sarah's commune, Jane, and Tory. The first day was pure hassle—too many people, not enough tools, no idea how to proceed. I find that I am just not able to cope with all the chaos of a lot of people; I have so little tolerance or patience these days. The next day, no one from Sarah's came, Tory was late, and Jane wanted to plant strawberries. So I went to work alone, working off my tension and frustration in slow, solitary labor. I made two trips by myself: four hours of pitching, packing, hauling. Then Tory and I made two more trips, working together quietly and easily. Today, Tory and I and Sharon and Alan finished the big room ourselves—twelve more loads. I was really flying when we finished. Skinny, filthy dirty, my jeans caked with manure and car oil (the clutch went out and I fixed it), stinking of sweat and the same T-shirt for three days, driving the truck easily and like a pro: I felt so strong, so real, so *good*. I am beginning my ranch and I am loving it. The garden will be superb, and I will have sheep by summer. I love the days I spend working, how quiet and slow yet full they are. The pleasure of being totally tired at night yet not aching or really exhausted—pushed just up to my limits but not beyond them.

few weeks locked in their house, they will return to it at night to roost. Feeding them inside will likewise encourage them to roost inside.

Ideally, you should build two or more runs per house so that each run can be rested and disinfected (with lime, or a planting and tilling under of mustard). This resting period will keep the soil from getting sour or worm-egg infested. The runs may also be used as "scratch runs." Put litter (straw, hay, leaves) and feed some grain out there —the hens will scratch and turn the litter, breaking it up and making beautiful compost.

If you let your chickens out to range daily, you won't need to build runs. It is essential that you first give them a long period of house confinement. They will form the habit of laying eggs and roosting in their house. You can then begin letting them out. The first few days wait until an hour before sunset. This will allow the hens a short period to roam and scratch before they become anxious to be back in the house and on their roosts for the coming night. After that, we've found the best ritual is letting our hens out at about two in the afternoon and locking them up just after dark. They lay all of their eggs in their house, are safe from raccoons and such at night, but still have enough time each day to range about in search of insects and wild green feed. The only disadvantage in this "free

range" method is that new birds can't be easily introduced into the flock—and some birds will insist upon laying their eggs in secret caches outdoors. When new chickens are introduced, you must keep your entire flock in the coop for about a week before letting them range again.

Some final details for your chicken house include perches or roosts, feeders, waterers, and nest boxes. Perches may be made of poles or finished lumber. Poles should be peeled and about three inches in diameter. Lumber (2×4 or 2×3 material) should have its edges rounded off (a small hand plane does this nicely). Place the perches twelve to fifteen inches apart and support them with crosspieces where necessary. We built our first chicken house with graduated perches which seemed to encourage late-night feather pulling. Our new chicken house has perches all on the same level and no feather pulling! We also hung these perches quite high—about three and a half feet—and the hens seem to like them up there. Perch space should be from eight to twelve inches per chicken, depending on the breed.

Feeders may be built or bought. They are necessary for mash and must be designed so that birds won't alight on or attempt to roost on them. Trough-type feeders have a rolling bar along the top to discourage roosting. Hopper feeders have

The Very Basic Chicken House For 25 Laying Hens

Showing Pole Construction & Board-and-Batten Siding

Overall Dimensions
Floor Area: 10 ft. by 7 ft.
Height: 6 ft. sloping to 4 ft.

Roof:
three sheets of plywood (4'x8')
covered with roll roofing.

Note:
This house presupposes that you let your chickens out to run for part of the day.

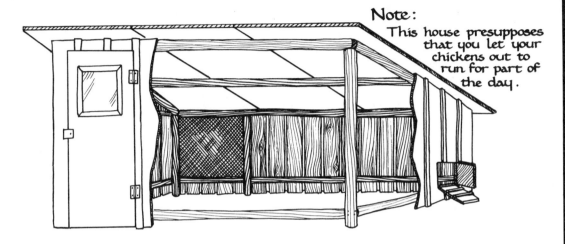

A Bird's Eye View

South Side
Exposure is very important for light and ventilation. This is a chicken wire window.

Nest Boxes: should be shaded from direct light. One box (12" square) per every seven birds. Put them at least a foot off the ground.

Deep Litter: (1 foot) on floor, straw or pine shavings, turned occasionally, will keep floor dry and odors down.

Trough Feeder: for feeding mash without waste.

Waterer: 2-gallon automatic waterer available from Sears for $5. Place on a box so litter doesn't get in the water.

feed scratch (grains such as corn, milo, wheat) on the ground to encourage turning of litter.

door

Roosts: should be 2" diameter poles. Each bird needs about 7" of space. Poles should be a foot apart on the same level and high enough to clean under.

protective coverings for the same purpose. We bought a few "automatic feeders" from Sears—they are large, round hanging feeders which let mash drop down as the hens eat it. Simple, inexpensive and easy to clean, they seem a good investment. They should be hung so that the feeding part is about level with a hen's back. Some type of hoppers should be made for oyster shells and/or grit. We use a large flat pan which the hens tend to get in . . . not a good idea.

Waterers may be of many types. We use the covered Sears type because they keep the water clean, can be moved in and out easily, and don't tip over. The hens can't walk in and foul the water, though they do tend to scratch litter into it. Any water pan or pail should be elevated on a sturdy block of wood and cleaned and refilled daily.

Nest boxes are easily made of wood. For each five to ten hens, you will need one nest. If you have heavy-type birds, make your boxes 12″×14″ and 12″ deep. For Leghorns and Banties, a nest 12″ square will do. Build your nests a foot or two off the floor and make hoods to keep out light and discourage hens from roosting there. Some chicken houses have nest boxes that open to the outside of the coop, which facilitate egg gathering. Nests should be cleaned periodically and filled with fresh straw.

One useful addition to your house would be a broody cage: a small coop or box built up off the floor of the house. It should have a screen or slatted bottom to allow air to circulate freely and to keep the hen's body temperature lower. This cage can be used to confine a broody hen (one who is trying to set eggs you don't want set), or as a temporary hospital for a sick hen.

Use of electric lights is another thing you should consider for your chicken house. It is a common practice on small farms as well as in large commercial poultry houses to use "artificial illumination" to stimulate egg production. The idea is to provide a longer day which encourages the birds to eat more and thus to lay more. It is especially useful in fall and winter, when egg production normally drops off. As little as fifteen minutes a day of extra lighting can affect egg production. A couple of hours of extra light in the early morning and again in the evening increases production but allows the hens plenty of sleep.

FEEDING A LAYING FLOCK

Laying hens must be fed a diet that is adequate for their particular needs as egg producers. Their basic ration will go to maintain their bodies, providing the nutrients for replacing tissue cells as well as the energy necessary to keep up body functions. The balance of their feed will then go to producing eggs. There is an art to feeding a flock for maximum egg production and healthy, active hens.

Hens need a diet fairly high in protein, with carbohydrates and fats in lower proportions. Fiber is needed to keep the digestive system well toned. Ash or mineral matter (lime, salt) is also necessary. Green foods are added as a source of vitamins and to keep the digestive system stimulated and healthy. Grit may be needed if the hens are confined without access to sand or soil. Around these basic needs, three systems of feeding are developed. One is the all-mash system; one is the grain-mash system; the third is the grain-milk system. If you follow the first system, you simply buy a ready-mixed commercial chicken mash that is intended as a complete diet for the laying flock. This type of mash contains all necessary nutrients and is available in a pelleted, "crumble," or mash form. It is fed free-choice in a trough-type or automatic or hopper feeder. The hens must be given water and oyster shell in addition to this diet, but nothing more is necessary. A mash that is intended as the sole food for a flock will state so clearly on the label. It is balanced, and you should not buy one of these mashes to use with grains. The second and third systems of feeding are probably better suited to the natural needs and instincts of poultry—they provide for exercise in scratching and seeking out feed, and a more varied diet. I feel that our flock has done best on a home mixture of grains and the

197

addition of either mash (system two) or a mixture of milk, linseed meal, and free-choice alfalfa leaves (system three). There are many ways to feed your flock. In the paragraphs below, I've tried to discuss most of the elements you should consider so that you can adapt your feeding program to what you have available. The price and availability of different grains, the productivity of your garden and other animals, and the preferences of your hens will all affect what you actually choose to feed.

Mash: Mash is a mixture of finely ground feeds, usually made largely of mill and slaughter house by-products. It is easily digested by poultry because it is so finely ground. Mash provides a high percentage of protein readily eaten and used by laying hens—its digestibility and concentration stimulates heavy egg production. It may contain alfalfa meal, wheat bran, wheat middlings, corn meal, and/or soybean meal. In addition, it will usually have animal-based protein of some sort. Animal protein is considered absolutely necessary in egg-laying rations. Tankage (the by-product of soapmaking and packing houses) and fish meal are common ingredients of commercial mashes. You can substitute dry milk as your animal protein source.

Dry mash should always be fed in troughs or boxes. If it is thrown on the floor or ground, a good deal of it will be lost. Usually it is fed "free-choice" (kept before the birds at all times). If your hens don't seem to like mash, try mixing some with milk —this will increase its palatability and should increase consumption.

Milk: Milk and milk products are an excellent source of protein for poultry. Liquid milk can be fed in pans or troughs; most hens will drink it readily. Sour, clabbered, or skim milk is all equally palatable. Dry milk should be mixed with mash or fed in troughs. Milk and milk products provide vitamins which other forms of animal protein lack. The lactose (milk sugar) is also considered beneficial to the growth of healthy intestinal bacteria.

Grains: Most grains are rich in carbohydrates and fats which supply heat and energy for the hen's body. Grains vary in palatability and food value and should be fed in mixtures. Yellow corn is one of the most popular poultry feeds, being high in food value and well liked. Wheat and milo are similarly desirable. Oats are commonly fed in combination with any of these three—the oats providing fiber which the others lack. Peas are a good protein source when fed as part of a mix. Beans should be cooked or ground and fed in a mash. Our hens are very fond of leftover rice—a very starchy and therefore fattening feed that should not be overfed. Other grains available in different parts of the country can also be fed.

Commercial "hen scratch" is a mixture of corn, oats, and milo—or wheat, corn, and oats. It is usually cheaper to buy these grains separately and mix them yourself. Grain may be scattered in the hen house or yard rather than fed in troughs. Hens enjoy the exercise of scratching and pecking for their grain. Don't overfeed grain as the extra will be left to mold and spoil. To determine how much your hens will actually eat, feed them measured amounts in a trough for a few days. Then feed slightly less than they have cleaned up. Overfeeding grain can also lead to excessive fatness of your hens—and lower egg production. If you are feeding mash and grain, most books suggest feeding scratch grain twice a day. One third of the day's ration is fed in the morning—this will encourage the hens to scratch around and exercise, but will leave their main appetite for mash. The remaining two thirds should be fed in the late afternoon. If you are feeding grain and milk, feed two thirds of the ration in the morning and one third in the late afternoon.

Minerals: Laying hens need lime, phosphorus, and salt. These are usually provided in commercial mashes, though more lime is usually needed for production of normal, hard-shelled eggs. The best source for this extra lime is crushed oyster shell. It may be fed with the scratch grain or free-choice fed in a box or trough. Another source for lime and calcium is egg shells—save your shells, dry them out, crush, and feed back to your hens. Bone meal is commonly fed as a source of lime and phosphorus. If you are feeding milk or meat scraps as a main source of animal protein, bone meal is unnecessary. Salt helps digestive processes. It is usually part of a mash mix—or can be added (eight ounces of salt per hundred pounds of mash).

Hens confined to yards or runs must be given free-choice grit, which may be purchased at a feed store (about a pound per hen per year is estimated to be enough). You may substitute with sand or with a regular shovelful or two of good rich earth.

Tory and I went to get more manure today. On the way back she asked me to drive slower and I said I would "because she was scared." She got angry with me, told me that I didn't have to protect myself by making her feel small. There I was, caught in the Macho patterning of defending myself by attacking others. Slowing down because she asked me to is different from slowing down "because she's scared." I knew she was right, but I couldn't give in—felt my stomach in a knot, could hardly talk for holding in the tears, kept asserting I knew the road and the truck and what I was doing.

When we got here, I walked away. I didn't want to see her, be near her. I wanted to cry and be left alone and build my defenses in private. But she wouldn't let me go—instead of arguing, she hugged me and I went from rigidity to a willingness to acknowledge what she was saying.

She is the first person who has dared to criticize me, to be real with me in a long time. I feel threatened and vulnerable, hurt, and scared. I also love her—for pushing us to accept and understand each other, for not politely accepting walls.

Seaweed is a rich source of natural salt, iodine, and other minerals. If available, it can be dried and powdered, then mixed in mash or fed in milk (a pinch per hen).

Green Feeds: Ideally, your hens should range freely part of the day and be able to pick bits of grass and herbs and weeds. Fresh greens are tonic, furnish vitamins and succulence, and stimulate the appetite. You can grow special crops for your hens —kale, chard, lettuce, rape—or feed garden excesses and kitchen scraps. Any pasture crop (alfalfa, clover, rye) can be cut, chopped fine, and fed to confined poultry. Ours appreciate grass, mustard greens, comfrey, even handfuls of weeds. Once you begin feeding greens to your chickens, try to be consistent about it. They grow used to this additional feed and look forward to it.

Water: Daily provision of fresh, clean water is a necessary part of feeding poultry. Water is used in all body processes, and a lack of water can make a hen unhealthy and unproductive. Using an automatic waterer will make this chore a little easier but check it often to make sure it is clean and full.

The amount you should feed your hens will vary according to their breed, their housing, their ages, and the season. It is hard to suggest an exact ratio or formula. For instance, very active breeds, such as Leghorns, need proportionately more feed per bird. A flock that is let out to range will pick up a lot of live animal protein in the form of insects and thus need slightly less protein in their mash. In winter, flocks need more grain to provide extra body heat. Different poultry books suggest different amounts and balances. One formula we found that can be adapted to a smaller flock is this: one hundred laying hens (Leghorn) need twenty-four pounds of grain and mash daily to maintain 50–60 per cent egg production. In winter, twelve to sixteen pounds of this should be grain; in summer, eight to ten pounds should be grain. You can juggle this around and experiment with your flock until you reach what seems like the proper feeding. Then try to be consistent—always make changes gradually as radical changes will immediately lower egg production.

If you want to get really scientific, the Poultry Nutrition section of the *Merck Manual* includes a table of "estimates of the total feed required per chicken per year for maintenance and the production of the indicated number of eggs." This table tells you, for example, that a five-pound hen (average size) laying very well (300 eggs) will need 92.9 pounds of feed. A hen of the same size laying poorly (100 eggs) will need 75.2 pounds. This same hen needs 66.3 pounds just for maintenance (0 eggs).

If you want to make up your own feed, use the following table in balancing your feed:

grain (at least two kinds)	40 per cent
ground feeds (at least four kinds)	30 per cent
animal feeds (at least one kind)	10 per cent
green feeds (large variety)	15 per cent
mineral feed, grit, and shell	5 per cent

Your ration should contain 15–16 per cent protein (5–6 per cent of this should be animal protein). This should be as varied and well-balanced a feed as any commercially available, and one you can adapt to your own farm or homestead.

Here is a feed ration worked out for a flock of twenty-four laying hens (and one rooster) which we have used for years on our homestead:

24 ounces whole or rolled oats
24 ounces cracked corn
22 ounces milo
6 ounces linseed meal
6 ounces wheat bran
6 ounces (minimum) alfalfa leaves
2 ounces (minimum) oyster shell
a quart or more of skimmed or whole goat milk
with a pinch of salt added

This ration seemed to maintain our flock well. We weighed everything out first in a small can so that daily feedings were easily done by the half can or quarter can and didn't have to be weighed. When we were low on milk, we substituted with a high-protein mash. When we had lots of milk, we fed extra. Usually the hens were let out each day to range in search of insects and greens. They were also given table scraps and garden surpluses.

keeping a healthy flock

My introduction to poultry health care came one morning when I found half of our flock of hand-raised White Plymouth Rocks with what looked like giant abscesses on the sides of their throats! We had raised these pullets from day-old chicks with never a problem and assumed they would go on being as healthy as they'd been. The "abscesses" were a shock and jarred such blithe assumptions. I ran for the *Herbal Handbook* . . .

Fortunately, what I was facing was not an enormous epidemic but a simple management problem. The hens were crop bound. Food ingested by the chicken passes down the gullet into a sort of storage pouch, the crop. The crop can become distended with coarse, undigested food before this food passes on through the digestive system. They had either eaten too much green grass or too little grit; their crops were stuffed full and not functioning properly. A fresh bowl of grit was greedily

pecked at. Some of the worst hens I took in hand—force fed them a teaspoon or two of olive oil and massaged the crop. By the end of the day the oil and grit had worked, and the crops had shrunk down to normal functioning size. Since then, I've had more experience with taking care of many flocks of hens, and have had relatively few health problems. My feeling is that it is easier to keep a flock healthy than to treat one that has become sick. Chickens that roam free part of every day, getting sun and fresh air and exercise, seem to stay healthy. They pick and scratch for bits of this and that, getting lots of protein in the form of insects, lots of greens in the form of grass and weeds, plus extras in the soil and sand. With a spacious house that is kept clean, and with proper food and plenty of fresh water, a chicken flock has few problems. It is really important to keep floors of houses and pens dry—moisture harbors all sorts of bacteria and disease. Lime should be regularly used as a disinfectant; it also deters lice and mites. In the following paragraphs, I'd like to discuss some of the simple and fairly common poultry problems I've had experience with. For a more comprehensive reference, you should get a copy of the *Merck Veterinary Manual* (see Resources), which has an extensive poultry section, or the *Herbal Handbook* (see Resources), which has a fairly good poultry section emphasizing herbal and dietary care. You may also be able to find a good poultry book dealing with other problems.

Lice and Mites: These are very common to poultry and usually manifest themselves in loss of feathers, irritation of the skin, and general lowered condition. There are more than eighteen types of mites that infect poultry—some are visible if you part the feathers, but many are not. Lice are usually visible as tiny grayish insects crawling in the feathers or attached to the skin. Preventive treatment for lice or mites consists of regularly liming floors and runs, and cleaning all wooden surfaces with a good disinfectant. Another treatment is to paint roosts and the lower parts of the hen-house wall (and the floor, if wooden) with used motor oil. The hens may be dusted individually with rotenone powder (if available) or a commercial insecticide. Hold the hen by the legs with head downward so that feathers will open and dust can be sprinkled in easily. Be careful, though, not to get it in the bird's eyes. Flocks that show signs of lice or mites should be dusted carefully and their house and roosts treated as mentioned. Ten days or so later, dust all the birds again. This second dusting is to kill all lice or mites that might have hatched from eggs not affected by the first dusting. After this, your flock should be free of these infestations. Whenever you bring home new birds, isolate them from the flock and give them a dusting (followed by a second dusting ten days later) to make sure they don't introduce these parasites into your clean flock.

Scaly Leg Mites: These are a particular variety of mites that cause rough, enlarged, dry scales to form on the feet and legs of the bird. They are very common, but very easily treated. You can simply dip the legs and feet of the infected bird in a mixture of one part kerosene, two parts *raw* linseed oil. Another treatment is to scrub the legs and feet with a mild ammonia-and-water solution and then rub in a garlic-and-vinegar mix. Repeat each treatment in a week, and again if necessary. Also paint the roosts with a mixture of motor oil and creosote. Again, any new birds should be checked for this condition and isolated and treated if they have it.

Worms: Poultry can be infected with a number of different types of worms, or internal parasites. All of these can cause variable symptoms, most common of which are diarrhea, emaciation, weakness, and lowered egg production. Free-ranging poultry are less likely to become severely infested than those kept in runs or confined to buildings. Keeping all poultry houses clean and disinfected is the best preventive. Garlic is a natural worm inhibitor and can be routinely and safely fed to your flock (use powdered or finely chopped, mix in milk or mash, and feed). For specific symptoms and treatments, consult a poultry book or the *Merck Manual.* There are several general-purpose worming preparations available commercially.

Prolapsed or Dropped Vent or Oviduct: A retained, soft-shelled egg, an inflammation of the oviduct, or overfat condition can cause a bird's vent or oviduct to prolapse. A mass of tissue will be hanging from the vent—it may be slightly ruptured and bloody, and is usually very dirty looking. We have had two hens with this problem. Both recovered with a simple treatment and later began to lay normally again. The bird should be removed from the flock to prevent cannibalism, and kept in a dry, clean cage by herself. The protruding mass should be washed with warm water and a gentle antiseptic or mild soap. It should then be lubricated—carbolated (medicated) vaseline works well, as does the broad-spectrum antibiotic mastitis ointment. If possible, the prolapsed mass should then be pushed very gently back into the vent. The bird should be very lightly fed until she recovers. Greens are especially good for her. She must have fresh water each day. The prolapsed tissue should be cleaned and lubricated twice a day (or more frequently). One book I read recommended giving the bird a dose of Epsom salts (if many birds show prolapses, the entire flock should be dosed with Epsom salts). The two birds I treated took about a week each to recover—the prolapsed tissue was drawn back into its normal place, and the birds seemed fine and healthy.

Coccidiosis: Watery, blood-stained droppings are characteristic of this disease, which is most common in young chickens three to eight weeks old. It is highly infectious and often fatal. Feeding milk and/or garlic is supposed to control this disease. I once treated some affected chicks with a mixture of garlic, powdered ginger, honey, and warm water given by eyedropper four or five times daily. Most of these chicks recovered. Birds that have recovered from coccidiosis develop an immunity to the particular species of coccidia they have been infected by. The *Merck Manual* discusses this problem at length, and suggests control programs for it once it has appeared in your flock. Generally, raising all young birds in a clean new (or very well disinfected) house is a good preventive.

Egg-bound: A hen that is egg-bound will strain, squat down, and assume what is called a "duck-like" position. She may be found sitting in the nest all day with a very preoccupied look. The treatment is to pour or inject some olive oil into the vent and after a few minutes, to try and work the egg down toward the vent opening. When the tip of the egg appears, the shell should be broken with a nail and the contents removed. Wash the vent out with cold water and isolate the bird for a few days.

Colds and Respiratory Problems: These problems are usually caused by overcrowding in damp, drafty houses, and are best avoided by better attention to housing. The birds affected by most respiratory illnesses will have mucus-clogged nostrils, watery eyes, labored breathing, and so on. They will be disinterested in food and usually sit huddled on the perch or in a corner. They may sneeze a lot. These birds should be isolated in a warm, draft-free house and given special attention if they are to survive. Clean the eyes and nostrils with a damp cloth. Eucalyptus oil may be applied around the nostrils or used as an inhalant in a steam tent. Garlic and cod liver oil are said to help. Epsom salts may be given in the drinking water. Chronic respiratory problems that don't respond to the above treatment may be treated with a drug known as Tylan or Tylosin. This may be given orally to the individual bird or put in the drinking water. I have used it on some very valuable pure-bred chickens that had some sort of sinus infection, with accompanying sneezes and discharges from beaks and eyes. They recovered after about four days on this drug, with no subsequent relapses.

Broodiness: This is not an illness, but is a problem—more severe in certain breeds than others. Broody hens are those who are ready and willing to hatch their (or anyone else's) eggs. They will be found in the nest box long after the other hens have gone to roost for the night—and by clucking, ruffling their feathers, and even pecking at you, will indicate their determination to stay where they are. A broody hen goes into a distinct physical state: her body temperature rises and she becomes almost comatose. Once or twice a day she will leave the nest to eat, but always in a clucking rush to be back. We have had a lot of Aracauna chickens (a wildish breed that still has all of its mothering instincts) so have had plenty of experience with broody hens. Our more domesticated hens—Plymouth Rocks, Rhode Island Reds—rarely go broody. Those few who have usually desert their nests before the chicks hatch. If you have a broody hen and don't want her to set on eggs, you will have to confine her in a light, airy cage for about four days to a week. She may be lightly fed (mash, but not grains) and given water. Usually by the fourth or fifth day, she will be out of her broody state and ready to lay again. Hens that are persistently broody should either be kept as brooders (you can give them other eggs that you want hatched if you do so at night, and with as little disturbance as possible) or culled.

Egg Eating: This is usually a symptom of some dietary deficiency, though it may just be a bad habit developed by accident. A lack of calcium in her diet will cause a hen's eggs to become very thin shelled, and they will break in the nests. The layer or another hen may then eat the egg—which is the start of a bad habit. Or a hen may be driven to break and eat the eggs intentionally out of some mineral, vitamin, or protein deficiency. Broken bits of shell in the nest, or eggs smeared with yolk mean egg eating! It is difficult, if not impossible, to catch the culprit unless you stumble upon her in the very act! Remove all broken eggs and bits of shell and replace the nesting material. Feed more oyster shell or dried, crushed eggshell and be sure you are feeding a balanced, varied diet. Some hens will become obsessive egg eaters and (if you can catch them with the evidence) must be culled. Give them the benefit of the doubt, though—sometimes a week or two of proper feeding will be a cure.

Moulting: Once a year, the chicken flock will go through a heavy moult, in which the birds lose their old feathers and grow a set of new ones. Feathers will fall off in patches and spots, leaving a

This has been a wonderful, empowering week—living out my fantasies, testing my strength. Jim Thompson came last Thursday and Friday to dig the postholes, and I walked beside his tractor to help mark each hole. It was so exciting to see the auger bore into the earth bringing up the blackest of loams, and red or gray clay—from one hole to the next the color could change completely: soft sand, to pure loam, to thick wet clay. It gave me a new understanding of the meadow and a new awareness of its contours, its minute undulations. I felt such joy at seeing the demarkation of the pasture-to-be, for the first time believing in sheep. Working with Jim was a pleasure; we became really comfortable, and he told me many stories of his family and the early days here.

"Lordy, lordy, lordy," he said when he first arrived. "This sure takes me back. I remember one summer when I was about sixteen, my dad was loading hay for Mr. Perry—this here was the Perry place then—and one day I was driving the hay wagon and helping Mr. Perry load it. We heaped that wagon as high as we could with hay down in the lower meadow below the cypresses and was driving back to the barn when we got hung up on a low hanging branch. Perry gave me his axe and told me to chop that branch off and oh boy was that ax dull! It was so dull Perry fell asleep on top of the hay before I was done chopping. Well, when that branch finally fell, it was with a crack that was like a rifle shot and those horses just took off with Perry asleep in the back of the wagon. He was half way down the canyon when he woke up!"

ragged looking bird until the new feathers come in. Most hens will stop laying for part or all of their moulting period, though a few good ones will lay right through. The moult period usually lasts for a month or two, and is usually in the fall of the year. In spring, hens may go through a light moult but usually lay through this one.

BUTCHERING CHICKENS

One of the first things I did when I moved to the country was get some chickens. I was lured, like so many others, by the prospect of fresh, fertile, unamphetamined eggs, and the bucolic pleasures of overseeing the daily pomp and circumstance of my own flock. Also present was a vivid remembrance of picking out a real, live chicken for Friday dinner with my grandmother, in a New York where the ragmen still made daily rounds in horse-drawn wagons.

It was awhile before I could bring myself to butcher any of my chickens, despite the fact that I had enough practical experience and reading behind me to know that four-year-old hens are a financial liability and at least half of those appealing fuzzy chickies grew up to be blustering roosters that fought constantly, wore out the hens with their non-stop sexual advances, and rarely could be given away, let alone sold. I think the ultimate deciding factor was buying a chicken for dinner which tasted transparent and had no dark meat on it, due, no doubt, to its never having set foot on the ground.

Several years and as many unsavory experiences later, I have learned the easiest ways to butcher and clean chickens, which I'd like to pass on to you. Killing a chicken isn't a pleasantry, but performing it in the most efficient way possible makes it easier for both of you. The chicken should not have been fed for twelve to twenty-four hours prior to butchering. This will eliminate a lot of mess should you accidentally cut into the crop or intestine.

There is one thing I can say about the time-honored head-on-the-block method, and that is, it works. It is also messy and upsetting to see a deheaded bird flopping around. If you plan to butcher your chickens this way, I suggest you put a paper or burlap bag over its head, to minimize the mess. The freshly deheaded body may also be put quickly into a bag (burlap) to minimize flopping and catch the blood.

The way I prefer to kill is by the "English" method, which breaks the neck of the bird. To do this, grasp the legs and tip of the wings in one hand to keep the bird from flopping around, and the head between the thumb and index finger of the other hand. Pull down on the head, stretching the neck, and simultaneously bend the head back sharply to dislocate the neck at the base of the skull. The bird will flutter some, but can be held. When it stops moving, you can behead it and let the blood accumulated in the neck run out.

There are several ways to pluck. I prefer to pluck dry, but it has to be done quickly and immediately, before the feather follicles contract and the feathers get hard to pull. First, pluck the tail feathers, then the primary wing feathers, one or two at a time, and then the body feathers. The big feathers come out easily by jerking, and the body feathers by a snappy rolling kind of a pull. You should pluck these latter feathers in handfuls and concentrate on getting the most feathers out in the shortest time, while they are still easy to pull. On some birds the skin will tend to tear more easily than on others—try to tear the skin as little as possible. I've found that tying the bird upside down to a 2×4 nailed between two trees so the bird is about shoulder height is the easiest way to pluck. It's a comfortable working height, and leaves both hands free to pluck.

For wet-plucking, or "scalding," you need two two-and-a-half-gallon buckets—one with hot water (about 180° F) and the other with cold. Immerse the bird in the hot water for a few seconds until the feathers are loose, then dunk it in cold water to avoid cooking the skin. Then pluck. You can dunk the bird again to loosen feathers that are still hard to pull—again, be careful to dip quickly into hot water and follow with an immediate dipping in cold. You can singe the fine hairs and pin feathers from the bird with a gas flame or with matches. You may also skin the entire bird, removing feathers and all in one neat package, if you don't care about losing the skin.

To dress or draw the chicken, cut the feet off at the joint, cut the wings at the joint. The neck may be sliced close to the base and then broken off. Make a slit beneath the breastbone about one and a half inches toward the vent. Make a circular cut around the vent, trying not to cut the intestines. Pull them carefully out of the body through the first cut. Next remove the windpipe, gizzard and crop through the neck. Each of these may be severed and pulled loose gently. The gizzard should be carefully slit open and the inner sack forced out and discarded. You should now be able to reach into the body cavity and remove the

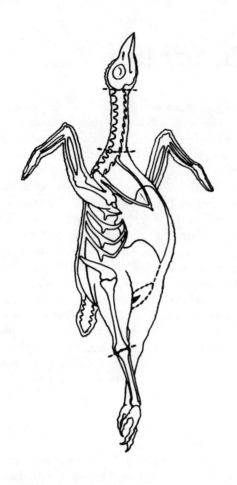

remaining viscera in one intact group (tricky, but possible) by working them loose first and then pulling out. You will probably want to save the heart, liver, and gizzard for cooking (these are called the giblets). Scrape thin veins and blood away from the heart. The liver will have a small green sack attached to it. This is the gall bladder, which is full of a very bitter fluid which you will not want to get on any of the giblets or meat you intend to eat. Cut this sack away carefully, leaving a small piece of the liver attached to it rather than risking cutting too close and rupturing the sack. Then remove the two oil sacks, one on the back and the other at the upper base of the tail. Rinse the bird by running cold water through the body cavity. The bird should then be chilled until it is ready to cook.

turkeys

In one of the new homesteading books, the chapter on turkeys consists of a single word: "Don't!" Everything else I have ever heard or read about turkeys stresses their stupidity, inability to function on the simplest level, and susceptibility to diseases. My own experience raising turkeys from tiny poults convinced me that turkeys are, on the contrary, fairly bright, personable, and resourceful birds. Like other creatures, if they are given space, fresh water, and good food, they thrive. If you are interested in low-cost, home-grown meat, a small flock of turkeys can provide you with delicious, high-protein meals for very little investment.

Our turkeys were purchased on impulse, with neither plan nor place prepared beforehand. We went to buy feed one morning and there in the brooder usually reserved for baby chicks were about thirty miniature, ostrich-like birds. They were Bronze turkeys, a breed that resembles the wild native American turkey and adorns Thanksgiving cards more commonly than tables. The commercially raised turkey usually found in grocery stores or on large turkey farms are usually the White Holland and Beltsville Whites—two breeds developed specifically for the market. The Bronze poults were all toms (males) and had been hatched with the holidays in mind. They would be slaughtering age (about six months) right around Thanksgiving. Without really thinking about what we were getting ourselves in for, we bought six. The feed-store manager cautioned us not to handle them at all—supposedly they are very delicate and easily upset, and can quite literally die of fright if you scare them. He also told us emphatically not to raise them anywhere near chickens, nor on ground where chickens had run. Young turkeys are susceptible to a disease called "Blackhead," which is transmitted by chickens and can remain in the soil where chickens have been for long periods of time. By the time we'd picked up the Purina turkey raising booklet and been convinced to buy medicated "Turkey Grower" (we never buy medicated feed as a rule, but the manager said flatly: "They'll die without it!"), we were thoroughly intimidated. We were sure the poults would all die before we even got them home.

All the way home, though, they hummed and chirped quite merrily. We worried over where to put them and where they could live when they were big enough to range. Every inch of our farm that wasn't brushland or forest had been trodden by the Blackhead-laden feet of generations of chickens! We decided we would build a pen somewhere far from the main barn and chicken houses and hope for the best. The immediate problem was where to put the poults when we got them home: our house was *full* of baby chicks we were raising through the brooder/light stage, including a hundred Rhode Island Red pullets we'd started a week or so earlier. We decided to put the poults in a clean box in the back room and be extra-careful to always feed and water them *before* we tended the chicks. Meanwhile, the warning never to handle the poults was fading from our ears: far from being terrified of us, they were actually, obviously very interested in people . . . before we even reached home, there was a lap turkey being groomed!

Our turkeys spent a week or so in isolation, eating their turkey grower and eagerly downing greens from the yard and garden. Turkeys are great grazers and young poults appreciate all kinds of greenery. We kept them under a brooder light for warmth, even though they were pretty well feathered. Like baby chicks, they seemed to thrive on special daily care—clean feed, clean litter, and fresh, clean water. We were careful to disinfect the waterer we used as it had previously been used with baby chicks. All boxes, feeders, and so on used with turkey poults should be brand new or disinfected before use. This simple precaution may be what keeps your poults healthy. In a few weeks, we began cutting the poults medicated feed with the unmedicated, soybean-base mash we raise our chicks on. Turkeys need a very high-protein diet, so we began supplementing this mash with milk from our goats and with linseed meal. They grew noticeably every day and began to look a little crowded in their little box. At this point, we went through some serious discussions about introducing them to our baby chicks. Obviously the poults hadn't been frightened to death by our handling; on the contrary, they *loved* being handled and would hop out of the cage onto our laps in a second. There they would pick and preen happily, chattering to themselves and us in their soft turkey voices. So, we reasoned, the feed-store manager had been wrong about this—maybe he was also wrong about the potency and omnipresence of Deadly Blackhead.

. . . We borrowed an old poultry book from a friend and looked at its section on turkeys. There was the same stern warning: *Never raise turkeys with or near chickens!* It had elaborate setups for revolving poults from pen to pen every few weeks to keep them from becoming diseased. We pondered more and decided to trust our own intuitions. Besides, if the turkeys were going to live on our farm, they would sooner or later be bound to come face to face with a chicken or its droppings. We resolved to begin their exposure and see what happened. On the big day, we moved the poults in with several young chicks. We had a good solid week of Deadly Blackhead Paranoia, but nothing happened. Chicks and poults thrived. They scratched, ate, preened, and slept together in total disregard of all dire warnings. By now, our turkeys were weaned from their special turkey grower (and its medication) and sharing with the chicks a diet of mash, finely ground grains, and plenty of greens and milk. When our poults were a little older, they moved outside to a pen with a small house for shelter. They seemed none the worse for their exposure to the chicks, and had begun what was to be a very communal existence with many chickens and other birds.

When the poults were large enough to intimidate our cats rather than become their prey, we began letting them loose every day. They loved to graze and nibble at grass and weeds and seemed perfectly secure wandering around the farmyard. At night we would round them up and walk them back to their little pen and shelter. They would dive into their feed bowl, almost inhaling their mixture of mash and grains, then settle down cozily for the night. Turkeys are great snugglers! When they are young, they sleep huddled close to each other, probably for the warmth and comfort of the contact. As they get older, they still sleep snuggled together, not only at night, but in the afternoon when they're napping and daydreaming. If you lie down in the sun in the presence of our turkeys, they will stroll over and settle down with you—preferably leaning up against your arm or leg. There they'll sit and doze contentedly, or pick gently at shoelaces or buttons. Turkeys are capable of great, gobbling excitement as they get older: a sudden noise or movement and their tails fan out and simultaneous gobbles pierce the air. Children and turkeys seem to have great fun conversing together—one shrieks, the other gobbles in a raucous and uninhibited dialogue! Despite this excitable part of their nature, our turkeys have never swooned when

a plane passes overhead, or died of fright when a horse galloped by. They seem to be fairly relaxed most of the time, pursuing the sun or leisurely grazing. Anything that happens around the farmyard is usually subject to their attention and inspections—a goat to be bred? they are present outside the pen . . . a gate to be mended? they are parading just outside . . . and so on. As you may guess from these stories, some of the turkeys are still with us . . .

After a month or so we stopped locking the turkeys up at night (at this point they ceased being "poults" and came into their own as turkeys). They established their own sleeping spot—first on an unused roll of fencing, later on the railing of the back porch. When the weather began to turn colder and rainy, we began a practice of herding them into the barn on particularly bad nights. If you live in a very cold or damp climate, you will have to treat your turkeys similarly; they will need shelter through the winter months or summer storms. Besides this simple consideration and a source of fresh water and some food, turkeys seem to need very little care. Their feed requirements will depend largely upon the type of grazing you have available, the number of birds you keep, and so on. Our turkeys have always foraged freely on land that stays green most of the year. They pick up scraps of grain from the chickens and from the goat mangers or horses—but rarely demand more. When they were in their first few months of rapid growth, they had free-choice of mash and grain. One long-term feed estimate I read said that it takes three to four pounds of grain to produce one pound of turkey (meat). This source said nothing about grazing, however. Even at the maximum four pounds of grain, the conversion rate is very good and makes the turkey a reasonably economic food source. I am sure that ours cost a great deal less to raise.

We lost two of our original six birds when they were still quite young. One was killed by an aggressive peacock, and the other died after exposure to an unusually early rainstorm. The other four stayed healthy and grew rapidly. We butchered two of them for food when they were about four months old. To be as humane as possible with these very humanized birds, we got up before dawn and took them still drowsy and barely conscious from their roosts. They knew no fear and were killed instantly. The traditional way to slaughter these birds is to hang them upside down and slit their throats, letting them bleed to death. We decided that this was much too slow and painful a death, so we used a sharp, heavy ax to strike the heads off, killing them at once. They were then hung to bleed out. When the bodies were a little cool, we dry-plucked and drew the birds much as you would a chicken. One of these young turkeys dressed out at about fifteen pounds; the other at thirteen. They provided us with many delicious meals—which we ate thankfully, in full realization of the lives they had lived. It is never easy to slaughter an animal or bird you have raised yourself, but somehow it makes a very tangible, real connection between your needs and your surroundings and how you choose to integrate these. For many years I was a vegetarian farmer, selling or trading or giving away the inevitable surpluses of lambs, kids, roosters that the farm produced. Being very much into raising and breeding all sorts of livestock, I was never able to bring myself to kill and eat the unproductive extras and culls. Now it seems that something has settled in my mind, al-

lowing me to accept and take part in the cycles of life and death, feeding and growing, that surround me. This realization—or resignation—or reawakening—came in part from my experience raising these turkeys. Of the two that remain, we plan to use one for meat and sell the other to a neighbor who has a small flock of females of the same type and wants a breeding male. We plan to buy or trade for some more poults in the spring and raise them as we have these. My advice about turkeys, then, is: "Do!" Try, as we have, raising just enough for your own needs. If you have any health problems, consult the fairly comprehensive poultry section of the *Merck Veterinary Manual.* Chances are if you are reasonably careful, buy your poults from healthy stock, and treat them well, you will have as few problems raising turkeys as we did. Besides being practical and economic, turkeys are beautiful, interesting—and bright.

RAISING RABBITS

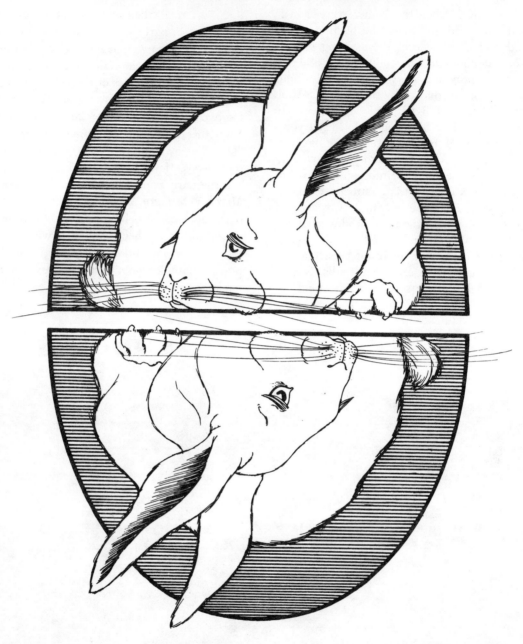

I don't know how to begin an expository article on raising rabbits—whether to stress the ease of keeping such pretty animals or to emphasize all the tasty dinners and warm furry garments they can be made to provide. But actually, if you are interested in keeping rabbits, these advantages need no further promotion, and it's time to plunge into the practical aspects of turning these possibilities into reality. This is complicated by the fact that I don't treat my animals like meat machines and don't

want to encourage anyone to do so. My practices, as a result experiences, are significantly different from those promoted by makers of the pelleted rabbit feed or by the agricultural agents of the United States Government. They result in less "efficiency" and less expense.

Before getting any animals consider whether the "beefiness" of the commercial New Zealand Whites will be offset by the fact that for home tanning of furs their pelts will be of a boring sameness. Our intentionally crossbred rabbits produce litters containing a cheerful combination of white, light and dark gray, chinchilla, red, "Wild Rabbit" (agouti), black and in betweens, but of course they don't have the feed conversion efficiency and "growthiness" of strains selected by commercial rabbitries.

The number of does (females) needed will be in relation to the frequency of eating rabbits. To produce one litter (6–10) of "edibles" a month, we keep three does and breed each one every three months. The does then spend one month on gestation and two months nursing bunnies. It's possible to breed more frequently than this, but I've found it requires constant consultation with the calendar and the does tend to lose condition after a year of high production.

We try to mimic Wild Rabbit life, so the doe lives with the buck (male) for the first twenty-five days of her pregnancy rather than the bare twenty-five seconds required to get her bred. (By the way, it is normal for the buck to scream and fall over in a faint upon ejaculation.) The buck's cage is surrounded on either side by does and their young in a semblance of his centrality in a natural rabbit warren.

One doe is bred on the first day of each month. On the twenty-fifth she's returned to her regular cage with a clean nest box to kindle on the first. The average litter size is 7–9. The bunnies are born blind and furless and won't come out of the box until they're about eighteen days old. They nurse for two months and are usually eaten in the month that the doe is being rebred, though an extra cage is available for the occasional "edibles" not butchered by the time the doe is ready to kindle again. With more does we would breed on the first and fifteenth. The buck would still have a doe with him at all times, but there'd be twice as many edibles per month.

There are two ways of feeding rabbits—either pelleted mixtures (containing ground grains, alfalfa meal, vitamins, minerals, and various chemicals) or a homestead blend of fresh greens, roots, hay, whole or cracked grains, and salt. The first involves no more thought than finding the nearest feed mill but considerably more expense than the latter method. Making up your own ration takes ingenuity, but much less cash—and most rabbits prefer it.

Alfalfa and clover make the best hay, but clover-grass mixtures will do. Pure grass hay is too low in protein and palatability for rabbits. Fresh greens are the rabbits' natural food but don't change over-night to this diet or severe scours may result. Just keep putting in a little less hay and a little more grass clippings, weeds, and vegetable garden wastes (especially carrot tops, old radishes, cabbage leaves, sweet potato vines, and bolted lettuce) until the rabbits are accustomed to a wetter diet. We feed greens from early spring to late fall but piece them out with hay on rainy or lazy days. Comfrey (a fairly easily grown herb) makes a good feed and has a long growing season, but some rabbits won't eat it if they can pick and choose so I give them comfrey in the evening (their normal feeding time) and a little hay in the morning if they've "cleaned their plates." As each litter grows, the amount of feed they consume will double and triple, so if they've eaten all the hay you gave them in one day increase the amount until they leave a few scraps.

Rabbits will eat any grain except whole corn, which is too hard and should be cracked. Ours prefer rolled barley and whole oats. No cereal grain has enough protein for the hard-working doe or fast-growing young so a supplement of soy meal is

211

RABBIT HUTCH

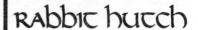

hay and grain storage

worm pit

Inner part of hutch has roof and walls, but its wire floor is actually "outside" the building. Outer compartment is all wire. Thus the animals can go in or out at leisure while the feed (and the feeder) don't get rained on. Size of both inner and outer parts together: 6' x 3' x 2'± (18-square-feet floor space as opposed to 10-square-feet "standard" cage). Worms are raised in the manure under the cages.

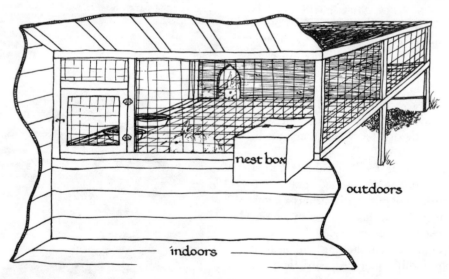

nest box

outdoors

indoors

The nest boxes are 12" x 16" x 12" and hang on the front of the cage. There is a hinged lid on top for viewing the babies and a 6" x 6" opening in the upper corner for the doe to enter. The high threshold will keep baby bunnies from falling out until they are ready to leave the nest.

added. I use 3 parts whole oats, 1 part chopped corn, and 1 part soy meal by weight, but any other grains can substitute in like measure. There are fancy metal self-feeders on the market, but for pure traditionality and a sense of heritage nothing beats the $.89 crockery rabbit dish to feed grain and water.

Water crocks should be kept full all the time. A doe and her eight-week-old litter of seven can drink a gallon of water on a hot day. In freezing weather give warm water at least once a day.

Rabbits need salt too. A circular "Bunny Spool" of iodized salt hung in the cage is usual, but chunks broken off a fifty-pound salt block are cheaper.

If the floor of your hutch is made of wire, provide each rabbit with a small board or piece of plywood to sit on.

Most rabbit diseases are easier to prevent than cure. Coccidiosis, a protozoan parasite causing white spots in the liver, is spread by dirty drinking water in rainy weather. Enterotoxemia, with fatal diarrhea, can result from a sudden change to wet green feeds. Correcting these causes will not eradicate germs, but the rabbits will be able to handle them without getting sick.

Slaughtering is the least pleasant part of my life's intersection with rabbit lives, but it comes right before all those delicious dinners and furry clothes so I try not to lose heart. Killing animals before they've been fed makes the work easier because the guts are emptier. Hold the rabbit upside down by the hind legs and hit it where the skull joins the neck with a short piece of iron pipe. Then hang it up by one hind leg and cut the throat.

Make a shallow cut through the *hide* only around the heels and down the inner part of the hind legs to the anus. Cut off the front feet, tail, and head. Now the pelt can be pulled off over the body inside out in one tubular piece. Rinse off any blood on the hide and with a little more work removing fat it'll be ready to tan.

Open the rabbit's abdomen with a shallow vertical cut. Pull out the intestines and save the kidneys, liver (take off the greenish gall bladder without breaking it), and heart. The meat can be jointed in the same manner as a chicken except that on a rabbit there is no dark meat and no "breast"; the tenderest part is the lower back. Any recipe for chicken will work with rabbit. Try it southern fried, in sweet-sour sauce, with dumplings, or however you choose.

So that's the story of how to be a successful rabbit raiser. But to be a compassionate rabbit raiser, well, try taking the animals on walks through the grass, call them by name, remember always to thank them for being born and for allowing themselves to be eaten.

RAISING A WEANER PIG

If you have a little land and want to raise your own food, be sure to consider a "mortgage lifter": the weaner pig. The initial investment is small, and the pig can· eat a varied diet (making use of what might otherwise go directly into the compost bin). It can live in a fairly small yard with a minimum shelter and can be home-butchered and cured. We were lucky enough to be "found" by a stray weaner pig whom we fed about thirty dollars worth of commercial feed and much free garbage in the six months we kept her. We spent twelve hours butchering and packaging one hundred and sixty-two pounds of meat for the freezer, and rendered another thirty-eight pounds of lard for cooking and soap. The total cost of the edible product was about fifteen cents a pound. Interested?

Buying a Pig: There are two basic types of pigs: those bred for bacon are long, lean, and leggy; those bred for ham are stockier and shorter bodied. A well-known bacon breed is the American Land-race, a white pig with erect ears. A good breed for larger hams is the heavier Red Duroc. Each

type produces both ham and bacon. If you buy your weaner pig from a hog farmer, you have a chance to look closely at her/his selection, and can tell general conditions of comparative size, weight, and health. Choose the lean, thrifty one over the fat, flabby type. Assume that for this privilege of selection and comparison you will pay a little more than at an auction, as you will probably be buying "by the head" rather than by the pound. Read the paper's livestock market report on current hog prices or talk with your neighbors to determine a fair price per pound before you go to buy your pig. Don't buy a little weaner of twenty pounds for twenty dollars because it is so cute. That is a dollar a pound!

Auction yards are a good place to get a fair price and sometimes an outright bargain. Get there early to look at the hogs in the lot pens, pick out a few good healthy lots and keep them in mind when they start running the hogs through. Don't buy obvious runts, lethargic pigs, or "razor backs" (those so skinny their backbones show). None of these animals will make normal gains. Most lots are sold "by the head" or per pound for the entire lot in the pen—which means you must take them all if you bid (there are usually ten or more). If you want just one or two, wait until they are offered "by choice," which means that the price is bid up to the top figure and top bidder is given her/his choice of any of the animals in the ring. Then bids are open again—top bidder gets choice—and so on. After the very choicest animals go for the higher prices, you can usually bid on a nice healthy, normal-size pig and get it for a good price. Prices of pigs at auction vary considerably depending upon the area, availability of stock, price of feed, and so on. I recently got a bargain on a seventy-five pound gilt (young female). Other pigs were selling from thirteen to seventeen dollars at about twenty-five to thirty pounds. I got her for twenty dollars because no one else wanted her—she was lame in one hind leg. She is getting better now and gaining weight. Usually, you will get more for your money buying a feeder pig between forty and seventy-five pounds, by the pound, rather than one just weaned (at five weeks), by the head. One more caution: look for hernias (they appear as a swelling at the navel or where testicles would be) and don't buy the animal with this condition. You would be buying trouble, as that animal must either have an operation or be raised only to a light slaughter weight of one hundred and ninety pounds.

Housing: Adequate housing requires a large pen (at least three hundred and fifty square feet per pig) enclosed by good stout fencing (preferably welded wire). Pigs need exercise and shade, and love a damp wallow in warm weather. If you are keeping them through the winter, their house needs a dry, warm floor raised above the mud of the yard. I'll never raise another pig through the rainy season without at least a partial cement floor in the pen. That is the best way to control flies, odor, and mud. Pigs are susceptible to sunburn and pneumonia, and need a well-built (though not fancy) shelter that allows them to get out of the sun or bad weather. A word about fencing—as the pig gets older and stronger, it may begin to root under the wire fencing. The traditional way to handle this problem is to run a discouraging single strand of barbed wire along the bottom of the fence.

Feeding: First, consider what extras your homestead or farm is producing that might go into your pig-raising project. Surplus milk—up to a gallon or two a day—is wonderful hog food. Eggs, kitchen and garden wastes, alfalfa scraps, and unwanted by-products of chicken slaughtering will all be relished by your pig. Corn, wheat, barley, and to a lesser extent oats are all fed to hogs. I investigated all edible throw-aways in my area and came up with a "stew" for my pig—a mixture of restaurant garbage (go through it carefully for glass, metal, and paper scraps), meat trimmings, fish scrap, and so on. The stew was cooked to kill any unsafe bacteria which might be found in unrefrigerated food. *Raising Small Livestock* (see Resources) estimates

that a pig will need about six hundred pounds of feed from weaning age to slaughter age (six months). With a bit of effort, I managed to find most of my pig's food for free. A pig should be fed regularly two or three times a day. It will also need a source of fresh, clean drinking water.

Butchering: We killed our pig by shooting her behind the ear toward the brain. This method is quick and painless and doesn't damage the meat. Don't attempt to shoot a pig through the forehead because the skull is so thick it is impossible for the bullet to penetrate it. The other common method of slaughtering is stunning the animal and then cutting the jugular vein (the animal bleeds to death). Shooting seems more humane to me.

Immediately after slaughter, the animal should be hung head down to bleed thoroughly. The carcass can then be scalded or skinned. We skinned ours out and sent the hide to a tannery to be made into leather. That process takes about three months and costs under twenty dollars for a 4′×4′ hide. The entire butchering process can be done at home with the help of some friends and some good knives. Detailed butchering instructions would make a book in themselves, so I'll recommend Mor-ton Salt Co.'s booklet: "Home Meat Curing" (see Resources). It has excellent illustrated instructions on killing, butchering, and curing. Curing, by the way, is what makes a roast a ham, and what transforms fresh fat with a little meat into bacon. Included are several recipes for sausages, as well as for other kinds of meats as well.

You can salt-cure your hams and bacons in your refrigerator or cold room, and smoke them in your fireplace or homemade smoker. I designed a smoker from two garbage cans. Though it did the job, it was too small to hold all of the hams and bacon at once. *Joy of Cooking* has some good ideas on smoking and on making your own salt cure. *The Foxfire Book I* is also valuable—it has the mountain folks' description of hog raising and some old-time recipes. I got a real sense of the terrific value of *all* the parts of a pig from reading those old mountain anecdotes. As they say, "it's all good except for the eyeballs."

Your home-grown bacon, sausage, roasts, and hams will be so succulent you'll never want to be without a pig again. Pound for pound, a pig can be the most versatile, thrifty food producer on the farm.

FENCING

Fencing, like all of farming, is a continual process. Not only do you constantly have to fence for new animals, new fruit trees, bigger garden space, or different use of existing pens and pastures, but fences seem to have a life cycle of their own. Fences fall down, are broken through, pushed over or apart. They sag, are battered down or run down, and dented in by pickup trucks sliding in the mud. Fence wires snap, ease out of their U nails, rust, spring loose, and coil themselves up around the nearest post. Posts age—sometimes amazingly fast —and lean, wobble, rot, corrode, or just plain fall down. You enter the life cycle of fences as creator or restorer. With a lot of energy and materials, you can create a marvel of straight, taut wire and sturdy vertical posts. With a lot of ingenuity and determination, you can patch, renail, restretch, and sustain old meandering farm fences. Your fences will grow and change with you, your farm, your animals.

Our fencing experiences began with some barbed wire fiascoes and progressed to woven wire and wooden posts. We (three women brand new to country farm) fenced an area for our half-dozen sheep, using a brand new roll of "heavy field fencing" from Sears and our brand-new fencing tools. We had a mixture of old redwood posts and fir poles, and a logical idea of how to proceed. We learned as we worked and learned more after we'd finished. The first part of our fence is a stretch that boasts nine-foot-high posts that are set one foot in the ground and lean at interesting angles to one another. Wandering from post to post is 36″ field fencing—the effect is impossible to convey in words. Joined to this is another section that is creative fencing at its best. Odd-shaped pieces of variously sized chicken wire are laced together by fragments of barbed wire, old rusty fencing, and some bits of baling wire. There are hastily thrown up boards ("half the sheep are out!") that careen from one post not quite to the next and have an extra piece nailed on to make those last few inches. One corner is secured with a large scrap of plywood. The whole effect of this part of the fence is that it is trying to —and about to—rejoin the forest (or perhaps create a new dump location). Next comes a fairly respectable line of field fencing that is nailed tree to tree with a few set-in posts where necessary. The last line we did has posts set in every twelve feet— and the posts are only a foot or so above the top wire and are sunk well into the ground. The wire there is fairly tight and keeps our goats out of the sheep and vice versa.

This first experiment/experience in fencing taught

us a lot of basics. For the eye bothered by relative proportions or enamored of orderliness, the fence we created is a continual shock. For anyone interested in proper fencing technique, it's a how-not-to lesson complete and eloquent. For the sheep who live within its confines, it's a fence!

Most fences consist of vertical posts placed every so many feet with horizontal fencing or filler of some sort. The posts may be metal, wood, or even concrete. Between them you can use woven wire, plain, barbed, or electric wire, rails, poles, boards— or any combination of these and other materials. Most of our fencing has been woven wire with wooden posts, though we've experimented with almost every other type. The principles are the same for most materials: posts should be strong and well-set; fencing should be tightly stretched and well-fastened; materials should be the best you can get (or can afford). The art of building a good fence isn't difficult and doesn't require a great deal of physical strength. It does take a lot of time and energy—so you should make the best initial effort (materials and construction) possible. Basically, a homestead/farm fence is for keeping various creatures in or out of various areas. The height, strength, and spacing of a fence is determined by the creatures it proposes to control. The life expectancy of a fence is subject to the quality of materials and work that goes into its making, and the wear it takes from weather and animals. A good permanent fence will last indefinitely. A poorly built or temporary fence may straggle on a few years.

Planning: The first step in fencing is to decide what type and how much fencing you need and where the lines will run. The best pastures are square (the most area for the least perimeter) or rectangular. Yours should at least have straight sides. We learned the hard way that circular fences, though very organic and aesthetically pleasing, collapse quickly because of the strain placed on each post. A straight line of fencing lets each post help support the others. We also learned that there's an art, which we have never totally mastered, to this straight-line business too. We've found that a really

Susan has come to live with me for the summer, to help me build the sheep fence. She says that she is worried about our trying to live in the same place. We have been friends since junior high school, the longest friendship in either of our lives and neither of us wants to jeopardize that with the petty hassles of sharing the same house. I feel confident that we can learn to be honest with each other and to give each other space. So far, being with her has been comfortable, loving, and good. What I worry about is her leaving—wasting energy on a months-from-now event. It's nice not to be so all alone.

Learning to work together, though, is harder than I would have thought—a delicate, silent dance. After years of unconsciously adjusting and accommodating myself to Peter's standards, pace, decisions, I have been luxuriating this last year in the freedom of discovering my own. Now with Susan, I find myself altering my pace to make room for us both. I want to teach her what I know, share my skills, but am reluctant to let go of control; find it hard to know how much information to share. I am pushing myself to meet the imperative deadline of the arriving sheep but don't want to force those pressures onto her. I find myself wandering off to work alone so as to give her freedom to choose what she wants to do; then I realize, as she asks a hesitating question, that leaving her alone is helping to perpetuate her sense of incompetency. We will work it out, this learning how to share work, because we do not compete or threaten each other and because we trust each other. But working together is not always a simple or natural process.

long string, staked down at ground level is the best way to get the line straight. Where this gets tricky is when the ground drops quickly and the string disappears over the horizon; we just try to sight as best we can. The easiest way we've found to get the corners square is to use the old 3–4–5 triangle from geometry: if one side is 30′ long, the other 40′ and the diagonal between them is 50′, you have a right angle. So you just stretch two pieces of string, 30′ and 40′ each, along your proposed fence lines and adjust their positions until the diagonal is the right length. We've been pretty lucky because most of our land is open, grassy meadow, but in places we've had to fence through brush or woods. There, you have to clear a path for the fencing, and we've found that one about three feet wide makes it easy to unroll the wire and get the fence up. After experiments with machetes, axes, hatchets, adzes, and knives, we strongly recommend the use of a small chain saw or bow saw and pruning shears when clearing brushy areas. (Goats are helpful too, especially in poison oak.)

Once your lines are laid out, you need to mark where your postholes should go. We've found a shovel does this easily and the marks will last for months. The closer you space your posts, the more strength your fence will have. For small pens and pastures where animals will give the fence constant wear, we space posts eight feet apart. Larger areas mean less-concentrated wear, so posts are eleven or twelve feet apart. Our garden fence, which is not pushed, rubbed against or stood on by animals, has posts every fifteen feet or so. If you are fencing a large area or over really steep terrain, you may want to use metal posts. These set really quickly

(you don't need to dig holes) and are not too expensive ($1.00 to $2.00 depending on the height); they will last about fifteen years. You will still need to set wooden posts every three to five metal posts however, as the wire is actually nailed to the wood posts and only sort of propped in place by the metal ones. We prefer to use wood posts because they are stronger, last longer and are slightly cheaper (in this area), but a neighboring rancher recently told us that she always uses some metal posts so she'll never lose her whole fence to a grass fire.

Wood posts can usually be bought in lengths of five to eight feet and in diameters of two and a half to six inches. The height you need is determined by the height of wire you will be using, allowing one third the length of the post to be set in the ground. (If you're using 36″ fencing with one strand of barbed wire above, you should use 6′ or 7′ posts, set 2′ to 2½′ in the ground.) It's a good idea to think about all the possible uses you may have for that pen and set the tallest posts you will conceivably need; once you've gone to all the effort to put those posts in the ground, you won't want to pull them out again.

Posts for straight, woven wire fences are sometimes as small as two and a half inches in diameter, but a minimum diameter of three and a half is recommended. Four- or five-inch-diameter posts should be used for barn lots, pens or corrals, and sandy or wet soils. Posts five inches or larger in diameter are usually used for anchor posts (gate, corner, end, and braced line posts).

The type of wood you use for fence posts is critical. Hard wood and especially the heartwood of hard woods and certain soft woods (redwood and

So many too's: too tired, too much work for too long, too many visitors, too few intimate touchings, too little money, too little time.

I feel low and unclear. Knowing I am pushing myself so hard I can't enjoy what I'm doing, feeling so pressured about finishing this fence that I eliminate slowness and fun from my life. The money pressures are terrible, so much needs to be done here that the money is swallowed up long before it is earned. My job as a carpenter is beginning to feel crazy; it literally saps all my energy and leaves me hardly any richer. Manual labor is really labor. It is hard working for friends, too, too much labor goes unaccounted for and unpaid; in my scrupulousness I cheat myself.

cedar, for example) will last much longer than saplings, sapwood, or soft wood. For comparison: posts (mostly heartwood) of red cedar would last fifteen to twenty-five years; white oak five to ten years, southern pine or willow, two to seven years. Untreated sapwood of any wood species will usually rot in one to three years (Farmer's Bulletin ≠2247, U. S. Department of Agriculture). The best post is the one least likely to rot above or below ground. You can greatly lengthen the life of any post by treating it with wood preservative. Soaked in preservative, sapling posts will last twelve to fifteen years while hard or heartwood will last almost indefinitely (forty to eighty or more years).

Posts and Postholes: What type of posts you buy (or make yourself) will depend a lot on your plans and on what you can afford. All that I can say is that once you've spent weeks or months building a fence, that five- or fifteen-year lifetime for the posts will feel much shorter than it did before you began.

Soaking the part of the post that will be in the ground is the best home method (you can also buy commercially pressure-treated posts). Use creosote, Penta (mixed 1 to 10 with used motor oil or diesel fuel) or any other good preservative. We use Penta and get our used oil at the local gas station. We use a fifty-five-gallon drum half full of Penta mix to soak most of our posts (it will hold ten good-sized posts). For smaller projects, you can use an old honey tin or small garbage can. We once lost thirty gallons of Penta mix when the wind blew our fifty-five-gallon drum over, so ours is now set several feet in the ground. A sheltered spot or some sort of prop will do as well if you have them. The absolute minimum soaking time is twenty minutes. We soak each post at least overnight, and we have friends who soak theirs each a week. Soaking is the best method because it allows the preservative to fully penetrate the wood. Warming the preservative (even the sun helps) speeds up the process and allows maximum penetration. Be careful not to let rain water get into the soaking bucket. The water will sink to the bottom and your post ends will be sitting in water, not preservative. If you have to paint the preservative on (which is *not* as good), use a soft, wide paint brush and warm preservative; keep applying coats until the wood won't absorb any more. It's important to remember that these preservatives are poisonous and will burn eyes or mouths. So keep cans, buckets, and brushes away from children. And be careful when you're working not to drop posts into the bucket, splashing Penta. Always wash your hands well before smoking or eating.

On our big projects, we have as many holes as we can dug by a neighbor who has a tractor with an auger attachment. This is possible only on fairly level and clear terrain. This is a great saver even at the rate of $20 an hour, as the tractor can do as many as a hundred holes (loose soil, open level land, and good conditions) in an hour. Along with the tractor, too, comes Jake who is full of old stories about the days when his "dear old Mom and Dad" settled this ridge, or of how he picked huckleberries in our woods during the Depression —"six cents a pound and pay your own freight."

For one large area where Jake couldn't get the tractor in, we bought a gas-powered hand auger (a one-person tool supposedly but we find working two women together is easier). The auger has definite limitations and I doubt we'd buy one again, knowing what we now know. It lacks the necessary horsepower to dig in wet or hard-packed soils and so is really only useful in loose, sandy loam. We have found it faster and less tiring to dig by hand on our heavy clay hillside. The carburetor also goes out often on our digger, and we spend a lot of time keeping it in tune.

We have done most of our hand digging with a clamshell posthole digger. It is an easy-to-use, efficient tool. With the handles together, you pound

the digger into the ground several times (lift and drop, lift and drop), then with handles open you lift out the dirt. The next time, you face the scoops in the opposite direction to complete the circle. It is also important to enlarge the size of the hole at the top so that you will be able to open the handles when you get to the bottom. Recently a friend introduced us to his hand-auger posthole digger which bores into the ground. We've found the two types make a good combination. The clamshell digger is good for breaking through sod, roots, hard clay or rocks. The auger is good for loose soil once you're through the sod.

Whichever method you use for digging the holes, the time of year you choose to work in makes a huge difference. It is worth waiting literally months in order to have the soil easy to dig. What you want is loose but not really wet (heavy) soil. In most areas, this means digging postholes in the fall or spring when the ground is neither frozen, saturated, nor dried out.

Setting Posts: Metal posts are a joy to set if you do so when the ground is soft. They set really quickly (about four times as fast as wooden posts) and are one of the few things in fencing that give you a definite and immediate sense of progress. To set metal posts, you use a large iron pipe that is sealed at one end. This slips over the top of the post before you stand it in place. Then line the post up as vertically as possible and in line with any preceding posts (this is hard to do on a steep hillside). Lift the pipe and slam it down on the top of the post (two people can work well together at this— one holds the post steady, and one works the pipe). Five or six good solid bangs should set the post deep and firm.

There are lots of methods of setting wooden posts. Extra time spent tamping them firmly in the ground will give your posts long-lasting strength. Tamping is the packing of earth in around the post. It is best done with a heavy pole or pipe whose weight will join with your strength to pack the earth. After some experimenting, I've settled on a 5' bar of ¾" galvanized pipe capped at one end with a ¾" cap, and at the other end with a 1¼" cap. Use the larger end for most tamping, since it packs the dirt well; the smaller end is for tight spaces. The different methods of setting posts depend mostly on how much money or energy you want to spend on them. The proper method is to mix sand or gravel with your dirt for a firm hole. Start with a layer of sand or gravel to create drainage, then al-

ternate layers: dirt-sand or gravel-dirt (each layer two or three inches thick). The trouble with this method is that it takes money or energy to buy or haul the sand or gravel. On most of our land, the loamy soil packs very well by itself, but where the ground is very wet (near springs and in low-lying areas), or is composed mostly of soft clay, we use the sand-gravel method. What is most important of all is to tamp really well and often. We put a few inches of soil in the hole, pack it until it is very firm, then add a few more inches and pack again. The post should be solidly in the ground when you finish. A post that you can move the day you finish setting it will wobble or sag very quickly when animals start pushing on it.

If you are setting posts in an area where the ground freezes during winter, you may have to take special precautions to keep your posts from being weakened or pushed up out of the ground by the changing freezing and thawing of the earth. Check with other homesteaders and old-timers in your area to see what is usually necessary to set a firm, permanent post.

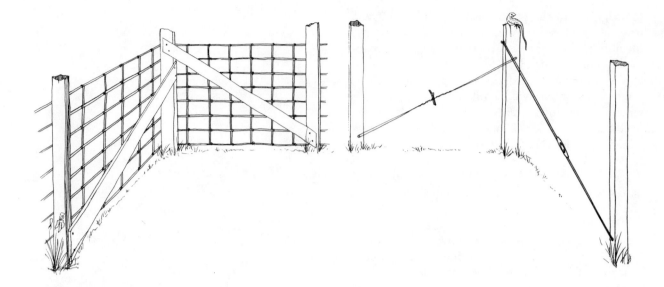

When you come to a corner, you should space your posts closer for extra strength in your fence line. The distance from the corner post to the next post in each direction should be from four to six feet. You can then run diagonal braces for even more strength (see illustration).

Setting posts is a nice two-person job. Take turns tamping and shoveling in the dirt, gravel, or sand. The rhythmic circling of the tamper, the touching of earth, metal, stone, and wood can become a beautiful ritual, another part of farm work that blends music and purposeful work.

Choosing Wire: Woven wire fencing comes in many heights, weights, and spacing. Heights range from thirty-six inches to six feet. Thirty-six-inch wire will do for sheep and some goats. Thirty-six-inch wire with one or two strands of barbed or plain wire above it will do for cattle or horses; thirty-six-inch wire with a strand of barbed wire along the bottom will do for pigs. Goats that like to jump will need forty-eight-inch wire. The six-foot-high fencing is for poultry and gardens (it will keep most deer out).

Weights are light, medium, heavy, and extra-heavy. Weights refer to the thickness and coating of the wire. We bought lightweight fencing for a kid pen once. As soon as our kids were a few months old, they began to break the wires by leaning against and standing on them. Heavyweight will last well and extra-heavyweight, of course, lasts the best. Another consideration besides the weight itself, is whether the wire is domestic or imported. American steel is of better quality and

wears much better for its weight, but it is also considerably more expensive. Of the imported brands, we've found that Belgian steel is far preferable to Japanese.

The spacing of fencing is talked about horizontally (the line wires) and vertically (stay spacing). General purpose fencing is available with line wires spaced as closely as one or one and a half inches at the bottom and four to five inches at the top, with stay wires every six inches. This will keep chickens and rabbits out of your garden—or chickens in a yard and safe from predators (it is much stronger and more durable than "chicken wire" netting). For our sheep and goats we use fencing with the stay wires every twelve inches so the animals don't get their heads caught (lambs and kids are especially prone to head catching). More closely spaced stay wires (six inches or even two inches) will keep dogs out of your pens but you have to be aware of hooves and heads getting stuck.

A roll of field-type fencing is usually 330 feet long, except for the six-foot varieties that come in 150-foot rolls. A roll of wire can cost anywhere from $20 to $50 so do some price comparison before you buy; prices for basically the same product can vary a lot.

There are very few good things that can be said for barbed wire. It goes up really quickly—one person can pull and hold sixty feet of wire with a pair of fencing pliers while her partner nails it to the posts. And it will discourage people or dogs from going over your fence and just about anything from digging under it. But other than that it is evil stuff: it will tear at you while it's going up, it's dangerous once it's up, and it's a nightmare to take down. Always be sure to wear heavy clothes (boots, leather gloves, levi jackets) when you're working with barbed wire. If you have long-eared goats, avoid using barbed wire along the tops of fences and never use it where goats can tear their udders. One tip if you're removing barbed wire: take a short length of stick and wrap long pieces of wire carefully around it. Secure the end of each wire well. This snaggy mass can be more easily handled than coils.

Stretching the Wire: Tightly stretched wire is just as important to a good fence as firmly set posts; the two support and sustain each other. Like everything else, there is an art to stretching wire, and we've tried many different methods since that first sheep fence. Having the right tools seems to be even more critical in this case than having the right technique. Essentially you need something with which to pull the wire (a fence stretcher, block and tackle, or a low-geared truck or tractor), something to pull from (a truck, tractor, or the next fence post), and something to keep the wire evenly spaced (a homemade fence spreader).

Our truck doesn't have low gear, and we have found it is hard on the clutch to pull directly with the truck. Where the ground is clear enough to drive the truck along the fence, we attach a fence stretcher to the truck at one end and the wire at the other and pull. Where we can't get the truck in, we attach the fence stretcher to the next post (which has been braced to prevent its pulling loose). We prefer the truck method since there's no danger of knocking the fence posts loose, but sometimes there's just no choice.

Fence stretchers come in two types, both of which are relatively inexpensive (around $10). They are absolutely indispensable tools if you are doing any large pastures or several smaller pens. (You can also "borrow" a stretcher, posthole digger, and fencing pliers from Sears for a returnable deposit if you bought your fencing from them.) One type of fence stretcher is a metal clamp that attaches to the top or bottom wire of the fence and has a ratchet pulley to get the wire tight. With this type you pull one edge of the fence, then move the clamp and pull the other (after you have nailed the first, of course). The other type of fence stretcher is just a small block and tackle with a ratchet to keep it from slipping loose. If you have a good fence spreader (to catch all the stay wires)

you can stretch each section in just one pull. This is the type we have, and we've been really pleased with it.

On the end of the block-and-tackle stretcher is a hook that must attach to the fence somehow. At first we used a two-by-four board with nails driven into it at the height of each line wire. It had a rope tied onto it top and bottom and the fence stretcher hooked onto the middle of this rope. This board was simple to make and it worked relatively well. The biggest problem we had with it was that the nails would gradually work loose and the fence would "pop" off the board at the top or bottom just as it was getting really tight. For our latest fencing project, we have taken a length of galvanized pipe the same height as our fencing and welded a hook onto each end of it. A piece of chain attaches to each hook, and the fence stretcher hooks onto the chain. The pipe is threaded through the line wires of the fencing, and it pulls tight against the nearest stay wire (see illustration). We used to stretch twelve feet of wire at a time, nail one post, and then move the whole apparatus. With this pipe now, we are often able to

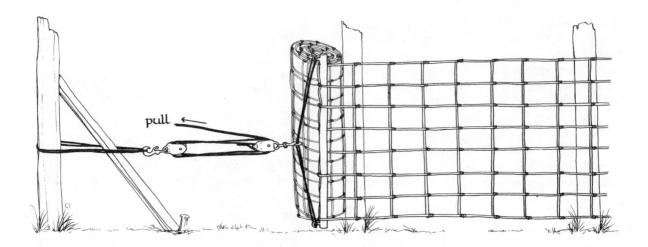

stretch sixty or seventy-five feet at a time (provided the ground is level) and then nail five or six posts before we have to move the stretcher and set up again. The pipe gives a good even pull along the entire width of the fencing, and it can't come loose at the top or the bottom. We've found that stretching the wire just until the stay wire begins to bend around the pipe gets the wire sufficiently taut. If you pull too much, the stay will be bent all out of shape, distorting the line of fencing, or the wire will be too tight to pull in against the posts. When you're going down a slope pull only a few sections at a time, or you will lose tension on the wire. On a particularly steep slope, you may want to pull first the bottom, then the top of the fencing for better tension. Wherever it's possible, we try to lay the roll of wire on the ground and unroll it that way. Then we stand the wire up (prop it, is more like it) and begin to stretch it. Where there isn't room to roll it flat, we tip it on one edge and roll it that way. Put your wire on the inside of the posts, so animals' weight will be on the posts, not the wire.

Once the wire is stretched, it must be nailed. Use one or one and a quarter inch galvanized fence staples (U nails) to do this. Be careful to drive the nails slightly crossways to the grain of the wood so that the post won't split. Wherever you can, set your staple across a stay wire so that the wire can't slip in either direction. We've found that nailing the top and bottom two wires and every other one in between is plenty; it's possible you could get away with even less.

If you're using metal posts, the process is only slightly different. Stretch the fencing and nail onto the wooden post (wooden posts should be set at least one in every five metal posts). This will keep

the wire from slipping loose when the stretcher is released (the clips hold the wire against the metal posts but don't keep it from slipping). Once the wooden post is nailed, go back and fasten the wire to each metal post. This is done with the clips that usually come free when you buy the posts. You need three for each post. Bring the bottom line

224

wire of the fencing against the post *under* the nub on the metal post that you can reach. Fasten it with a clip and then bring the top line wire *over* the highest nub on the post. Fasten that and set a clip at the middle. It took us awhile to figure out how to fasten the clips (halfway down into the canyon we suddenly figured out how to hold the pliers!). You need a good pair of fencing pliers, the kind with a curving hook on one end. One side of the clip is an open hook; the other is a kind of loop. The hook fits over the line wire on the left side of the post (if you're standing behind it). Then bring the clip around behind the post and push the looped end under the line wire on the other side. Bring your fencing pliers over the top of the wire and hook them through the loop. Then bend the loop of the clip up over the wire until it clamps firmly in place. If you can't get the hook on the pliers through the loop, you will find that the tip of the pliers is also made to grip the loop and will bend it to a postion where you can get the hook in. This is hard to describe in words and often tricky to do in practice (especially when your free clips are too small for the posts or the nearest stay wire prevents your getting the pliers in position) but gradually you will become quite dexterous with this awkward tool. We have found that it is sometimes easier to work the pliers partly from the front side of the post and then finish from the back, so having one person on each side is an advantage.

With most fencing, it is easiest to work in pairs. One person pulls the stretcher, the other watches the wire. One person braces the post while the other one nails. One person pulls the wire against the post while the other fastens the clips. And then there is all the just plain moral support when you're halfway down a canyon slope and it's getting dark or when the wire just plain won't get tight in that gully. But I have also fenced a ten-acre pasture almost entirely alone, and there isn't a single job that can't be done with just one person and a little ingenuity.

Retightening and Bracing: A woven wire fence that has begun to sag because it wasn't stretched tightly enough or because the posts are loose can be saved! You can place sapling poles as vertical braces every few yards. Notch the poles and push the top wire into the notch. If you have goats, running poles or boards along the tops of fences protects the fence from a lot of wear, especially in areas where your animals *frequently* jump up on the fence—near shelters, gates, feeding places. A strand or two of plain wire stretched above the fencing will take a lot of the leaning of horses off the fence. Sears makes something called a "fence and wire tightener," a little gadget that twists the wire tight again—a box of these costs $8.25. They are put on with a carpenter's brace.

POULTRY AND ANIMAL FEEDERS

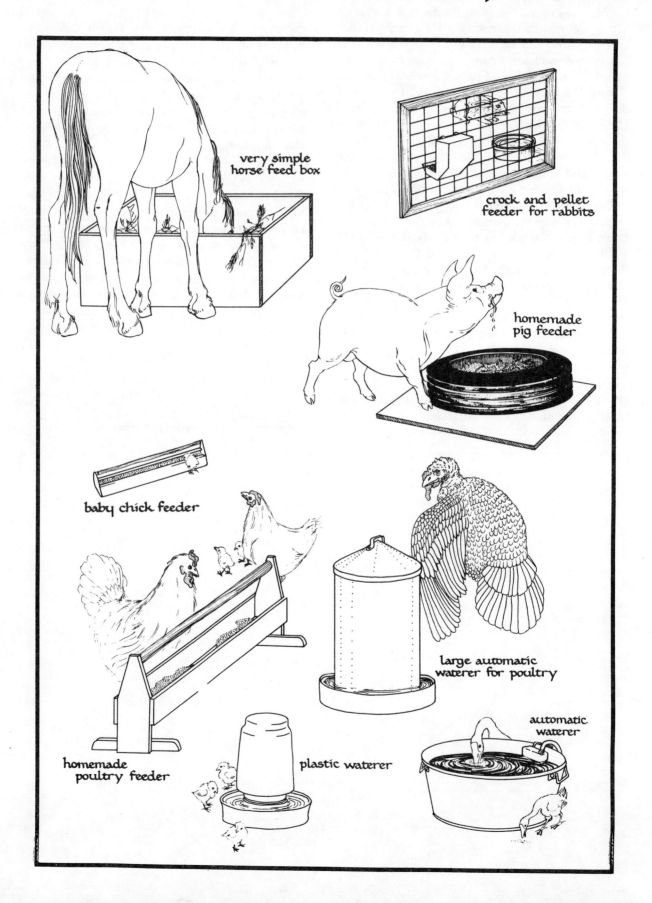

very simple
horse feed box

crock and pellet
feeder for rabbits

homemade
pig feeder

baby chick feeder

large automatic
waterer for poultry

homemade
poultry feeder

plastic waterer

automatic
waterer

Very Simple Horse Feed Box: Until you are inspired to build a beautiful corner manger for hay and grain feeding for your horse, build a sturdy wooden box to keep the feed off of the ground and out of the dust. Use ring-shank nails and heavy boards to make a strong enough box.

Crock and Pellet Feeder for Rabbits: These are two pieces of equipment you will want to buy rather than trying to duplicate with a homemade effort. They are very inexpensive and proven in design. The crock water bowl is heavy enough so rabbits won't tip it over. It must, of course, be dumped and washed and refilled daily. The pellet feeder is metal, attaches to the side of the hutch and may be filled from the outside with a few days' worth of feed.

Automatic Waterer: After you've carried a million water buckets, you may find the automatic waterer an attractive device! It works on a simple float principle and fills the water bucket as the animals drink. There are plain and fancy models, ranging in price from under ten dollars to over fifty. The one pictured is the inexpensive model. It must be attached to a hose or water pipe and then screwed tightly onto any suitable bucket. This one is enclosed so that nosy goats can't play with it. Every few days, the bucket must be dumped and cleaned—otherwise, this little device makes morning chores a bit easier.

Homemade Pig Feeder: The base of plywood makes it impossible for the pig to root and tip over this feeder. It is the simplest construction imaginable: an old tire nailed tightly to the piece of plywood. It can be filled with water or very liquidy food or milk, and the rubber and plywood swell together to make things leakproof. Very handy and brilliant! The pig can have a most enjoyable time rooting around in its food without wasting or soiling it.

Baby Chick Feeder: This tiny metal or wooden feeder may be purchased or homemade. It keeps the chick scratch from being lost in the litter, and allows all of the chicks equal chance at the feed. A wooden dowel should be run across the top of the homemade version and nailed with a single nail at each end so that it can pivot if a chick tries to roost on it (preventing droppings from ruining the feed). The commercially made metal type has a cover with tiny feeding holes to achieve the same protection.

Homemade Poultry Feeder: An enlarged version of the feeder discussed above, the wooden trough-type poultry feeder was a standard for the family farm. It should be built to the size of your flock, with a wooden dowel used as described above. This trough is mainly used for mashes that would get lost in the litter.

Large Automatic Waterer and Plastic Chick Waterer: Both of these waterers must be purchased rather than made, but they are essential equipment for your poultry. The larger version is made of galvanized steel, and comes in eight-gallon size for the large flock—and on down to a two-gallon size for the small flock. The plastic chick waterer shown holds a gallon of water and is useful for breeding pairs of birds as well as for watering up to sixty chicks. There are other types of waterers available commercially too. You may choose a water bowl with automatic attachment, a dome-type waterer (also automatic), or a metal pan with a guard to keep chickens from roosting on it. There are also trough-type waterers available. My choice would still be the two illustrated. They are inexpensive, easy to handle, clean, and eliminate the need to run a water line to the poultry house.

sheep and goat feeders

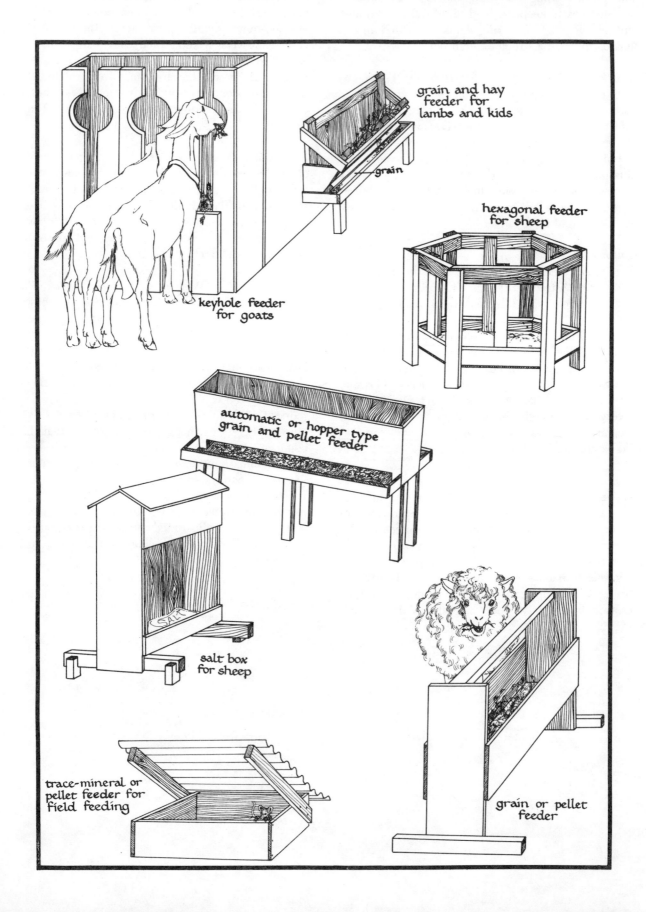

grain and hay feeder for lambs and kids

grain

keyhole feeder for goats

hexagonal feeder for sheep

automatic or hopper type grain and pellet feeder

salt box for sheep

SALT

trace-mineral or pellet feeder for field feeding

grain or pellet feeder

Keyhole Feeder for Goats: This type of feeder is designed to save hay by preventing the goats from dropping it on the ground. It should be about four feet high. The slot is about four inches wide, and the diameter of the round or keyhole part should be seven or eight inches. The keyhole part should be made high up, as illustrated—this forces the goat to lift her head up each time she wants to leave the feeder and prevents her (usually) from taking a mouthful of hay along to drop on the ground. About twenty inches from the bottom of the feeder you can nail a board, as shown, that will help keep the hay in the feeder and also give smaller animals a toehold for using the feeder. The keyhole feeder may be built with boards or with a sheet of heavy plywood. One sheet of 4×8 (standard) plywood will comfortably fit four keyholes.

Grain and Hay Feeder for Lambs and Kids: This simple feeder I designed for use in a special lamb feeding area. It was attached to the sheep house so the back part is solid wood. It is basically a tray made of a 12″-wide board with a 2″-high edge running around it, and a traditional hay manger attached to the tray. The manger slants back to allow a small area for grain feeding. The manger itself is made of one by two material nailed vertically every six or eight inches. This feeder is sturdy and versatile, and can be made in a morning with simple hand tools and scrap wood. Dimensions should be adapted to the animals you will be feeding—i.e., mine was built for very small lambs so it is only about a foot off the ground. You could enlarge this feeder for your goats, horses, or cows.

Hexagonal Feeder for Sheep: This feeder was actually designed for goats, but I liked the idea for feeding our ewe flock supplemental hay and grain. It is not hard to build once you have cut out all the pieces and can see how they fit together. I used some old 2×4s for the structural part of the feeder, and a piece of plywood for the floor or base. The head spaces on the one I built are about 8″ wide. In retrospect I think this is a little too wide—our Angora goatling squeezes through the spaces and tramples the hay while getting her choice of the feed. The major advantage of this type feeder seems to be that the larger and more aggressive animals can't keep the smaller and meeker ones away from the feed. Jostling simply results in a revolving game of musical spaces (with the traditionally designed feeder, one particularly bossy individual can keep almost everyone away by smashing into and pushing them).

Automatic or Hoppertype Pellet and Grain Feeder: This feeder may be filled up with alfalfa pellets and/or grain for free-choice feeding. As the animals eat from the feed trough more pellets/grain drop down into the trough. This feeder should have a hinged cover to keep the animals out of the reserve feed and to keep the feed clean. If used outdoors, it should have a gable-type roof that overhangs the feed trough enough to shed the rain.

Salt Box for Sheep: This is a range-type salt feeder that is designed to protect the salt from erosion by rain or snow. To keep animals from tipping it over, it is secured to stakes driven well into the ground. The opening is made high enough to discourage the sheep from trying to rub their backs on the top board or roof.

Grain or Pellet Feeder: This is a feed trough that is easy to build and is very durable under heavy feeding conditions. It is simply a long box with a top piece that acts as a brace, and horizontal braces on the legs that act to stabilize the whole feeder. This feeder will not tip over even when a half dozen goats are jamming around it. I used the one I built for grain and pellet for feeding our sheep—then liked it enough to build another for our goats. The one disadvantage of using this feeder for goats is that they will stand up on it and jump in if they can—soiling the feed. Very agile smaller goats have a wonderful time balancing on the top brace board, too. Otherwise, it's a useful and well-designed feeder!

Trace-Mineral or Pellet Feeder for Field Use: A simple long box with a corrugated tin or aluminum cover to protect it, this feeder can be put out in the fields for your sheep and other grazing animals. The one I have is 6′ long and about 18″ wide.

TAKING CARE OF YOUR ANIMALS

Developing an alternative to rushing a sick or hurt animal to the vet has been a slow but necessary process. I have lived in the country for many years now, with a small homestead that has included goats, horses, peafowl, sheep, chickens, turkeys, and so on. When I first moved here, I knew almost nothing about veterinary medicine: a sick cat or injured dog meant a phone call and visit to the vet, reliance on her or his word, and a big bill. Moving to a farm and coming up against our first animal emergency, I acted accordingly—rushed down to the nearest phone and called the local veterinarian. He wouldn't come out to the farm and wasn't interested in seeing an injured sheep in his office. I learned then to start relying on my own abilities to diagnose and deal with routine illnesses in our animals, to act in emergencies, to have the basic confidence to take care of our own. It was a process begun out of necessity, and one that has taken many many years to develop. With the help of some very good books, a lot of talking and exploring with other first-time (and old-time) farmers, and some advice and lessons from willing veterinarians who aren't out to mystify their profession, I've come to a place where I can trust my understandings and intuitions about the animals I live with. I believe that you *can* be your own veterinarian a lot of the time if you are willing to really work at learning, and if you have an active love for your animals. There will be times when you will

Pits of despair. This whole summer seems to be either a test of my endurance or a lesson in limits. Everything that could possibly go wrong has. Any one incident would have been a major crisis; the sum total is about to break me. The latest is that the truck won't run—a fuel line clogged or something. It's not just the frustration of having no vehicle with a full-time job that gets me but also the frustration of not knowing enough to fix it myself and not knowing where to go to learn what I need to know.

I'm trying to cut back now. Recognizing that this pace is too hard. The women's work weekend gave me some perspective—seeing "how little" twelve women could accomplish in two days helped me to know "how much" Susan and I have done this summer. Part of my problem is my image of the job: I talked myself into believing that fencing ten acres was a task we could finish in a few weeks; I still haven't readjusted to the reality that it is a huge amount of work. So instead of feeling satisfaction at our accomplishment, I feel frustration that we haven't done more. Now, I'm trying to rediscover pleasure in what I am doing instead of just feeling enormous pressure. We've just built an emergency sheep pen to deal with all the breeding complications; so now the ten acres can take as long as it has to. The next major projects are the winter garden and the hay shed, both of which are fairly immediate pressures. I said to Susan last night, "Just one more bad year, and then it will be done." Even before I caught her ironic smile, the greenhouse, ram pen, new orchard, barn, passed before my eyes, and I knew it would be fifty more "bad years" unless I slow the pace and change the process to make them fifty good ones. I have dreams enough to last a lifetime on this land.

need professional help and the valuable alternative of a competent veterinarian, but much of the time you can take care of things yourself and do very well at it. Initially, it will take some financial investment in books and supplies, but mostly it will take determination to actually begin, and courage to trust yourself.

Learning to Look: The first step is to assume a special responsibility for your animals. This means developing a daily consciousness of each animal. Familiarize yourself with the animal when it's healthy: how does it look and act? What are its feeding habits? Some animals have voracious appetites, others are slow or picky. Any obvious variation from normal habits is suspicious. Look at the animal's fur or feathers or wool, and know how it should look when the animal is healthy. A goat's hair should be smooth and shiny; a sheep that is losing clumps of wool may be wormy. Notice how the animal stands when it is relaxed, how it walks, how it carries its head, and how alert and energetic it is normally. All of these things can become signs of some disease or mechanical problem. A goat that rolls its head back may have bad teeth, but some goats do this out of habit, and it is "normal" behavior for them. Look at the color inside the animal's mouth (gums changing color from pink to gray may indicate poisoning) and notice its teeth (abscessed or broken teeth may be infected; young animals losing teeth may go off feed; older animals with badly worn teeth may need special feed). Animals' eyes also reflect their state of health: they should be bright, clear, and free of discharges. The inner eyelid is pink to red in a healthy animal; paleness or a grayish cast is often a sign of severe worm infestation. Body temperature is also critical. Passing a hand over the animal's body will teach you how warm it feels naturally, whether it has been standing in the shade or out running in the pasture. You can pick up fevers or subnormal temperatures this way, though you will have to use a thermometer to get an accurate reading of the body temperature. Finally, notice how fast the animal breathes when it's quiet and when it has been exercising. You may want to check this respiration rate with a watch someday, but this is enough to give you a rough idea of what is "normal" and what is cause for alarm and further investigation.

You should familiarize yourself with the look (texture, color, consistency) of the normal feces of each animal or bird. Many illnesses and digestive problems make a dramatic showing in scouring (diarrhea), blood in the feces, and so on. Be aware, too of the natural or normal look of the animal's skin so that you can recognize infestations with external parasites or skin diseases. Finally, you should check each animal regularly for swellings under the skin that might indicate abscesses forming. Taking a critical look at each animal every day will develop a kind of sensitivity in you that becomes second nature. It is this sensitivity that will help you recognize and catch diseases before they become major problems.

Any animal that is sick or hurt should be isolated if possible, kept quiet, and watched. It is best to keep the animal somewhere close to its usual companions so that it doesn't suffer further from loneliness and the trauma of separation. A "hospital" or observation stall in the barn—close to, but separated from the other animals—is a wonderful asset. Once you've noted any obvious changes in appearance, body temperature, and so on, you can begin to analyze the subtler levels and work toward a diagnosis. Eliminating possibilities, being sensitive to similarities in diseases, and being fairly cautious about leaping to conclusions are the best guidelines. Reading widely really helps—often you will find yourself with a vaguely familiar paragraph in mind that relates to what you see before you. This often gives you a beginning base for diagnosis. Most diseases that will affect your homestead animals are common, have distinctive symptoms, and may be fairly easily distinguished from one another. Most accidents on the small farm—animals gorging on grain after breaking into the barn, tears and scratches on the udders of milking animals, and so on—may be handled simply and efficiently if you know what to do in advance and have the supplies on hand. Hopefully, you will be able to start and deal with smaller problems before you come up against a real emergency or a serious illness. What is important is to trust your own intuitions, your own knowledge about these animals you see every day. From there you can rely on yourself to diagnose and treat, or you can call in a veterinarian and offer some concrete observations that may help her/him treat the animal.

Keep accurate records of anything that goes wrong with your animals and birds, including how you noticed the first symptoms and eventually treated them. These notes can become invaluable references. Not all animals act the same when sick, and not all will respond in the same way. There are, however, some very valuable specifics to record and make use of again. These records should include all routine immunizations, wormings, hoof trimmings, and so on. They should include observations about the individual animal's lambings or kiddings or foalings. In a few years, you will find yourself with the most incredibly valuable homestead reference work!

Basic Supplies: What you will need in terms of veterinary supplies will depend upon the creatures you keep, your seriousness about this aspect of their care, your economic situation, and your own priorities. I've found the following items to be pretty much indispensable: a stock thermometer (heavy duty, especially for use with animals), plain vaseline for use with thermometer, hydrogen peroxide (for cleaning out wounds), wound ointment (any general-purpose antibacterial cream or salve), cotton (sterile), bandages and adhesive tape, and tincture of iodine (for multiple uses, from disinfecting navel cords of newborn lambs to treating abscesses). Rubbing alcohol is also very useful for cleansing the skin before giving injections. Mineral oil (a quart or more) is invaluable for treating bloat and other digestive system problems in all kinds of livestock. Baking soda is similarly useful. If you keep dairy goats or milk cows, you should have mastitis infusions on hand, as well as a kit for detecting mastitis (see Appendix article). As you learn to give your own immunization shots, you will want to have a supply of disposable or glass syringes and needles on hand. Depending upon the animals and birds you keep, you will want to have a small stock of commonly used antibiotics, injectable vitamins, and so on. Almost all of the supplies noted are available from country pharmacies or feed stores, or may be mail-ordered from animal supply houses (see Resources). As you learn more about your animals, your basic veterinary cupboard will expand and become more complex and more useful. I like to keep on hand: injectable calcium salts (for treating milk fever in dairy animals); Ketostix (for detecting ketosis in sheep and goats); injectable electrolyte and amino acid and B-complex vitamin solution (for use with animals suffering shock, severe dehydration, etc.); boluses for disinfecting the uterus and for stimulating passage of afterbirth; Kaopectate (for scours); an obstetrical lubricating powder for difficult deliveries of lambs and kids. I keep on hand Combiotic (a penicillin/streptomycin combination given to animals with shipping fever, pneumonia, mastitis, and other infections); Tylosin (used for pneumonia cases, respiratory infections

in poultry, and so on), Sulmet (sulphamethazine in oblet, large pill, form, used for bacterial scours, pneumonia, and certain clostridia-type infections), and a few other useful antibiotics or sulfa drugs. I've also kept on hand certain antitoxins (for tetanus, for example) and antidotes for allergic shock (antihistamines or epinephrine). All of these things are fairly inexpensive and may save an animal's life. They must, of course, be periodically replaced as they pass their expiration date and become ineffective. What you need may differ greatly from what I've listed. You will find yourself making up just such a list, little by little, and stocking up according to the needs of your particular animals and farm.

Using Books: Most illnesses of farm animals and birds are common and may even be prevalent in certain areas or under certain conditions. You can learn a lot by talking with other people who keep the same types of animals, by talking with local veterinarians, by going to your farm adviser (check with your county agricultural extension service). You can also learn a great deal from certain books. In the Resources section there are specific and general livestock books listed. These are invaluable as reference works on a certain species of bird or animal. In addition, I have found having a copy of the *Merck Veterinary Manual* to be indispensable. This book is technical (read it with a dictionary handy) and tends to make things rather mystified (intentionally?), but once you become comfortable with it, it has a wealth of information on everything from animal nutrition to specific treatments for uncommon diseases. It explains how to make an accurate diagnosis, how to administer all types of medication, how to treat parasites, how to recognize and deal with poisoning, and so on. It is well worth the rather high initial investment as it is a book that will be useful for years and years.

Another book I would wholeheartedly recommend is the *Herbal Handbook for Farm and Stable* (also listed in Resources). This book discusses the health and care of animals from a natural, sensible viewpoint, stressing herbal and nutritional treatments for all types of animals and poultry. I use it constantly and continually come across new and useful information.

If you live near enough to a college of veterinary medicine, you can have your choice of the many very fine books available. Some of those I've used are listed in the Resources section, with publishing information. You may be able to order them through a bookstore or through your library. Besides these books, you may find government pamphlets and agricultural extension service booklets useful. Extension courses (from the University of Pennsylvania, for one) are also available. Other sources of excellent, up-to-date veterinary information are the magazines put out on certain species of livestock. The *Dairy Goat Journal*, for instance, has a regular veterinary column dealing with all the problems you might ever encounter in your goat herd—plus special, more detailed articles on specific problems and diseases. If you can, subscribe to the magazines listed that deal with your particular animals. Save all back copies—you may find yourself referring to an invaluable article that was printed, say, three years ago! (I have—many times.)

Giving Liquids and Pills: If you are giving a tablet, pill, or capsule, give the animal or bird a little water before and after. Large tablets and oblets can be dipped in oil or butter to lubricate them. Hold the animal's mouth open and drop the tablet as far back on the tongue as you can. Holding the mouth closed and stroking the animal's throat will stimulate swallowing (be sure the animal *does* swallow—cats and goats are especially adept at rolling pills into the corners of their mouths and spitting them out later!). A good technique for dosing a goat or sheep is to stand with the animal's neck gripped between your legs, holding the head up with one hand under the jaw, and opening the mouth with your other hand. It requires a bit of dexterity or a friend to help! If you can't get the animal's mouth open, you may need to use a

Things were on the up for a few days. I was feeling easy, loose, and happy. Put some hard days into steady work here and it felt good to be deep down, really tired from work done. It seems like forever since I spent whole days working here, and it's what I want to do. Susan and I cut all the poles for the hay shed with the chain saw—the first time since Peter and I bought that saw that I have used it to cut trees. Yesterday we stripped the poles and set all the posts in the ground. Then I dashed off to town to get the 2×6s for the rafters and the parts to fix the Volvo. Finished the laundry at six and discovered the truck starter had gone out leaving me stranded with a loaded truck. Tory showed up an hour later to tow the truck to where we could push start it, and she pulled half the bumper off my truck and broke the tail light. But the truck started and we headed home. I was stopped by the Highway Patrol and got a ticket for no tail light. Got home exhausted and helped Susan finish chores by moonlight. Got up this morning and discovered the freezer had broken down and all our food was thawing. Found Jane after a long search and got permission to move everything to her freezer. Borrowed a car, packed up all the food and moved it to her house. Got out the rented wheel puller for the Volvo and discovered it didn't fit. Susan borrowed a car and took the wheel puller back to town to exchange it, and I began framing the hay shed. I framed two walls and put up the rafters, feeling skilled and happy, moving along steadily and not thinking of much except cuts and measurements. Susan came back with Michael and they started work on the Volvo. An hour later, the wheel still wasn't off and Michael was talking about taking a cutting torch to the brake drums ($20 each!). Susan decided to let them sit. The truck is still broken, no time to fix it. Susan came to help me and we finished framing the shed before dark, except for the bracing. Concrete sign of a little progress. We're sitting here tonight drinking brandy by the fire and trying hard to laugh. Seems like things can't get much worse.

balling gun or a mouth speculum. Both of these instruments may be purchased from an animal supply company—neither is expensive, and they are easy to use and make dosing simpler. The mouth speculum is also useful for checking teeth if you suspect an animal has broken or infected ones. Birds may be held firmly in the lap and their beaks forced open by a gentle pressure on either side of the beak where it attaches to the bird's face. A pill may then be dropped down into the throat.

Giving a liquid is a little trickier, though you can use the same basic approach to handle the animal/bird and open the mouth. You must be very careful not to get the liquid down the windpipe—this can choke the animal/bird or cause inhalation pneumonia. Give only a little liquid at a time, and be careful that you are not holding the head at too severe an angle. A drenching bottle (the force feeding of a liquid is called drenching) may be made using a lamb nipple (available at feed stores) and any sturdy bottle. This may be used on a cow, horse, sheep, or goat. The hole may be cut bigger to allow thick liquids to run out faster. A large animal such as a horse will have to be held by the halter, and the bottle slipped into the side of the mouth and tipped up. Be prepared for a drenching yourself, as much of the liquid will end up on you in many cases! For small animals and birds, an eyedropper may be used in place of the bottle.

Powders may be concealed in the animal or bird's feed. Strong-tasting or -smelling powders, such as wormers, may have to be further concealed

by the addition of molasses or other strong-tasting but palatable feed. My horse can sniff out a worming medicine in anything and refuses to touch it until his nose is well lubricated with Vicks, and he can smell only the pungent odor of that! Cats, dogs, and chickens will usually accept powders mixed into their food; sheep and goats are fairly easily tricked by the molasses addition.

Giving Shots: Injections are the quickest and most efficient way to place drugs directly into the animal's body or to give immunity to common diseases. They are given three ways: subcutaneously (under the skin); intramuscularly (into the muscle); and intravenously (into the vein). All routine injections I've given have been of the first two types; intravenous injections require special skill and conditions. I have learned to give an intravenous injection and to take a blood sample in case I ever need to do these things, but for general purposes you will probably be giving intramuscular and subcutaneous injections.

Most drugs and vaccines are given at room temperature and should be well shaken before used. Many are stored in a refrigerator and must be kept at that cool temperature to remain viable. Take out only what you need and return the rest of the substance or solution to the cool environment. Use a sterilized needle of the proper size for the animal or bird you are treating. Needles come in different lengths and thicknesses. In general, you would use the medium and larger gauge needles (18 to 21) for large animals or very thick medication and the finer gauge (21 to 23 or 25) needles for smaller animals and birds. Needle length varies from five eighths of an inch to an inch or two. For general use, the inch or inch-and-a-half needle seems fine. In animal supply catalogues, needles are usually listed by gauge and length together (21 × 1, for exam-

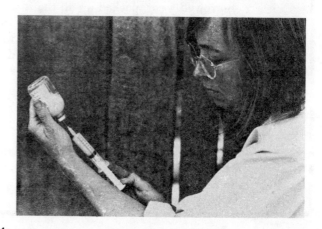

ple, or 18×1½). A needle may be sterilized and reused once or twice, but it tends to get blunt or ragged with use. Syringes also come in various sizes, and should be chosen according to the amount of medication or vaccine you will give. I've found a 5 or 10 cc. syringe to be the most useful, though I keep a few 3 cc. syringes and a few 20 cc.'s on hand for special use. Syringes, too, must be sterile. They may be reused as long as they are handled carefully and sterilized thoroughly. You may sterilize by boiling needles in distilled water for at least fifteen minutes, or by keeping them in alcohol. Syringes may also be boiled (plastic ones will melt if boiled at too high a temperature for too long) or stored in alchohol. Needles must be rinsed well before using (syringes too), particularly if you are giving penicillin (alcohol is antagonistic to penicillin and will interfere with its action).

Shake the medication or vaccine well (unless it specifically states on the label not to) and hold the container up to draw off the drug/vaccine. If you insert the needle and pull back on the plunger and nothing flows into the syringe, you may have to inject air into the bottle. Do this by withdrawing the needle, pulling the plunger and filling the syringe with air, and then reinserting the needle and pushing the plunger in. This time the vaccine or drug should be able to be withdrawn into the syringe. Holding the syringe needle up and tapping it gently once it's been withdrawn from the container will cause air bubbles to rise to the top. They may then be ejected by gentle pressure on the plunger. The injection is ready to give when it is at room temperature. Be sure to check the expiration date on all vaccines and antibiotics to make sure they are still viable.

Restrain the animal as well as you can. A goat or sheep may be thrown down and one person should hold the animal's legs and press a knee against its neck to hold it. A kid or lamb will be more comfortable and less frightened if it is held in the lap rather than thrown. A horse or cow may be tied or held by a friend. Cats and dogs may have to have their claws clipped or be muzzled—being overcautious may save injuries or struggle.

Clean the area to be injected with rubbing alcohol or a special antiseptic stain made just for this purpose. The subcutaneous injection is best given in the loose skin of the animal's body—the skin on the neck or shoulder of a sheep or goat is fairly loose; the skin at the back of a cat's neck is very loose. For the subcutaneous injection, the needle is held parallel to the animal's body at the

injection site. Pinch a little tent of skin between your thumb and first two fingers and slide the needle into this tent, still being sure that the needle stays parallel to the animal's body. You will be able to feel the needle moving freely just under the skin.

At that point you can press in the plunger and inject the contents of the syringe. If you are injecting a large volume of fluid, it helps to move the needle in an arc as you make the injection—this spreads the material out over a larger surface. You may also want to inject the material in two or more areas if you are giving a really large volume (as when you give an electrolyte solution to a dehydrated animal). When the syringe is empty, pull the needle out with a swift, sure motion and reclean the area with alcohol. Under certain conditions, you may then want to massage the area to spread the vaccine or medication; in certain instances, though, you are not supposed to do this. Read carefully the directions on the insert or bottle of the material you are using before you give this shot and follow them!

The intramuscular injection is given into a large muscle, such as that of the thigh or shoulder (or the chest of a horse). Again, clean the area of the injection with alcohol before and after the injection. The needle should go directly into the muscle, more or less perpendicular to the animal's body. It should go in a quarter to a half inch or so, depending upon the size of the animal and the muscle. The needle should go directly into the body of the muscle but not beyond it—you will be able to feel this. Before you inject the contents of the syringe, pull back slightly on the plunger. If traces of blood appear in the syringe, you have hit a blood vessel and *must* withdraw the needle and reposition it. When the needle is properly positioned, inject the material and then withdraw the needle quickly and cleanly. If an animal struggles a lot when the needle is in place, it is sometimes better to let go of the syringe or withdraw it rather than risk breaking the needle off or tearing the animal's skin. Staying calm and understanding that an animal will often struggle against being held rather than against any pain will help you.

Injections are not difficult to give if you are careful and considerate of the animal involved. It is a skill you will make use of many times in caring for your animals both routinely and when they are ill.

simple immunology

In six years of homestead farming, I've often found it necessary to use vaccines, antitoxins, and so on to protect and treat our animals. For a long time the terms and materials—toxoid, antisera, modified live virus—were unfamiliar and I was completely mystified by them. This spring, in trying to work out a protective routine of vaccinations for our goat herd, I decided to unravel the mysteries. Using some veterinary textbooks (including one on applied pharmacology), discussions with our veterinarian, and other sources, I came up with the following. It is a very simplistic approach to immunology, but I hope complete enough to give a usable overview. It feels good to read through our supply catalogues or face treating a sick animal with a basic understanding I've never had before.

To begin with, some definitions. I have always had to read my *Merck Veterinary Manual* with a dictionary at my elbow—the language is technical to an extreme. My first step in demystifying immunology was to list all the words I understood vaguely or not at all, and look them up. I found as I worked I continually had to refer back to Webster.

The healthy plant or animal body is in a smoothly functioning state. When that state is altered by foreign organisms and body functions are disturbed or interrupted, the individual is said to be diseased. Pathogenic (disease-causing) organisms or their toxic (poisonous) products must be dealt with if the individual is to be restored to a healthy condition.

Immunity is the ability (state or power) to resist a given disease. This may mean resisting an attack by the infecting microorganisms or resisting their toxic products. The animal body resists by producing antibodies, which are substances or bodies in the tissues or fluids (blood or serum) that "act in antagonism to specific foreign bodies" (the toxins or bacteria). In the case of one type of antibody, agglutins, this "acting in antagonism" means causing the bacteria to clump together. Another type of antibody is an antitoxin, which is capable of *neutralizing* the toxin that stimulated its production. Thus, antibody production is a means of restoring the animal body to a healthy or normal state by destroying and removing the foreign bodies that threaten that state.

If an animal contracts a disease and you want to help stimulate the antibodies with which to fight that disease, you will be practicing a therapeutic immunization. If you want to give your animal protection against a disease that is prevalent in your area or very common and widespread, you will be using prophylactic (preventive) immunization. In either case, you will be making use of the animal body's ability to produce these miraculous, necessary substances called antibodies . . .

An individual may be naturally immune—certain species of animals are immune to certain diseases. This may be called "inherited immunity." Dogs, for example, are immune to foot-and-mouth disease that plagues cattle. Goats are nearly immune to tuberculosis. If an animal is not naturally immune, it may *acquire* immunity to a certain disease. It may do this on its own, or it may gain temporary immunity from its mother's milk. Or you or your veterinarian may intercede to vaccinate your animal— another example of acquired immunity. The ways in which an animal acquires immunity or acquires extra immunity become quite complicated. It may become actively or passively immune, and each of these states may be acquired or conferred artificially or naturally. For clarity I would break these down as follows:

Active Immunity
A. Natural. Any animal that recovers from an attack of a naturally contracted disease will develop this type of immunity. An animal exposed to a low level of the disease over a long period of time may likewise build this up. This type of immunity usually gives the animal lifetime protection against the disease. An example of this would be true Goat Pox —a goat that has had this disease once will never have it again (there is a similar, common, but less severe disease also called Goat Pox which may reoccur again and again in an animal).
B. Artificial. For this type of immunity, the animal is injected with a material that will stimulate it to produce specific antibodies that protect it against a certain disease. The material injected might be a toxoid—by definition a toxin (soluble poisonous substance produced by a pathogenic organism) that has been treated to destroy or almost destroy its poisonous properties without affecting its ability to stimulate antibody production. If you keep sheep, horses, goats, and so on you should be developing their active immunity to tetanus with tetanus toxoid.

Or the material injected might be a bacterial vaccine. This would be used to create an immunity to a specific bacteria, and the vaccine would be a suspension or culture of inactivated or killed bacteria. It may be called an anaculture or a bacterin. For instance, when you vaccinate your cattle or goats against enterotoxemia, you might use a bacterin prepared from cultures of clostridium perfringens bacteria.

The third type of material used to produce artificial active immunity is the viral vaccine. These vaccines consist of suspensions of killed or live virus that have been reduced in their pathogenicity. Or they may be suspensions of virus that would be pathogenic to one species of animal but when injected into another species stimulate antibodies without being pathogenic in effect. When you vaccinate your puppy against canine distemper, you are using a viral vaccine.

Live viral vaccines generally give a stronger immunity than dead ones, but must be handled with utmost care. The live organisms can contaminate the environment and infect other susceptible animals. Living organisms may be excreted by the inoculated animal for several days and can be a source of infection. Used empty containers and syringes should be burned or immersed in a very strong disinfectant.

Active immunity—whether you use a toxoid, bacterin, or vaccine—usually takes from ten to fourteen days to develop. Often you will have to give two inoculations two weeks to a month apart. Immunity from the second inoculation develops quicker (two to four days) and is stronger. You will usually have to give the animal an annual "booster" shot to maintain the animal's immunity.

Extremely young animals are often not able to produce antibodies properly in response to vaccination. They may be temporarily protected against disease in other ways (section following). Before you vaccinate a young animal, be sure it is old enough to produce antibodies in response to the injection. If it isn't the injection will be useless (though probably not harmful). Most toxoids, bacterins, and vaccines will come with an insert stating the minimum age for the animal. If not, you should consult the list in the *Merck Manual* ("Common Immunizing Agents" in the Addendum) or talk with a veterinarian or pharmacologist.

Passive Immunity

A. Natural. This is gained by the young animal during its first few days through the mother's colostrum (first milk). Available antibodies will include those acquired naturally and artificially by the mother. The best way to protect young kittens from distemper, then, is to make sure the mother cat has received her regular inoculation. Another example of natural passive immunity is that gained by a newborn lamb against enterotoxemia when it nurses its vaccinated mother. Antibodies present in the colostrum will protect the young animal against diseases for variable periods of time, depending upon the strength of the mother's antibodies, the growth of the young animal, and so on. As the passive immunity wears off or becomes weaker, the animal should receive its own vaccination. You will have to research the proper age for vaccination of any particular type. Generally, young animals can be vaccinated at four to six weeks with a second injection a month later. Revaccination at weaning time is sometimes necessary.

B. Artificial. When immediate protection against a disease is needed, an animal may be given temporary artificial immunity. This may be given to an older animal who has been exposed to a disease or to a very young animal who is susceptible to a disease the mother has not been vaccinated for. This same immunity may be given to an animal who has already contracted a disease. If given in time, it can effect a dramatic recovery.

Depending upon the disease in question, you will use one of three types of sera to give artificial passive immunity. Where a toxin is involved, an antitoxin is used. Tetanus antitoxin is a commonly used example. An animal that has been injured and is susceptible to the toxins produced by tetanus bacteria should be given antitoxin. Even an animal that has had a regular series of tetanus toxoid and is fairly well protected should be given antitoxin if the wound is severe and tetanus prevalent in the area.

Antigens are the substances that stimulate antibody production. When the antigen is a suspension of bacterial bodies the sera used is antibacterial. When the antigen is a vaccine prepared from a live virus, the sera is called antiviral.

Use of these antiserums gives an immediate but very temporary immunity, usually lasting ten to fourteen days. If an animal has been exposed to a serious disease, it can often be given antiserum for immediate protection and vaccine/bacterin/toxoid for long-term protection at the same time. If an animal has contracted the disease, antisera may be given and other therapeutic measures taken. The vaccine/bacterin/toxoid should not be needed in this case—if the animal survives, it will have developed its own natural active immunity.

I decided last night to list all my troubles so I could look at them and deal with them—couldn't do it. It was too completely depressing. Only a virgo would try such a thing in the first place!

The starter motor went out on the truck again two days ago, so we pushed it to start it and I left for work. At Anderson's I met the school bus. Worried about the engine dying, I swung over to let it pass and got stuck in the mud on the shoulder. Didn't want to bother the Andersons at 7:15 so I walked back down the road and woke up Alan to borrow his car.

Susan just called here at work to say that when she got up to Anderson's to see about the truck, she found them pulling out their picket fence so it would be easier to pull the truck out. When she apologized for being so much trouble, Fred said, "When someone needs help, I'm glad to give it. I just figure I'm lucky it isn't me."

When administering any type of sera, you should watch for and be prepared for any allergic reactions. Sera are either homologous or heterologous. The former means the antiserum is taken from the same species as the animal receiving the serum. The latter means that the donor animal and receiving animal are of different species (for example, lamb dysentery antiserum is prepared from horses). Allergic reactions are more common when the antisera used is heterologous.

One severe allergic reaction is anaphylactic shock. Signs include muscle shivering, difficult breathing, and anxiety. This type of reaction usually happens within a few hours of administration of the antisera. Antidotes—including antihistamines and adrenaline—must be given immediately, as the animal may die quickly once it has gone into shock. Another antidote is epinephrine. which is an antispasmodic and vasoconstrictor. It is easy to administer (may be given subcutaneously), effective, and inexpensive. Keep this on hand as it is indicated for shock caused by many common vaccines.

Another type of allergic reaction manifests itself in "increased irritability" and swelling of the head, ears, and neck. It will happen within a few hours of antisera administration. A third reaction is called "serum sickness." It involves the appearance of a rash, painful joints, and fever, and it may take days or a few weeks to appear. Both of these reactions may be treated with antihistamines.

I have yet to see allergic reactions in any of the animals I've treated, but just in case I keep epinephrine and antihistamines handy and watch the animals carefully for signs of reaction.

I learned the value of vaccinating our animals the hard way—by losing kids to tetanus and grown goats to enterotoxemia. Both of these diseases are common and very easily prevented by a regular, in-expensive, simple-to-give vaccination. I've learned to give our kittens and cats distemper shots, our horses tetanus and encephalitis shots, our sheep and goats tetanus and enterotoxemia protection. I've also learned the value of using antitoxin when an animal has the first signs of—or has been exposed to—a serious disease. If you are keeping animals—either as pets or part of your farm or homestead—you should give them the attention and care of routine vaccinations. Being domesticated, these animals are subject to all sorts of exposure and conditions that can cause diseases. You can also prevent their suffering or dying from diseases natural to them even if they were in their "wild state." Livestock books, other breeders, your veterinarian or farm adviser can all help you decide what disease you should actively prevent by vaccinating your animals. The materials you will need are available through your veterinarian, rural pharmacies and farm supply and feed stores, and through mail-order supply companies. Often pharmacies and feed stores don't display their vaccines and such (most must be refrigerated) but have them—you will have to ask. Laws about purchasing and using these and other veterinary supplies vary from state to state. Usually you are well within your rights as a livestock breeder (the definition is flexible; in one small animal supply catalogue I have a breeder is legally someone who owns one or two purebred animals with the intention of breeding or showing them).

Actually, giving the vaccinations is not difficult. Read the earlier section ("Giving Shots") in this book and/or have an experienced person show you and then watch you the first few times. Follow all instructions and precautions on the materials you use. And begin to practice this whole new area of animal care—it's well worthwhile!

I WANT TO LIVE TO BE
AN OUTRAGEOUS OLD WOMAN
WHO IS NEVER ACCUSED OF BEING
AN OLD LADY

I WANT TO LIVE TO HAVE TEN THOUSAND LOVERS
IN ONE LOVE
ONE 70-YEAR-LONG-LOVING-LOVE

THERE ARE AT LEAST
TWO OF ME

I WANT TO GET LEANER AND MEANER
SHARP EDGED
COLOR OF THE GROUND
TILL I DISCORPORATE
FROM SHEER JOY.

BUYING GOATS

Most people who buy a goat wind up keeping two, three, or more. This has to do not just with the sociable nature of goats (for they crave the company of others of their own species and usually don't take too well to being the only one), but with the special qualities of the goat. An animal that combines independence, cleverness, affectionate nature, and high productivity, the goat is a valuable member of the homestead. She (for most goats you keep will be does) has fairly simple needs—a small shed, a pen or pasture, a monthly hoof trimming, and daily feed and fresh water. You will have to be home twice a day if you have a milking goat—a serious basic commitment. Her small size makes the goat easy to transport in breeding season and easy to handle and house. The initial investment is not large and will come back to you many fold if you choose a good animal. The milk she produces (good-tasting and especially nutritious) will easily outweigh the cost of keeping her. Her kids (if you can bring yourself to part with them) will add to her productive value. The most important thing in buying a goat, whether it's your first or your fifth, is to choose her carefully according to your needs and expectations. Love at first sight is not to be discounted—but try to be sensible!

Before you buy goats, you should learn as much as possible about them and look at many different animals. People who raise goats are usually quite happy to talk with you and show you their animals. Goat magazines and books can teach you a lot, but it comes to life in the animals themselves. I learned more about conformation in the goat by attending one small local goat show than by reading two years' worth of *Dairy Goat Journal!* Learn the basics and then go out and look at goats . . .

There are five breeds of dairy goats and infinite varieties of crossbred goats. Saanens, Toggenburgs, and Alpines resemble one another in body type and general appearance, though they are quite different in coloring. All have medium-sized ears, which are held out erect from the head, and rather straight noses. Saanens are pure white and are usually large animals. Toggenburgs must be brown (shades vary) with a light or white stripe down each side of the face, white or light legs, and a white patch on either side of the tail on the rump. Alpines are multicolored, with certain established patterns such as *cou blanc* or *cou clair* (white, tan, or gray neck and shoulders shading to black hindquarters, black markings on head) and chamois (tan, red, bay, or brown with black markings on head and neck; black stripe down back and back legs). Nubians have pronounced Roman noses, with long floppy ears. They may be any color or combinations of colors, ranging from pure white to spotted gray and black. La Manchas may also be any color (red, black with tan trim, brown splashed with white), with straight or dished faces and very tiny ("elf") ears or almost no ear at all ("gopher" ear). All La Mancha bucks must have gopher ears to comply with breed standards. Crossbred goats may combine breed characteristics, so that a goat with Toggenburg markings, floppy half-Nubian ears and a slight Roman nose might appear. When you go to buy a goat, take into consideration the breeding bucks in your area. When you go to breed your doe, it should be a careful effort to produce *better* kids. You probably won't want to haul her two hundred miles to breed with a good buck of the same breed, but you should never breed with the buck down the road just because he's a male. Also, buying a goat of the type prevalent in your area will give you a wider choice of animals.

The basic vocabulary of the goatkeeper is simple. A female goat is properly a doe. A male is a buck. A castrated buck is a wether. "Kids" range in age from a day to a year, at which point they become "yearlings" (although a young female may be called a "doeling"). When a goat has her kids, it is called freshening—coming into milk—and her age may be discussed in terms of her number of freshenings. A doe is usually bred at the age of ten months, but this may vary from seven to eighteen months. A "first freshener," then, is anywhere from twelve months to twenty-three. "Grade" is the term used for a goat of unknown, or partially known, breeding. A purebred animal is one whose ancestors are all of one type. She may or may not be registered. The language of registered and pedigreed goats is more esoteric. Most of the terms —A.R. doe or star buck—refer to milk production tests and records. You should ask the breeder to explain any terms used, as they can be quite complicated.

Spring is usually the best time to buy a goat. Kids are abundant. Milking does are available. People are culling their herds (selling those animals they no longer want) to make room for special kids. A large number of available goats gives you your choice of animals.

Whether you go to look at a goat advertised on the local bulletin board or to pick one out of a

large dairy herd, don't be afraid to ask questions. The economy and pleasure of goatkeeping is related to the productivity, temperament, and health of the animal you buy. Because you are buying a dairy animal, you should first consider milk production. It varies tremendously from animal to animal and is largely a result of good or poor breeding. If you are looking at an older doe, ask for production records on her and if possible on her daughters. If you are looking at a doe kid, ask about her mother's (dam's) production and any milking-age sisters. Find out about her father (sire) too; he should be from an excellent doe. If a doe kid comes from a line of good milkers, you can expect her to be good, too. If she comes from a line of unrecorded milkers, you have no way of knowing whether she will be worth her keep (though it is a rare goat who doesn't at least pay her feed bill).

The most accurate milk records are kept by weight, in pounds and tenths of pounds. One pound equals approximately one pint; eight pounds to the gallon. A normal lactation is 305 days, so if someone says a doe gave "1,600 pounds," you assume it was over a ten month period. The average is important. Some goats will milk very well the first few months and then drop radically. Others will milk consistently, if not spectac-

PARTS OF A DAIRY GOAT

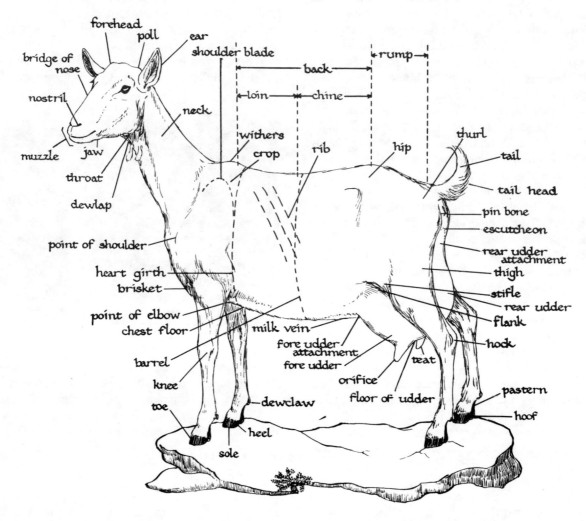

ularly. If someone says a doe gives "a gallon a day" or "this doe is giving three quarts," find out how long she's been milking and what her average production is. Ask if "a gallon a day" is actual weight or an approximation. The first month after kidding, a doe will gradually increase in her milk yield. After that, she should level and hold for as long as possible. Gradually the goat will slack off, and once she is bred (usually after milking seven months) she may drop a lot. Some goats will milk a year or two without being bred again.

Breed, age, and management will greatly affect a doe's production. Saanens, Toggenburgs, and Alpines generally give more milk than Nubians and La Manchas. The milk of Nubians and La Manchas had a much higher butterfat content so that it is excellent tasting and can be easily separated for butter, ice cream, or sour cream making. An indi-

vidual animal may be the exception to these general rules: there will be heavy milking Nubians and poor Toggenburgs around. The crossing of two purebred parents usually produces a doe kid of exceptional productivity and strength. A doe that has many crossbreedings in her past, though, is usually *not* a good goat. Size, vigor, and milk production are lost in casual crossbreedings. The uniformity of color and type in purebred animals carries with it a uniformity of other dairy characteristics. Good and careful breeding results in good animals. The practice of purebred dairy goats is not just someone's fancy or fetish.

A young doe, first or second freshening, is at the beginning of her milk producing potential. Until age five to seven, a doe increases yearly in the amount of milk she gives. She reaches a peak productivity for a few years then begins to decrease.

244

Does may kid and milk to age twelve, and some keep producing to eighteen! The value of an old, proven doe is in the kids she will give you.

The way a doe is kept will greatly affect her production too. A well-fed and healthy goat with an adequate shelter, access to salt and plenty of clean, fresh water, and space to exercise in the sun should be milking at her maximum. If the goat looks thin, has a rough coat, or appears depressed, look at her surroundings—and her—more carefully. She should be getting high protein hay and grain. If she's well fed and still thin, she may have worms (look at her inner eyelid; paleness indicates worms). If her coat is rough, she may have lice (severe infestation lowers milk production). If her shelter is drafty, she may be putting more energy into keeping warm than into milk production. A few weeks of good care (proper feeding, worming, or delousing) will restore her surprisingly.

All these things taken into consideration, how much milk does a good goat give? Fifteen hundred to sixteen hundred pounds a year is a fairly good average for a young goat. Translated into daily production, this is two to three quarts. Older does should be giving two thousand pounds or more to be considered equally good. There are many dairy goats on record who average ten pounds (five quarts) a day and more. Unless you want to put a lot of money into buying an outstanding doe kid, you should probably settle for the good average milker. By breeding her with an exceptional buck, you may produce those special kids. Also take into account that a doe for sale is usually not the best in the herd.

The goat's conformation is important. Look for strong, straight legs and a deep body (allowing for maximum feed consumption and, correspondingly, maximum production). The back should be straight. The slope from the hip bone to tail should be as gradual as possible; this allows for a large and well-attached udder. The udder itself should be shaped as much as possible like this to give maximum milk; a pendulous udder may have little capacity and is subject to injuries; a goat with very small teats is hard to milk (see illustration).

If possible, milk the doe you are considering buying. She should have large enough holes to milk easily. Don't be put off is she tries to kick you or the milk pail; she may just be reacting to you as a stranger. Even the chronic kicker (and some does are on sale for this very reason) can be cured with patience and gentleness. You should also check to see that there are no hard lumps in the udder. These indicate mastitis, past or present. Mastitis is

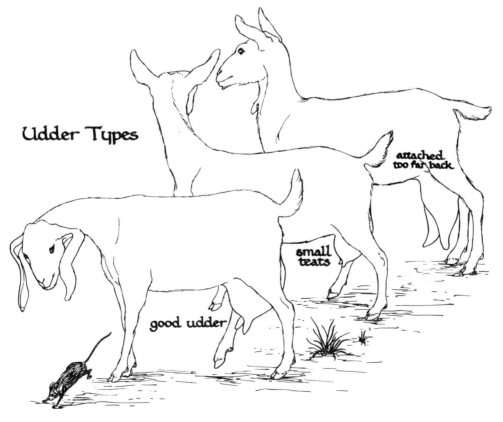

Udder Types

attached too far back

small teats

good udder

an infection of the udder which causes fever, lowered milk production, and other problems. It may be mechanically or bacterially caused. Once a doe has had mastitis, she is more susceptible to getting it again. She may also lose a good part of her udder tissue and never produce well. Generally, you should not buy a doe that has or has had mastitis.

If you are looking at a doe kid, you can't tell much about her udder without seeing her mother and/or sisters. Udder type is inherited; remember the influence of the buck too. You can check to see that the kid doesn't have double teats (an inherited fault). Widely spaced teats on a doe kid are supposed to indicate a good udder coming.

Prices, like goats, vary greatly. An excellent purebred doe kid may cost from a hundred dollars up. In our area, a grade kid sells for between twenty and thirty dollars newborn or a few months old. Registered kids or kids from very good does usually start at about fifty dollars and may be sold for a hundred or more, if they have exceptional breeding. A bred yearling will sell for anywhere from fifty dollars on up (and even cheaper if she is crossbred or of mediocre type). The price on milking does, too, varies tremendously. Depending upon their age, production, conformation, and so on, milking does sell for fifty or sixty-five to one or two hundred. There are "bargain does" around ("twenty dollars for a milking nanny"), but more often than not these does are not bargains. What you save in initial investment you lose in long-term keep. Consider the following:

(At the time this book is being put together, feed and milk prices are fluctuating pretty wildly, so it is impossible to guess figures that will be accurate by the time this writing reaches you. Use the following more as a sliding scale of values than an absolute—the conclusions are still quite valid.) Assume that it costs about $8.00 a month to keep a milking doe (feeding three pounds alfalfa hay and four pounds of grain daily, plus bedding, salt, and other incidentals). Add in $10 for the year (for uncounted extras). Therefore, it will cost about $106 a year to keep her. Milk is valued at $1.00 per gallon. The initial investment for a bargain goat is $30. She gives an average of three pounds a day or $114 worth of milk in a year. She earns her keep, pays you back half your investment. And you have her kids to sell. A good goat costs, say, $50 or $75. If she gives an average of six pounds a day, she produces milk worth about $280, which pays her keep, repays your investment, and then some. And her kids are valuable. The superior goat, chosen carefully for maximum production and giving at least an eight-pound average, may cost upwards of $100. In one year's milking, she produces $405 worth of milk. Pays for herself and her keep as much as twice over. And her kids you won't want to sell. So the higher price on the better goat really is the bargain, if you can afford her in the beginning.

Buying a kid or pair of kids is a good way to begin if you aren't in a hurry for milk, and can afford to feed them for the period of time it takes them to mature. The price of a really fine kid will equal that of a medium-priced milking doe, and your investment in raising her will pay off in long-term production. By the time she freshens, you will also know a lot about goats, their habits, and needs. Of course, if you are struggling to survive, the medium-priced milking doe or bred yearling would be a better choice. You can then breed this doe to an outstanding buck to produce exceptional kids.

Whenever you buy an animal, find out specifically what she's been eating, how often she's been fed, and so on. A goat taken from its accustomed place, people, and companions gets very homesick. You can minimize the difficulty by providing feed and a routine she's used to. If you want to change her feed, milking schedule, or bottle feeding in the case of a kid, do it all *gradually*. Also ask about any problems with her routine care (hoof trimming, extra sensitivity to cold or drafts, and so on). Get a record of immunization she has been given. If she's a mature doe, ask about any problems she's had in kidding. You might also want to find out what bucks she has been bred to, what type of kids each breeding has produced, and who has her kids. I advise strongly against moving a goat who is very close to freshening. The upset of the move could cause a case of ketosis (see "Taking Care of Your Animals"), and you might lose the doe and her kids, or at the very least have to go through a traumatic illness with her. It would be much safer to ask the breeder to keep the doe in her familiar home until after she has kidded. If you are buying a registered animal, have the breeder sign the papers over to you and mail them in (to the Dairy Goat Association) on the day you purchase and pick up the animal. You might want to check the animal to see if her tattoo (all registered goats must be tattooed) corresponds with her papers, and to see that everything is correct on her papers. Handle your new goat with extra patience and sensitivity—both of you will be happier for it.

KEEPING GOATS

Goatkeeping is not a simple matter of building a shed, fencing a pen, and buying a goat. You soon find yourself in a complex relationship with an animal whose nature is truly "capricious." A goat may be gentled into domesticity and fed and bred into productivity, but she remains of an independent and suprising nature.

My friend Catherine writes of Snowball, the Missouri Saanen, ". . . on top of that problem is her insane personality—I've never seen a goat like her; lying down while in the stanchion, trying to get the milk bowl so she can drink her own (or any other goat's) milk! (She doesn't self-suck; I think she was pan fed as a kid and is hung up still.) She has a food fetish: one day it's only pure oats, then she switches to corn and molasses, then mixed grains, then pure corn—everyday we run through this whole routine with her, offering her handfuls of this and that until she chooses her entree, fill the

bowl, start milking, bite her six or seven times (hard!) when she tries to lie down, try to rescue the milk at the same time, offer her dessert (usually the entree with molasses) cause she won't ever finish off a bowlful of grain unless she's really hungry. At the very least one has to stir it around to reattract her attention . . . And she's such a pretty animal too, very fine boned and high class in appearance. She has a water fetish also! She will only drink from a white pail or sink. The water must be from one of the three ponds (none of the other animals share her preference) and it must be immaculate, fresh and, in winter, warmed to room temperature . . ." All goats are not like Snowball (!), but each has her own distinct personality, peculiarities, and delightful individualism. It is this amazing capacity for *character* that makes goats so endearing to the people they live with, and turns many a casually begun relationship be-

247

tween person and goat into a prolonged love affair (or at least a refreshing duel of wits!). A goat can be the most charismatic personality on the homestead!

The wild goat is a nomad and browser. She travels as part of a flock led by the "queen" (an older doe who searches out the best feed and routes to follow), and followed by the buck (whose task is to protect the rear guard of the flock). Does of all ages, kids and younger bucks make up the rest of the flock. They travel together in quest of their subsistence diet of brush, leaves, bark, and so on—a diet that changes with the seasons. The goat uses her horns rarely, relying on her speed and agility to escape danger. She must be alert, sensitive, and hardy to survive. A whole set of behavioral patterns, natural skills, and needs carry over in the domestic animal. They create the particular problems and delights of goatkeeping.

Life as a communal, free-ranging animal doesn't make the goat well suited for solitary confinement. She likes the company of at least one other goat, though she might transfer her sociable affection to people or other animals. Goats do well in small flocks of various aged does. Some care must go to see that smaller animals get their share of food—at the hay feeder or grain trough, goats aren't prone to generosity. Generally, goats of similar size will live peacefully together. In larger flocks, the social order is a source of quarreling. Rank is pretty well determined by head banging jousts which try strength, endurance, and determination. Once established, a social order will hold until a new goat is introduced or until a young one begins to challenge her elders. The "higher up" goats will demand first choice at the feeders, pick and choose

the better (to their minds) sleeping areas, and so on. Perodically the queen of the flock will do a bit of jostling and head banging just to remind everyone of her position and her intention to remain in charge. This jostling will go domino-like down the hierarchy, until each member of the flock has asserted her superiority over the one(s) she considers inferior(s). Beyond these little skirmishes, the goat house stays fairly peaceful . . . goats form fast friendships with one another, and within these relationships there is a great deal of loving companionship. Friends will almost always sleep nestled up close to each other, bodies touching—they will graze together, play, and lie around chewing their cuds and enjoying the sun within reach of one another. Breaking up these friendships can be a source of great mourning. I have seen goats reunited after a separation of a year or more who greet each other with obvious joy and quickly re-establish their habits of friendship! These close friendships are almost always formed between doe kids who are born in the same litter and raised together. These sisters become literally inseparable—even to the point of having their heat periods together and kidding on the same day!

Some goats are very tolerant of and gentle with kids or small goats who are introduced to the flock. Others are dangerously rough, and may actually hurt the newcomer by smashing and banging at it. Even the most agile kid may suffer from not being allowed to eat properly. In general, it is a good idea to raise your kids separately from the older does. When they are well grown, they may be put in with the others. You can provide a "creep" area for these smaller goats—a section of the goat house that is fenced or partitioned off in a way that

It has been hitting me that everything is cumulative—that as my commitment to the land grows so does the magnitude of what needs doing and what can go wrong; as my self-awareness grows, so do the complexities of my relationships and the possibility of joy or pain. This summer has been the hardest of my life, unrelenting work, hard enough if everything had gone well. Last summer, leaving Peter, was the hardest summer of my life up till then, yet now it seems relatively simple and clear. It's hard to face the realization that my life will continue to become more difficult and complex—yet saying that helps. Much as I long for some peace and ease, I don't want to "mellow" into complacency.

allows the smaller animals in and keeps the larger ones out. Within the creep area, the smaller goats can have their own feed and a space where they can rest and sleep without being bullied. This is one good alternative to totally separate pens, but the kids may still suffer some jostling when they leave their creep. You will have to watch your goats to make the decision about creep area versus separate housing—if they are fairly gentle with the kids, the creep will do; if they are rough, it may not be worth the chance that one will really hurt a smaller newcomer.

Bucks should be kept separate from the does to prevent random breedings and most importantly to keep the "goaty" flavor out of the does' milk. The buck's odor (extreme in breeding season but noticeable year-round) is picked up by the does and will ruin the flavor of the milk. Buck kids at the age of three months are capable of breeding; they should be separated from the doe kids (who can be accidentally bred when they are as young as two months). Buck kids may also breed does that are much bigger than they are. I was once given a full-grown Nubian doe who presented me with triplet kids from the four-month-old buck kid she had been penned with! Bucks enjoy the company of other bucks and should be kept in proximity to one another if possible. If you are keeping just one buck, you might think of keeping a wether (castrated buck) as a companion for him. When it is not breeding season, bucks of similar size may be kept together. They engage in a lot of mock-fighting, so a smaller buck might be unintentionally injured by a much larger buck. The larger animals also tend to dominate the feed and even the housing, to the detriment of the young buck, who may be badly stunted from lack of sufficient feed. Once breeding season begins, the bucks' play-fighting be-comes rougher and more serious. There will be a lot of bloody heads and bruised knees if the bucks are left together. They may seriously hurt one another—I've read of bucks who have been paralyzed in "combat" with their pen mates, and have seen a buck knocked full on his side and stunned by his friend of a month earlier. The ideal living situation, I think, would be adjoining pens with separate small houses, so that the bucks could see one another and talk over the fence, but couldn't harm each other. A large common yard might be used when breeding season is over. Fencing between buck pens must be incredibly strong—our bucks have broken through the heaviest gauge woven wire fence, snapped 4×4 posts right off at ground level, and broken 2×4 boards that separated them from a doe in heat (even though she was hundreds of yards away, and completely out of sight!). The best buck fence I've come up with is one made of 5' and 7' high redwood pickets that are at least 2″ thick and 3″ or 4″ wide, nailed to 2×8 boards run horizontally top and bottom. Spacing between the pickets is about 2″—allowing the bucks to see through but not to get so much as a toe between. This fence sounds extravagant, but it is solid, permanent, buck-proof, and in this area it was free (wood salvaged from the winter beaches). Something similar could be built with saplings or recycled lumber. A final word of advice is to build a very separate breeding pen away from the bucks' regular pen(s). This will save a lot of wear and tear on the buck who is not being used with the in-heat doe, and save a tremendous wear on your fencing. It will make things calmer, too. Bucks will learn quickly where the breeding pen is and can be let out of their pens when wanted for service; mine usually beat me to the gate of the breeding pen.

All goats—does, bucks, kids—have similar basic needs. They must have a good shelter, sufficient food, clean water, salt, and enough space to exercise. Periodically, they need to have their hooves trimmed. The milking doe should be milked twice daily at regular hours. Kids may be on a three-bottle-a-day feeding schedule, which should also be regular and consistent. Goats love to go for walks, to be brushed, and just to visit with their people. Plan a few minutes or hours whenever you can to stop and scratch your buck's ears, or take your kids for a ramble in the woods, or play running and leaping games with your "old" does (you'll find they're still kids at heart!). This is one of the pleasures of goatkeeping . . . and of being in the country.

In good weather, goats will spend most of their time outdoors. When it rains, snows, or gets damp and foggy, the domestic goat flock moves indoors. Though I read praises somewhere of the "sturdy La Manchas standing knee-deep in running streams while they browsed on the stream-side vegetation," my La Manchas fairly fly to keep from getting drenched in a rainstorm, and they pick their way most gingerly over the muddy spots in their pasture! A simple shed will house a few goats. It should be as draft free and rain tight as possible. Two to four goats can live comfortably in a twelve by twelve shed. The floor of the house may be dirt, wood, or concrete. The dirt floor is my first choice for a small goat house: it is easy to clean, allows some of the moisture to seep away, and makes the building very simple. Wood is also easy to clean, but tends to get wet and stay wet (even when deeply bedded) if the flock can't be locked outdoors to allow the floor a good drying (in our rainy season, this is impractical). Concrete is the least desirable floor for the goat shed, though it is wonderfully suited for the milking room. Concrete holds the chill of even mildly cold weather, and transmits the chill up through the bedding to the goats. It is particularly bad for milking animals, as a chill to the udder can lead to a case of mastitis. It also seems that it would be hard on the goat's feet and legs. Whatever floor you choose or find yourself dealing with, use plenty of bedding and keep the bedding clean. Bedding soaks up urine, keeps the goats cleaner and drier, and makes them more comfortable. You can use sawdust, wood shavings, straw, or even uneaten alfalfa stalks. The bedding should be regularly cleaned up and replaced. Sprinkling hydrated lime on the floor before putting down fresh bedding cuts the ammonia odor of the urine and disinfects as well (the lime can be bought at most feed stores and is very inexpensive). Goat manure and soiled bedding make a wonderful garden mulch. The manure is "cold" (that is, it decomposes with very little heat), so that it can usually be put directly on the garden. If the bedding is very soaked with urine, leave it in a pile for a few days before applying it around plants—the ammonia may burn plants if put on too soon. In a day or two, it seems to evaporate to a safe point.

Along one wall of your goat shed you may want to build a sleeping bench. This is simply a sturdy 12″ or wider board from 4′ to 10′ (or more) long raised 12″ to 2′ off the ground. Many goats will sleep up on this bench rather than on the ground.

Windows will add light and a feeling of space to a small goat shed, but be sure to protect them from inquisitive hooves and noses. Sturdy, fine-mesh screen or chicken wire stretched over each side of the window will protect glass or plastic. Or you can place windows above the goats' reach. We got a tremendously light space in a small buck shed by using a panel of corrugated translucent fiber glass in the roof. It was the center panel, lapping over sheets of tar-papered plywood on either side. We drilled holes in the fiber glass with a small hand drill, nailed it down gently with standard roofing nails, then tarred all the seams generously. It is quite leakproof but tends to condense and drip some in cold, rainy weather when the animals' body heat rises and hits the cold fiber-glass surface. The light makes the house feel twice as big.

In bad weather, your goats will appreciate indoor water. Put the bucket up high enough to keep it clean, and secure it so that a playful romp won't dump it over. Change it at least once daily; empty and rinse the bucket. Goats like *fresh* water and milkers will produce more if you indulge them this way. In cold weather, goats like their water slightly warmed. We live in a mild climate and our water rarely gets icy cold, so we don't usually take this extra step. I'm sure that goats in cold climates would much appreciate it.

Most dairy goats have short coats and thin skins. They keep warm by eating and digesting roughage; the breakdown of this roughage acts as an internal heater. If goats are kept in a drafty shed, they are not only susceptible to chills (which can lead to diseases such as bronchitis and pneumonia), but they must use more energy to keep warm and therefore put less into milk production. Your feed bill goes up; your milk production goes down. Drafts are caused when air can enter at one point

GOAT AND HAY
STORAGE BARNS

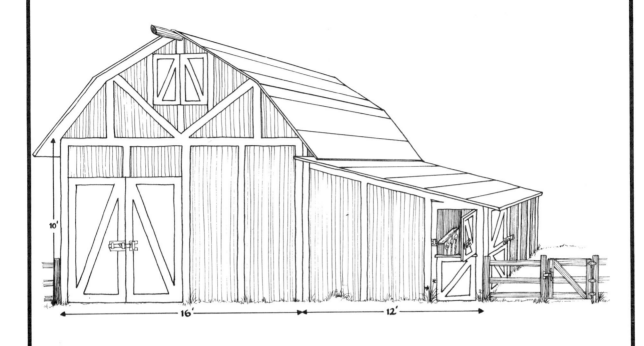

10'

16'   12'

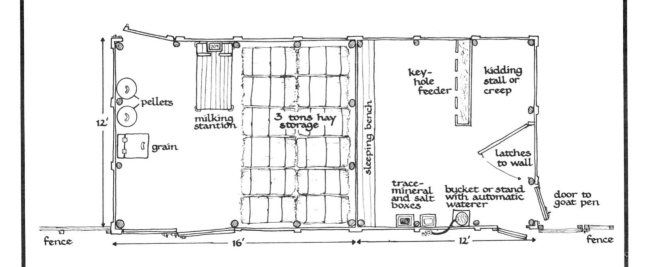

pellets

milking
stantion

grain

12'

3 tons hay
storage

sleeping bench

key-
hole
feeder

kidding
stall or
creep

latches
to wall

trace-
mineral
and salt
boxes

bucket or stand
with automatic
waterer

door to
goat pen

fence

16'   12'

fence

and leave at another, creating crosscurrents and up-and-down drafts. An open door won't make your goat shed drafty if the rest of the shed is tightly built. This is especially important if you're converting an old building to be a goat shed. Be careful not to keep your goat in too close or too warm quarters, either; she'll be chilled when she goes outside.

It's difficult to estimate how much outdoor space each goat needs. Generally, the more the better, though the goats will do well in a fairly small pen. Goats like to wander, investigate, and browse. A large area with brush, trees, fallen logs, and such would be ideal. If you keep the goats in a small area, taking them out for walks will provide them with exercise, supplemental feeding, and diversion. With exceptionally good pasture, you can graze about four goats to the acre. Tethering is a poor alternative to penning. I've heard of at least two goats in this area who strangled on tethers (one a kid, one a grown doe). The tethered goat is also at the mercy of dogs, sudden changes in the weather, and so on.

The smaller the goat yard or pasture, the better the fencing must be. Well stretched, medium- or heavyweight 36″ field fencing will keep most goats in. You can run boards or poles along the top to protect the fence when goats stand up on it (they do frequently) and to increase the height. Horizontal boards or poles make good fences if they are spaced closely enough so the goats can't squeeze through them. Avoid using barbed wire; goats can catch and tear ears, udders, or skin. Trees within the goat yard must be protected or they will be gnawed on and bark stripped. You can fence them or "paint" them with a goat manure and water mix-

ture to discourage chewing.

Many of the goat health problems you will encounter are covered in the sheep section, as much of the material in "Keeping a Healthy Flock" will apply to goats as well as sheep. Other problems are covered in the specific longer articles in the Appendix. The book *Aids to Goatkeeping* listed in Resources is one of the better reference works I've seen for this aspect of goatkeeping. I would also suggest a subscription to *Dairy Goat Journal* (see Resources) as a continually updated source of information on goat health care.

With her quick intelligence and insatiable curiosity, the goat is easily bored with life confined to pens and schedules. She learns the schedule easily and well, knows when you'll be coming to milk, feed, or water her. Beyond these routines, she needs something to do besides eat and lie about in the sun. Given a few simple elements—a seesaw, a pile of rocks, a large stump or log—goats invent and play a variety of games amongst themselves and with people. They also make use of any opportunity to create new games. The wheelbarrow you use to clean their house is perfect for leaping in and out of (and hopefully overturning). Complicated latches challenge the mind and mouth (goats can undo most any, given a fair chance). Breaking the fence brings you to fix it—with all manner of fascinating tools to be nosed about and bothered. Visitors are to be nibbled, jumped at, and generally exploited for affection. Goats are forever trying to understand things—what is it? how does it taste? what play can I make with it? Learning to communicate with them, share their romps and games and quiet contemplations, are the bonuses in this far from pragmatic relationship you have entered into.

FEEDING

A whole book, or at least lengthy chapters, could be written on feeding goats. There are as many systems and ideas as there are goat breeders. What you feed your goats will depend upon the climate and area you live in, the type of pasture you have, hay and grain available, and so on. An adequate diet of good-quality feed is necessary to keep any goat healthy and productive.

Like the sheep and cow, the goat is a ruminant. This means she is a cud chewing animal having a specialized four-compartment stomach adapted to digest roughages such as grasses, hays, brush, and silage. The goat is a browser, normally eating a large variety of weeds and shrubs, grasses, and tree leaves. Her basic diet should be mainly roughage, which can be provided by pasture, hay, hay and silage, or hay and pellets. She can convert this highly fibrous diet into usable nutrients for her body maintenance and growth. The addition of grain to her diet is necessary for maximum growth and milk production, and to balance the nutrients of pasture or hay. Salt and trace minerals will complete her dietary needs.

A goat can survive by browsing, provided there is an abundance of natural grasses, weeds, and brush, and her pasture is not overgrazed. Too many people tether their goats or confine them to small areas and expect them to live on almost nothing. The result is a thin, low-producing, and often stunted animal. The "scrub" or "mountain" goat is a small, rugged creature who can live in an area where other animals would starve—she is a far cry from the high-producing dairy goat you would like on your homestead. If you want your goat to forage for most of her needs, you must be sure she has plenty of space and a choice of year-round vegetation. If you expect to have a normal-sized goat who can produce healthy, normal-sized kids and who will provide you (and family or friends) with milk, you will have to supplement feed your goat even when she's grazing. Alfalfa pellets are an excellent supplement, containing high levels of protein. You will also want to feed her some grain—preferably a dairy mix with a good protein level. If you can plow and plant your pastures, you may be able to simply feed the grain supplement, leaving your goat to graze on a mixture of grasses and legumes that will meet her roughage and protein requirements. Each goat will need at least a quarter of an acre of good mixed pasture. At certain times of the year (when spring pasture is watery or late summer pasture is woody and indigestible or snow covers the fields), the pasture cannot provide adequate feed. Hay or pellets will have to substitute. Pasture should be regularly manured and put through crop rotation to keep it at maximum productivity. Pasture feeding is probably the very best system for the goats, providing them with exercise, a varied diet, and a more interesting life!

If you don't have pasture land or the acreage or climate to grow hay, you will probably end up buying hay to feed your goats. Legume hay (alfalfa, clover, soybean, etc.) is excellent goat feed. Legume hay mixed with grasses (timothy, sudan) is good as long as at least 50 per cent is legume. Hay comes in two-wire or three-wire bales, ranging in weight from eighty pounds to 120 or more. You can buy it by the ton slightly cheaper than if you buy it by the bale. You may also be able to make a work-trade with a local farmer helping to bale and load the hay in exchange for a small percentage of the crop. If you are buying alfalfa, try to get second or third cutting (the fields are cut five, six, or more times; middle cuttings are excellent nutritionally, with fine stems which the goats will eat more readily). Look for green, leafy hay. Avoid dusty or moldy hay; the goats may eat it, but it isn't good for them. Also watch for weeds in the hay. We've found foxtails, thistles, and all kinds of questionable plants in bales of alfalfa. Some plants may turn up that are unpalatable or even poisonous to goats. Store your hay in a dry shed, off the ground and not in direct sunlight (this will bleach out some nutrients). Hay will keep for a long time but gradually decreases in food value—so try to keep and buy only for about a year's time.

An average-sized milking doe will eat about three pounds of legume or mixed legume hay daily. A mature buck will need four to six pounds daily, depending upon his size and body metabolism, and how much he is being used for breeding. If your doe is being fed silage, she will eat about two pounds of hay and two to three pounds silage. These estimates are very generalized and you may find your animals eat a bit more or less. We feed our goats twice a day, morning and evening, removing any stems that haven't been eaten. These stems can be fed to sheep or horses, used as bedding or added to a compost heap. Leaves that sift down to the bottom of the feeder should be periodically removed (they make a good, protein-rich chicken feed). Our goats also enjoy eating some of their bedding straw or some oat hay. These both provide roughage, but aren't nutritionally adequate as a complete diet.

Many goat breeders are now feeding alfalfa pellets rather than alfalfa hay. The pellets are usually made with good quality hay and have the proper high percentage of protein that the dairy goat needs. They are easier to store and handle than loose or baled hay. A major advantage of pellets is that there is no waste of stalks—everything is rolled into pellet form and if hopper-type feeders are used, there should be very little loss or waste. Goats fed hay will waste as much as 30 per cent of their feed, so even though pellets seem more expensive compared ton for ton with hay, they actually save you money. A mature doe will probably eat about four pounds of pellets a day. If she has access to pasture, she will probably eat less. Feeding some low-protein roughage-type hay (oat hay, for instance) will give the goats extra bulk and something to do if they don't have pasture. Goats milk very well on pellets. Kids and bucks seem to like them too. Be sure all animals you pellet-feed have been given protective shots against enterotoxemia (see Appendix article) before putting them on this high-protein, concentrated feed. When you go to buy pellets, you should know that they come in two types: plain alfalfa pellets, and pellets with a mixture of alfalfa and grain. If you feed the grain-mix type, adjust your feeding of concentrates (grain mix) accordingly. Grass-mix pellets are also sold in some areas—these have a low percentage of protein and are not really suitable as a basic feed for the dairy goat. If you make a feed change from hay to pellets (or vice versa), do it slowly. Over a period of about a week, gradually replace part of each feeding with the new feed. Watch your goats to be sure that none go on a "voluntary starvation" program—most goats seem to relish pellets, but there are a few who just won't eat them.

Another alternative to hay feeding which our farm adviser praised highly was "Wonder Hay," which is the local brand name for hay cubes. Hay cubes were *not* popular with our goats, and I've talked with other goat breeders who have had similar experience. Hay cubes are, quite literally, two-inch cubes of chopped and compressed hay: neat, tidy, and very modern. Our goats had a fine time picking them out of the feeders, nosing them, tossing them in the air, and trampling them into the ground. After that, they were looking for their breakfast . . .

If you have extra space in your garden, you can supplement your goats' diet with a number of crops. Comfrey, kale, cabbage, carrots, and chard are all enjoyed by at least some goats. Most goat books suggest growing mangels or sugar mangolds as a good winter feed (don't feed to bucks; they can cause sterility in some males). If you feed any strong vegetable (cabbage, for example), give it directly after milking so it will not affect milk flavor. At least two garden products—rhubarb and beet tops—are poisonous to goats, so be sure to check one or two goat books before you try a new feed.

We have never fed silage to our goats, but have visited one goat dairy where home-grown silage was a main part of the goats' winter diet. Silage is "fodder, either green or mature, converted into succulent winter feed for livestock through processes of fermentation . . . [It] is cut fine and stored in an airtight chamber [silo] where it is compressed to exclude air; it undergoes an acid fermentation, which gives it an agreeable flavor and prevents spoiling." Silage can be made with corn, grass and beet pulp, sorghum, and so on. A suggested ration using silage is a half pound of grain mix, a pound and a half of corn silage or roots, and about two pounds of hay.

Besides her daily hay or pasture, each goat should be given a ration of grain. What she is given and in what amount depends upon her age, her size, whether and how much she is milking, and so forth. It is important to remember that grain is a supplement to a diet consisting chiefly of roughage. Grains are mixed to balance their content—fiber, fat, protein, carbohydrates, and so on. Oats, for example, are very high in fiber; corn is high in carbohydrates; linseed and cottonseed meal are high in protein. Milking does and growing kids need 14–16 per cent protein mix. Dry yearlings and bucks need 10–12 per cent protein. There are commercial goat mixes that combine grains, molasses, and trace minerals in a formula especially designed for goats. There are also dairy rations mixed for cows that come close in content and are usually slightly cheaper. Horse feeds (usually 10 per cent protein) may be supplemented with linseed meal to make a goat ration—or you may be able to buy "brood mare feed," which is higher in protein than other horse mixes. I would like to caution you about one thing if you are considering buying a grain mix: check the feed tag carefully for the presence of *urea*. This is a chemical which is produced synthetically from carbon dioxide and ammonia. It contains a high percentage of nitrogen, which ruminants can convert into protein. Urea is added to feed as a substitute for natural protein-rich oil meals such as soybean and linseed. Goats (and sheep) are particularly sensitive to urea and may be poisoned by it if too much is added to their diets or if it is introduced suddenly. Goat breeders who have lost animals after feeding a urea-containing formula caution against feeding *any* to your goats. Others suggest introducing it gradually, and not feeding too much. After reading a few articles and personal letters about urea, I decided it wasn't worth the risk—ever—and choose our grain mixes accordingly. Urea becomes toxic to the goat when it breaks down in the rumen, and one of its constituents—ammonia—cannot be utilized quickly enough by the animal. Symptoms of urea poisoning include bloat, muscle inco-ordination, uneasiness, and convulsions. The goat may die. I haven't heard of any treatment that works once the animal has developed these symptoms, so prevention is obviously the answer! To avoid such problems as urea and preservatives being included in most commercial grain mixes, you may want to buy grains separately and make up your own goat mix. *Aids to Goatkeeping* (see Resources) has a section of suggested mixes with protein balanced properly for milking does, kids, and so forth. The formula for yearling goats, for example, is a mixture of: four hundred pounds rolled oats; four hundred pounds ground barley; two hundred pounds wheat bran; fifty pounds linseed, cottonseed, or soybean meal. *Raising Small Livestock* (see Resources) also lists formulas you can make up yourself.

How much of what grain mix to feed your particular goat is an open question. There are so many variables that no formula is absolute. One to two pounds per day is a generally agreed upon ration for a yearling, a dry or pregnant doe, or a buck. The dry doe, though, may become overly fat on this ration; the buck may become thin if he is being heavily used. You will have to learn to balance your feeding to keep your goats in good condition but not overweight—this is an art that takes a while to develop. A milking doe will usually need about one pound of grain for each three to four pounds of milk she produces daily. Again, there is considerable variation possible. Some goat breeders feed their goats twice this much; others limit their does to three pounds a day regardless of milk production. Some goats will limit themselves to a certain amount of grain; others would eat until they burst (or so it seems). Until you develop a good eye for the condition and responses of your goat, hold her feeding to a tried formula.

In addition to any combination of hay, grain, pasture, and silage, a goat needs access to salt. Salt may be provided in block form or loose (placed in a box nailed high enough up to avoid contamination from droppings). You should also provide your goats with trace minerals, which may be purchased at a feed store in block or loose form.

The goats will self-regulate their intake of salt and minerals, taking what they need as their bodies dictate. Some goat breeders also keep a mixture of steamed bone meal and salt before the goats at all times, and/or a mixture of limestone and salt. Bone meal is particularly good for pregnant does. Don't forget that your kids and bucks need regular access to salt and trace minerals, too.

The final feed requirement of your goat is, of course, water. Fresh, clean water is an absolute necessity both to your goat's health and productivity. Kids and bucks likewise need fresh water daily.

BREEDING

We had three does by the time breeding season came. During our first year of goatkeeping, our goats were all pure Nubians and the man we'd bought them from agreed to loan us one of his Nubian bucks. For a month or so we let the buck run with the does. For the first time, our milk tasted "goaty" as the buck's odor permeated the barn and animals. Five months later all of our does freshened within two weeks of one another. During the does' "rest" period, the last two months before kidding, we were without milk. All at once we were overwhelmed with milk. It was the first of a series of breedings with whatever buck we could find, with little consideration for anything but getting the does milking again. It took us three years raising goats to learn that breeding good goats is a little more complicated than just producing milk. With two or more goats and a little planning, you can have a year-round milk supply. With some care as to breeding, you can produce kids that will be superior to their mothers in milk production and breed type, in conformation, and udder.

Choosing a buck with which to breed your doe takes time and energy, but it is usually an interesting and very rewarding project. Well before breeding season, you can begin to investigate the bucks in your area or within a reasonable distance. Any large goat herd or dairy usually has bucks at stud. You can also check bulletin boards at local feed stores for advertised buck service. If you can't find a good buck locally, check Breeders' Listings in the *Dairy Goat Journal* (see Resources; if you don't want to subscribe to this magazine, your library or bookmobile probably can get it for you). In spring and early summer, special "Buck and Kid" goat shows are held all over the country—again, check *Dairy Goat Journal* for listings of these events. This is an ideal place to see a large number of bucks, watch them judged professionally according to breed type and conformation, and to meet and talk with their owners.

Before you go to look at a buck (either at a show

3 A.M. Katie and Susan and I have been up all night canning—the house is full of steam and the sticky smell of too-ripe fruit, my hair is damp and frizzing all around my head, my hands feel like a nineteenth-century washerwoman's (me being the one with small enough wrists to wash the insides of the jars . . .). I am exhausted and happy and high. Katie's asleep on the living room floor now—surrounded by sixty-two quarts of pears and plums. Canning compulsion is a madness best shared. We've worked like automatons in a factory—I cut and clean the fruit, Susan packs it into jars and adds the water, Katie seals the jars and watches the kettle, all the while all of us singing, singing, working in rhythm, packing pears to a beat. One would think we were depression children, knowing the value of each bruised plum. Instead, we daughters of affluence begin at midnight to make huge pots of plum sauce and jam to use what cannot be canned whole. Living in the midst of America's waste, we even add the pits to the compost heap, mildly hoping to reap one sprouting tree from all this fruit. The country has taught us not to waste what is so generously given, and in our fever, it has given us, too, the joy of playing together while we work. Too happy to sleep, I sit smiling at the rosy beauty of these plums.

or at a dairy or herd), think about your doe and the kind of kids you want from her. What is the most important to you: high milk or butterfat production? show-quality kids? good udder attachment? an improvement in her body type? What are your doe's faults and weaknesses? If you have a doe who milks well and has a nice udder, but who has extremely weak pasterns and legs, you will want to look for a buck who is particularly strong on his legs and throws offspring who are similarly strong. If your doe is overly refined and/or very narrow in the chest, look for a buck who will broaden and strengthen her kids. If you are interested just in high milk production, look for a buck who has sired outstanding daughters (or, if a young buck, one who comes from top-producing stock). When you are fairly certain of what is important to you and the type of buck who will blend characteristics well with your doe, go and look at different bucks and talk with their owners. Explain what you are trying to correct or improve in your doe, and ask to see the buck's daughters. If he is old enough to have daughters milking, ask about their milk production as compared with that of their dams. Most people who keep bucks and have them at stud are willing to discuss their animals' strengths and weaknesses, show you the buck and his offspring, and help you make your choice.

Stud fees to a good, registered purebred buck usually range from ten to twenty-five dollars. Some breeders have facilities to keep your doe; others expect you to bring her in heat and take her home the same day. It's a good idea to make arrangements ahead of time if you decide this is the buck you want to use.

You will probably choose to breed your doe with a buck of the same breed if she is a purebred animal. If she is registered and the buck is registered, your kids are automatically eligible for registry. If your doe is not registered, her kids from a registered buck may be recorded with the Dairy Goat Association as "½ Grade Americans." When you breed the doe, you should be given a service memo by the buck's owner. When your doe has her kids, this memo will be needed to register or record them. Ask the buck's owner to explain how you may register the kids, or write to the American Dairy Goat Association (P. O. Box 186, Spindale, North Carolina 28160) for instructions and application forms. If your doe is crossbred, you will have to choose a buck of the breed you wish the kids to resemble.

Breeding season for goats begins in July or August and lasts until February or March. The season will vary according to your location: if the summer temperatures go too high, for example, the bucks

other does may mount her. Some does have a mucous discharge and/or a swollen, red vulva. The in-heat doe may fight with her friends, stand outside in the pasture and yell, pace a lot, and so on. If the doe is milking, her production will usually drop noticeably just before, during, and after her heat. In short—the doe in heat acts abnormal, and her behavior ranges from the subtle to the obnoxious (a case of this is our Alpine doe who spends almost twenty-four hours a day screaming her heat for all to hear!). Unless you have a buck nearby or are really familiar with your doe and her heat signs, it may be difficult to catch her heat. One trick that may help you is the use of a "buck rag." Keep a rag that has been rubbed on a buck's head (scent glands are located just back of where the horns would be) and smells fairly strongly of buck in a tightly capped jar. When you think a doe is in heat, present her with the rag. The degree of her interest in the buck's scent may confirm her heat. Bringing the doe to the buck will tell you for sure. Usually the doe is brought to the buck because a buck may not breed as readily in a strange, new place, and because the buck owner is not usually interested in letting her or his buck leave home. If the buck show no interest in the doe when she is put in with him, she is probably not in heat. You should leave them together for a short time to be sure. Also be certain you aren't putting the doe in at feeding time —a hungry buck may not feel particularly amorous.

A doe in heat will stand for the buck. They will sometimes arch necks and bang heads a bit, but usually he will mount and breed her rather quickly. The doe will probably wag her tail, stand by the buck and may nudge and rub him. If the doe runs from the buck, she is probably not in heat. If she runs and he pursues her for any long period of time, she is probably going into (or out of) heat and may be fully in the following day. Usually a buck will not continue to pursue a doe who is not close to her heat period. In a successful breeding, the buck throws his head back as he ejaculates, and the doe usually hunches her rear quarters forward. Semen can be seen dripping from the doe's vulva. This one breeding is sufficient to produce kids, but it is a good idea to leave the two together for a period of time and let them breed again. Occasionally, a doe will not stand for the buck even when in heat. This doe must be held to be bred— not a usual occurrence, but I have had one Nubian doe in my herd who would never stand willingly. Most goat books suggest breeding a young doe with an older (experienced) buck, and using a

may be temporarily sterile and the season will not begin until cooler fall weather sets in. Breedings late in the season may not take at all. If you want your doe bred for certain, plan on catching her late January or early February heat. On the average, a doe goes in heat every twenty-one days. Variations range from sixteen to twenty-six days. Young does (ours at least) seem to have their heats more often, and stay in heat longer. If you watch your does carefully and establish their cycles (keeping records year to year helps, as most does tend to have a fairly regular length of cycle), you can work out a projected breeding schedule. By breeding some does early in the season and others toward the end, you can have a continual supply of milk. Very early or very late heats seem to be shorter in duration and "lighter"—i.e., the signs are more subtle.

A doe's heat may last one to three days. The second day is the most fertile, and the heat signs are usually most pronounced then. Signs of heat include tail wagging (usually with the tail held flat against the body and "flagged" vigorously), excessive bleating, general nervousness, and a changeable appetite. The doe may mount other does—or

There is a war within me between farming and wildness. Yesterday we stretched the wire on the first side of the fence and felt very high: there it was—a fence! Today I walked to the pond and couldn't get there—a fence blocking my path. This place of open, golden grasses has become defined, a field. How I miss the wildness—the sense of land that lives free, belongs only to itself. And how I yearn for fields and fences and barns: white-board paddocks and a rose garden; for a haven of orderly, productive tameness among the wild. But there is not room enough here for both.

young buck on older does. I am not sure I agree with this idea—as many fine kids have come from yearling does mated with a six-month-old buck.

An alternative to this "hand breeding" method of bringing the in-heat doe to the buck is pen breeding. Here, the doe(s) and buck are allowed to live together for a month or longer. This method is easier and fairly certain, but has important disadvantages. If the does are milking, their milk will pick up the buck's odor and taste very strong. A young buck should not be used in this way, as his overuse during his first breeding season may make him partially or fully sterile. Most critically, you will not know when to expect your kids to arrive. One good use of pen breeding is in getting does settled very early in the season—particularly young does that have been held over from the preceding year because they weren't quite old or large enough. Does that are pen bred seem to go in heat earlier than normal.

If the doe has been bred but doesn't settle (conceive), she will go into heat again in her normal cycle. You will have to take her back to the buck to be bred again. Sometimes a doe will settle but will have what looks like a very light heat at the proper time in her cycle—brought to the buck, she will refuse to stand. There are pregnancy tests being developed for goats, but as of yet they are not considered particularly safe for the animal. One useful test you can do yourself is this: three weeks after the doe has been bred, check the area directly in front of the udder, or where the udder will be. In an unbred doe, this area is soft and gives to gentle

pressure; in the bred doe, the area is hard and taut. This test is not absolute, but it seems to be about 80 per cent effective—at least in the goats I've checked.

The doe that is bred repeatedly and will not settle may be suffering from a vitamin E deficiency that makes her temporarily infertile. You can try feeding her a vitamin E supplement or E-rich feed, or give her vitamin E injections. She may also have a much more serious condition—cystic ovaries. To determine this, you will have to have a competent veterinarian check the doe. If she is a really valuable animal and this is her problem, it may be worth having her operated on. She should subsequently be able to breed and conceive normally. Occasionally a doe will just have an "off" year and will not settle even when bred repeatedly with different bucks. The next year she may be bred again and conceive, carry and deliver healthy, normal kids. If you have a doe who will not settle but who has no apparent physical problem, it may be worth the wait to see if she will settle the following year. One other breeding problem is the overly fat doe—she will not settle until she has been slimmed down. Some does will also (some years) simply not go in heat. An artificial heat may be induced by giving the doe "Pregnant Mare Serum"—this stimulates the doe to ovulate and she may be bred a few days after this injection is given. This serum may be purchased from an animal supply house or be given by your veterinarian. Consider all of these possibilities before you decide you are dealing with a "barren" doe or a "hermaphrodite" kid!

kiddinG

If you recorded the date your doe was bred, you will have a fairly good idea when to expect her to kid. The gestation period for goats is five months, with a variation of about 144 to 156 days and an average of about 150 days. If you have pen bred your doe and don't know quite when to expect her kids, you will have to be extra-observant. Does "bag up" (begin to come into their milk) at various stages of pregnancy. I've seen a dry doe come into her milk a few weeks after being bred (this is unusual, but it does happen) and seen others that haven't really bagged up until a day or two before kidding. Usually, a doe's udder will begin to fill in a month to a few weeks before kidding. This is one of the most obvious signs of an impending kidding if the doe is dry and her udder has been shrunken, or in the yearling doe who has never been bred before. The size of the doe may also tell you something

about her state of pregnancy, but this can be very deceiving. Some does will grow enormous with their kids; some will barely show. Although this would seem related to the size and number of the kids in the womb, it isn't always so. It's not unusual that a *hugely* pregnant doe in our herd will give birth to two normal-sized kids, and a much smaller looking doe surprises us with triplets. I've learned to expect the unexpected!

Be prepared for your doe's kidding well in advance. If you have more than one doe, you should have some sort of separate space for kidding. Ideally, a small area of the goat shed sectioned off will allow the doe privacy without the trauma of suddenly being removed from her friends. Removable partition walls make a handy, temporary kidding stall. Make these walls low enough so that the doe can see her herd sisters. I've read that other preg-

What's happening with Susan and me is really good. Every time we get into difficulties I think will tear us apart—like my resenting her low energy or her being angry at being judged by my standards—we come through the other side stronger and clearer, forging a commitment to honesty from our shared experience. We are learning that being direct and honest makes us higher, that we trust and understand each other more. With Peter I always felt that if I was honest he wouldn't love me. With Susan I feel like the more honest we are, the happier we are together. I'm beginning to trust this way of being, to just be myself around other people and take the consequences. What surprises me is that the consequences usually feel better than the old ways ever did.

nant does should not be around when a doe kids, as they may deliver themselves prematurely—but my experiences with goats don't support this theory. Never have I seen one doe "excite" another into delivery! Our goats are usually very curious to see kids born, and will hang around the barn waiting, rather than going out to browse or sun as they normally would. If they are locked out, they will stand around outside, obviously waiting. All of our does have seemed perfectly comfortable with their friends peeking over the kidding stall—but they like the privacy of their kidding space, and will bang heads with an intruder to make this known.

Familiarize your doe with the kidding stall well before she needs it. Bring her in, let her sniff around and get to feeling at ease. I usually leave the doe in the stall a few nights before she is due. By the time she is actually ready to kid, the stall will be a place she knows and isn't upset by. The stall should be clean and deeply bedded. Straw is a preferable bedding. Pine shavings, which work so well in the general goat house, tend to get stuck all over the newborn kids and are a nuisance.

Besides preparing a kidding stall, you should have on hand a few basics: iodine, some string, clean towels, some sort of mild antiseptic, and a lubricating powder or jelly (this may be purchased from an animal supply house—it is very inexpensive). You might also want to invest in some uterine boluses, which are used to disinfect the uterus and vagina after any manual help in a kidding, and in some boluses which are used to stimulate passage of a retained afterbirth. Both of these types of boluses are available from the supply houses, are

very inexpensive, and may be needed. You should also have on hand a few lamb nipples (oversized rubber nipples that fit on most narrow-mouthed bottles) if you plan to bottle-feed the kid(s). Have a place ready for the kids in advance—a special, well-bedded, draft-free stall or box—if you plan to bottle-raise them.

A few days before a doe is going to kid, she may have a yellowish mucous discharge. Confusing this with the discharge that marks the actual breaking of the cervical seal before kidding, I've spent many a long night waiting patiently for kids that came days later! Enough experiences like this have taught me to differentiate between the two—the second, a true indicator, is much thicker and a deeper yellow. The udder of the doe will usually fill up tremendously in the last few days. Some goat owners pre-milk their exceptionally heavy milking does to save strain on the udder. This means that they begin milking the doe before she kids, saving (by freezing) the first milk (colostrum) for the kids when they come. I have never done this, even with my heaviest milkers, but it is a common enough practice. Usually on the day the doe is going to kid, her udder will become very tight, pink, and shiny. She will develop noticeable hollows in her flanks (just below the hip bones). At this point, she should be isolated in the kidding stall. She will often be found off by herself in a corner of the goat house or pasture, seeking this very privacy from the rest of the herd. You should be prepared to spend some time with her, to attend the birth of the kids and deal with any problems. Be sure to remove any water pails from the stall, as the doe might mistakenly drop a kid in one the few moments when you step out.

The thick mucous discharge described earlier will usually appear about an hour before the doe kids. She will probably be restless, pawing little hollows in the bedding and circling around a lot. She will lie down, get up, lie down again endlessly, it seems. Often she will appear to be listening for something with ears (should she have them) held forward and an intent expression on her face. Or she may look back toward her tail and call softly. Some does will press their heads against the stall (or your arm) and grind their teeth. Actual contractions begin irregularly. They will become more and more intense, more and more regular, and the doe will begin to strain. Hopefully she will be lying down by now (I've caught kids delivered into the air by confused first fresheners), and *not* with her rear quarters pressed against the stall! The doe's water will break and a fluid-filled sack will appear, stretching the canal for the coming kids. A kid is normally born with front feet first, and its nose forward, resting on its front legs. The first appearance the kid makes will be the tips of the hooves—usually white in color, and encased in a heavy membrane. Once the two front feet and nose appear, a few good pushes from the doe will deliver the kid. Sometimes the tips of hooves and nose appear and are retracted a few times before the kid is actually born. Don't be overly anxious and pull a kid who is partially out—if the front feet and head are coming in normal positions, the doe can probably deliver this kid unassisted.

A kid is born in one beautiful fluid motion: suddenly before your eyes is a perfectly formed, active little creature—the most incredible tiny goat, thrashing and shaking its head with surprise! The doe will usually move around to lick and nuzzle the kid, clearing its face and nose of mucous and stimulating it to move and breathe. This is a moment of intense delight and love for all concerned. Mother and kid communicate with soft bleats and murmurs that are quite unique to this time, much like the talk of the ewe and her new lamb. Your part in this is mostly to sit back and smile! If the mother has had a hard time and doesn't move at once to clear the little one's nose and mouth, you may help out with a towel or handful of straw which will catch up the mucous and slime. A second kid will sometimes follow the first immediately. The doe will get up, move about and paw another "nest." She may stop to nuzzle her first-born—but she may also step on it in her preoccu-pation with the coming one—so be prepared to move the kid to a safer place. Sometimes a doe will forget herself and lie down on this first kid—again, be prepared to move quickly! More often, she will arrange herself beside it and begin the groaning and straining to deliver its littermate. It is not uncommon for the second of twins or triplets to be a breech birth—born hind end first. I've seen many kids born this way, with no trouble for either doe or kid. Once the second kid is out, be sure the mother doesn't turn back to her more vocal, stronger first kid and start to clean it before the newcomer. This is a common mistake does make—common enough to warrant your presence at every kidding. This mistake can mean a chilled or even suffocated second (or third) kid. Moving it up close to the mother's nose will usually claim her needed attention.

If a doe strains hard for a long time and no kid appears, or if a portion of one leg appears and not the other, you may have to take a more active part in the kidding. Give the doe enough time to be sure she is really having trouble, but don't wait until she is completely exhausted. Most books suggest a period of half an hour to an hour of heavy straining before you intercede. I believe you should trust your own intuitions and your feeling for your animal at least as much as these figures. If your goat is really communicating pain and difficulty to you, don't wait too long to help her. Your own experience will teach you how to receive your goat's communications. Don't be intimidated into inactivity, either. If you are calm, gentle, and considerate of your goat, and if you are absolutely as hygienic as possible, you will probably be helping, not harming, this animal.

It will help to have a friend stand by to assist you. First, wash your hands and arms and the doe's vulva with a mild antiseptic and warm water. Rub a generous amount of lubricant powder or jelly on your hand. There are disposable plastic gloves most veterinarians use for this work, but I find they are more bother than help. They make it all but impossible to feel what's happening. With a large animal, such as a cow or horse, these gloves probably work better. Have your friend hold the doe if she begins to struggle a lot. If the kid is about to be delivered, you should feel first in the vagina a pair of tiny, hard hooves. If you follow the legs up, you should next feel the nose. If the nose is not within a few inches of the hoof tips, chances are that the kid has its head bent back and that is what is preventing the birth. Don't be afraid to reach in quite far to reposition or locate a kid. The doe's passage is quite elastic and your hand and arm will fit pretty easily. Think *through* your hands. Try closing your eyes and concentrating totally on what you feel. If the kid has its head bent back, you can push the kid up toward the womb a little and straighten the head. Often, one leg will be found bent back and must be straightened. Sometimes, both head and leg will be bent back. Work slowly, calmly, and above all gently. Don't force or rush anything. Repositioning a kid is usually fairly simple.

Now and then there will a be a really difficult situation, as when twins get intertwined and stuck in the passage. Confronted with a tangle of legs and heads you may have a moment's panic. Relax and remember that the doe can't deliver these kids without your help, and that you *can* help. Concentrate first on understanding what is actually happening. Locate each head and the positioning of the legs as well as you can. It helps to remember

that the front legs bend back at the knee and hind legs bend forward at the hock (or knee). Also the hind leg has a little bone which protrudes at the hock like this (see illustration). I learned this from an artist's anatomy book and put it to work the very next time I had to deliver a very *confused* set of kids. Once you understand the positioning of the kids, you can move to untangle them and to gently push one back toward the womb so that the other can come out.

If you ever have to pull a kid, do it very carefully, working *with* the doe's contractions. Pull down toward the doe's hocks and never pull hard. Usually once a kid is arranged correctly, the doe can expel it herself. Sometimes, though, a doe will be exhausted and/or her contractions will become so mild that they won't force the kid(s) out. In this case, you should take a firm hold of the kid's legs and pull it very slowly and gently. The head and shoulders are the hardest to clear—after they are out, the rest comes easily.

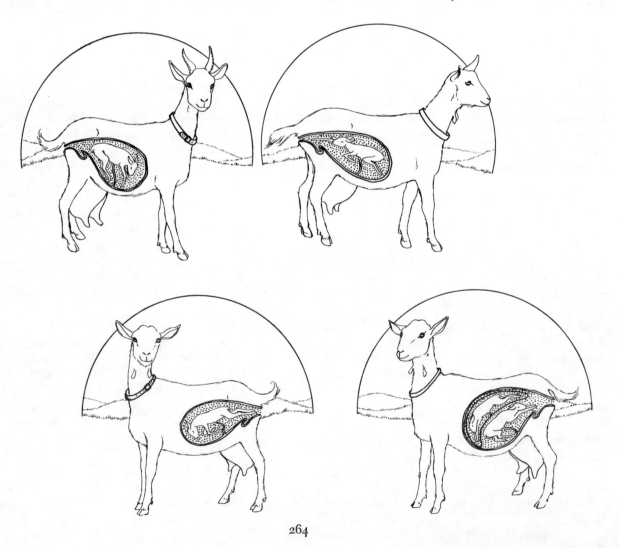

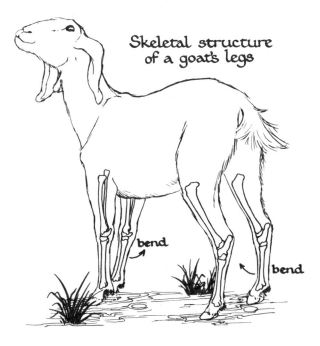

bend

bend

If a kid is coming rear end first (a common position in twins), you should be certain that the kid is not allowed to remain half-delivered for more than a few seconds. Pressure on the umbilical cord may cut off its supply of oxygen and suffocate this kid. A kid that is coming head first will not suffocate this way because it can open its mouth and breathe.

It may be hard to tell when a doe is finished with her kidding. She will usually deliver the kids all within an hour or so of one another, but I have seen at least two does who each waited four or five hours before delivering their final kid. The doe who is finished usually looks very sunken and will be on her feet, attentive to her kids and no longer "preoccupied" looking. If you suspect that a doe has another kid, wait a long time to see if it is delivered normally. I have had experience with one old Nubian doe who actually "retained" one of her kids. She had no further labor, seemed to be done after delivering two huge, healthy buck kids—but she had a strange look in her eyes. It took me a good eight hours before I decided to trust my intuitions and threw the doe down for an examination. She was still quite dilated and a half-arm's length up in the passage (still partly in the uterus, it seemed) was a pair of tiny hooves! I delivered this surprise—a beautiful little black doe kid—with no help at all from the mother! You should know that once the doe has delivered all of her kids, the cervix begins to close up and the birth canal narrows. If not—your intuitions may prove you right!

Whenever you introduce your (even carefully washed and disinfected) hand into the doe's vagina, you run the risk of introducing dirt and bacteria which will give her an infection. It's a good idea to use the uterine boluses mentioned earlier. When the doe has finished kidding, insert one or two boluses well up into the vagina or into the uterus if you can. These will break down and disinfect the uterus, preventing that infection. There are also the boluses used to stimulate the passage of a retained afterbirth. Normally a doe will pass her afterbirth within a few hours of kidding. There should be one placenta per kid, though *identical* twins may share one. You may see the doe pass these, or find them in the stall, or see her begin to pass them. Some does will eat all or part of the afterbirth. If you have watched a doe carefully and

Winona's kidding today was difficult and long, the second kid's legs bent back beneath her. I sat beside Winona aching with each contraction, empathetically sharing her labor and realizing the limits of my knowledge. Finally I realized that I would have to go in and help pull out the kid. There is such a vast gap between my book learning and the experience: between the drawing in the book and the actual feel of that confused, contorted mass in the vagina. Once I figured out which were the front legs and straightened them out, the kid came quickly and Winona seems to be doing fine. There is so much that I have had to learn by doing; no matter how much I read, it's never the same as the reality. But it's a hard way to learn when the life of an animal I love is at stake in the process.

know that she has not passed her afterbirth twenty-four hours after kidding, you should use these boluses. If you have had to help with a difficult kidding, there is a chance that the placenta tore abnormally, leaving pieces attached to the wall of the uterus. This is another case where the boluses should be used, and the danger of infection attended to. In over fifty kiddings, I have only seen one case of retained afterbirth—and this in a young doe who had had a very difficult kidding involving a partially reabsorbed fetus. The afterbirth tore and came out in shreds. Despite the use of the boluses, this doe retained a good deal of material that began to decompose and gave her a serious infection. She was feeling poorly, not eating, running a high temperature and discharging an evil-smelling combination of blood and pus. I called our veterinarian in to flush out this doe's uterus and give her intravenous antibiotics. With a lot of care and special treatment, this doe recovered. If you've had to help a doe with her kidding, watch her carefully for the next few weeks for signs of infection. She may have a normal discharge of blood and mucous for weeks after kidding—but any sign of pus or any bad smell indicates infection. She should be treated with antibiotics and possibly attended by a veterinarian.

Once your doe has delivered her kids and passed her afterbirth, you can give her a bowl of warm water and molasses (a quart of water to about a third of a cup of molasses). Most does will slurp this up eagerly. It is a ready source of energy and iron, gives extra body warmth, and tastes good! You can then proceed to milk the doe, if you are planning to bottle-raise the kids. The first few milkings, you should take out only enough for the kids —a few ounces for each. Taking too much milk from the does will predispose her to milk fever or mastitis (see Appendix articles) or result in a congested udder. Simulate the kid's nursing by milking the doe a little each time, but many times a day. After two or three days, you can cut down on the number of milkings and begin to draw off more milk each time. By the end of the third or fourth day, the doe can be on a normal twice-a-day milking schedule, and milked out fully each time.

As important as frequent milking is careful feeding. The doe should be on a light diet of good quality hay (or pasture or pellets and hay) and a very gradually increased supplement of grain. Too much grain in the first few weeks of milking can create udder congestion. Some breeders feed a special light grain mix (wheat bran and rolled oats with molasses) for the first few days after kidding. Extra attention to the feeding of this newly freshened doe will pay off in her health, contentment, and long-term production.

Yesterday was the Anderson's annual baseball game—a Norman Rockwell piece of Americana, complete with Kool-Aid, chocolate cake, watermelon, and all the neighbors. Jack Winters pitched and you could still see the lines of a fine athlete in his seventy-five-year-old body. I struck out continuously but loved the whole event. There was three-year-old Juli swinging her bat just enough so Jack couldn't hit it with the ball, Rainbow getting a home run on errors, Leila losing the ball between her feet.

After the ball game, we sheared Max. I discovered I've learned how to move swiftly and competently with the shears. I got a smooth cut without hurting him or having to cut twice. I've gained so many skills this year that self-sufficiency has taken on a whole new meaning!

RAISING A KID

If you are present for the birth of your kid, its care begins at once. As soon as it arrives, check to see that its nose and mouth are clear of the mucous membrane it has been encased in. This sack will usually break when the kid is born but may need to be torn and removed. Usually the first movements of the newborn kid or the movement of the doe will break the umbilical cord. The cord will tear, leaving anywhere from an inch or two to half a foot attached to the kid. Some blood will flow from the cord. As soon as the umbilical cord breaks, dip the end of the kid's cord in iodine. This is an important step in preventing disease: the end of the cord, like an open wound, can allow in bacteria and subsequent infection. If the cord doesn't break naturally, you can tear it. The safest way to do this is to tie a thread around the cord a few inches from the kid's body and another thread a few inches below. Tear the cord between the two threads. It is essential that you tear, rather than cut with a knife, for the tear is ragged and uneven like the natural break. The clean cut of a knife could cause hemorrhaging. Remove the threads and dip the kid's cord in iodine as soon as the blood ceases to flow. The cord will gradually dry up and fall off, leaving a

healed stub. Check periodically to see that the cord is shriveling up, and that the navel has not become infected. If an infection occurs, the area right around the navel will swell. Yellow pus will probably leak from the skin just around the cord. This whole area should be cleaned and drained as well as possible, and treated with iodine or antibacterial cream. Repeat treatment until the infection subsides. If the infection has become severe and the kid has systemic symptoms (loss of appetite, fever, and so on), you may want to also give this kid a series of broad spectrum antibiotic injections (such as Combiotic).

The newborn kid is slimy and wet—susceptible to chills in even the nicest weather. Usually the doe will lick her kids clean and dry. If she has more than one, or the weather is especially cold, you may help her out. If the doe has had abscesses, you should not let her clean her kids. Some abscess-causing bacteria can be passed in the saliva from the infected animal's mouth into the cord or the mouth of the kid (even if the doe is not showing signs of abscesses when she kids—a history of them makes her a possible carrier). The kid cleaned by this doe may harbor the bacteria for these abscesses, and get them as an older animal. Rub the kid you are cleaning all over with a towel, crumpled newspaper, or even straw. This not only dries the kid, but stimulates its circulation. If a kid becomes really chilled or has been born out in the rain (I rescued one kid at a friend's house who had been dropped in a puddle!), you may need to dry it well and place it under a heat lamp or by the kitchen stove. Whenever you move a kid from place to place, wrap it in a towel or jacket to prevent chills. Avoid moving kids from really warm to cold areas. Try to make temperature changes gradual. If you are living in a really cold area, you may need a special stall with heat lamps to keep your newborn kids from freezing. Be sure you are there for the birth of all kids to prevent such a tragedy—even the kid that has been licked dry by its mother may freeze in severe weather.

Most new kids are on their feet—if somewhat unsteadily—within ten or fifteen minutes. They begin to search for milk almost as quickly, bumping with their noses at whatever is closest. Occasionally a kid will be born with really weak legs and be unable to stand, though it will try. You can make splints out of cardboard or balsa wood and tape. Usually weak joints correct themselves in a week or two. They can be the result of poor nutrition during the doe's pregnancy, or they may be hereditary. Give the kid's legs ample time to strengthen before you make any conclusions. Sometimes a very large single kid or an overdue kid will have spongy cushions of overgrowth on the bottom of its hooves. These cushions may make it difficult or impossible for the kid to stand and walk. It helps to stand the kid on any surface that will wear down the cushion—our wood floors have worked perfectly. In a few days that "deformed" kid may be bounding around perfectly normally.

A kid should receive its first feeding within an hour or so of its birth. And depending upon how you plan to raise your kid, you should begin a consistent feeding program. My choice for kid raising has been to bottle-feed the kids from the very beginning. It takes a lot of time and energy but seems worthwhile. A bottle-fed doe kid is more than a pet —her relationship with you grows and deepens until as an adult milker she is perfectly at ease, trusting and calm with people. The bottle-fed buckling or wether, similarly, will grow up liking and trusting you. A kid that is left to nurse its mother seems to remain a little wary of people forever—even when it has been handled regularly and been part of your herd for years. This, at least,

has been my experience and the experience of other people I've talked with. This basic wariness may make the goat difficult to catch, hard to manage, never quite trusting. Bottle-raising kids is also a way to protect your doe's udder. A nursing kid butts and punches the udder to make the milk flow. As the kid grows, its punching is more and more vigorous. A doe who has been bred for high milk production has an overdeveloped udder that may be damaged by this punching. Mastitis (an infection of the udder that can harm the doe, decrease milk production, and destroy udder tissue) may be the direct result of the natural act of nursing a kid. The well-bred dairy goat is no longer really a "natural" animal. And, finally, if you are bottle-raising your kids, you can keep accurate production records on your does. You are also very aware of each kid's growth, health, and personality.

We've tried some alternative methods before settling on bottle-raising as *the* way. Leaving the kids with their mother for the first three or four days allows them to nurse often and encourages the doe to come into her milk slowly. The separation can be traumatic, though, and it is harder for the kid to accept the bottle. A kid that has been allowed to nurse may also try to nurse its mother or other does when it is put with the herd—this can cause injuries and mastitis, if not just annoyance. Pan-feeding is another alternative we tried. It is faster and somewhat easier than bottles, but the kids tend to gulp down their milk (bad for digestion) or step in the pan.

However you decide to raise your kid, it is essential that it have the first milk from its mother. This thick, sticky, yellowish milk (called colostrum) is rich in protective antibodies, minerals, and vitamins. It is also slightly laxative and acts to clean out the kid's system. If a doe dies during kidding or freshens with a bad case of mastitis or congested udder, you may have to use substitute colostrum for her kids. If you have a freezer, keeping some colostrum frozen from an earlier kidding doe is a good precaution. If you can't do this, you may have to make up a formula (see "Raising a Bummer Lamb" for a formula that may be given to the kid as well). If you are bottle-feeding your kids, milk out just enough colostrum for each of the first few feedings. A few ounces per kid will be sufficient. Heat it in a bottle set in a pan of water (colostrum scorches easily, so be careful). A goat's body temperature (102°–105° F) is higher than a human's, so the milk should be quite warm. Use a lamb nipple —a large rubber nipple that fits on a small-mouthed bottle. The first few feedings can be very trying, so start out with your patience level high. Some kids will just grab onto the nipple and guzzle —others must be "taught" to suck. Dribbling a little milk on the kid's lips usually gives it the idea. Be sure that the milk isn't too hot or too cold— either of these mistakes will make it more difficult to get the kid to nurse. If the kid is really unable to get its co-ordination together enough to nurse, try seating it in your lap and guiding the nipple into the corner of its mouth. A gentle pressure of the nipple pressed up against the roof of the kid's mouth sometimes encourages the reluctant one to take those first few sips. I don't know why this works, but it does! When you are bottle-feeding a kid—from the first few times on through the months ahead—always be careful to hold the bottle at a sharp angle to prevent the kid from sucking air. Also take the bottle away as soon as the last drop of milk is gone. Inattention to this can be fatal (see "Bloating" below). Bottles and nipples should be kept absolutely clean, and milk should be fresh, given at the proper temperature, and given on a regular schedule.

For the first few days, a kid should be fed small amounts often. Five or six feedings of a few ounces each during the day and one or two (or more) feedings during the night is right. Keep the kid on its mother's milk for the first four days at least; the doe keeps producing valuable antibodies for that long. After that period, kids can receive milk from other does or be switched to powdered milk mixed with water. Be sure that all changes are made *gradually* over a period of days to allow the kid's system to adjust. If you are going to feed a powdered milk, be sure that it has enough vitamin D or your kid may get rickets (see "Lambing" articles). Feed stores carry special formula dry milk for raising lambs and calves. Buy the lamb-type milk as it is richer in fat and more suited to the goat kid. The fat in whole cow milk is indigestible for a goat kid, so if you must feed cow milk, be sure it is skimmed. We've raised kids on powdered milk, powdered milk and goat milk, and plain goat milk. The goat-milk raised kids have been noticeably healthier and have grown better. Of the various brand name powdered milks available, I would recommend Land O' Lakes Lamb Replacer. Kids do much better on this brand. A word on overfeeding: this is the quickest way to kill your kid! A kid that is not regulated as to amount will almost certainly overeat. Watch a ewe nursing a lamb or a doe nursing a kid someday—the mother allows the

baby to suck for just a few minutes, then moves away to prevent it from gorging. Kids that are overfed can die from enterotoxemia, bloat, and other digestive problems. It is always better to slightly underfeed the kid and to feed it small amounts often.

After a lot of reading, talking with other goat breeders, and experimenting with our kids, we settled on a fairly stable kid-feeding program. Kids are fed a few ounces every few hours the first two days; gradually the amount per feeding is increased, and the number of feedings per day decreased. After four or five days, the kids are cut down to three feedings a day. All kids are given a certain set amount per feeding—some may drink all of this and others leave a bit, but no kid is allowed more than the set amount. Over a period of about three weeks, the amount of milk given per feeding is increased slightly until each kid gets a maximum of a half gallon per day (divided into three feedings). When a kid is two months old, it is put on two feedings a day of one quart per feeding. At three months, the kid is cut to one quart a day and given that until it is from three and a half to four or even five months old. Some kids will wean themselves early; others act as though they'll never be self-weaned. Buck kids who are being raised for breeding stock are usually bottle-fed until they are at least six months old. Almost every breeder I've ever talked with has a different kid-raising schedule, but most seem to agree to about a half gallon per day per kid and *at least* three months of milk feeding.

Back to the newborn kid . . . it should be kept in a draft-free place with plenty of bedding (straw is fine) that is kept clean and dry. Change the bedding regularly, as it may look dry on top but be soaked beneath the surface layer. Kids may be kept in a barn or shed if the weather isn't too cold— otherwise, they may have to be kept in the house or in a heat lamp-warmed stall. Kids will sleep together for warmth, so a single kid may need extra consideration. I like to keep kids in the house for their first week—putting them outside on sunshine-warm days. A kid needs exercise, fresh air, and sunlight. It will sleep a lot during its first few weeks, but should be full of bounce, energy, and curiosity when awake.

Check each kid for physical defects soon after it is born. Does and bucks should be checked for double teats or double holes in teats. A double teat is actually two that have grown together. A kid may have one double teat and one normal. The

Susan has decided to stay for another year, to put off going back to school at least that long. There have been many subtle changes between us lately; the balance of who dominates in work situations keeps shifting all the time—whoever has more energy or cares more, leads. When one of us feels low or inward, the other just keeps everything going; the chores get done, the garden cared for. She has begun initiating projects, making dreams of her own. The horse pasture and the just-begun chicken coop are all hers. We are finally becoming partners in the land. I'm finding I'm not afraid to share control with someone who loves it and feels responsible and doesn't need to be in control either, and perhaps without realizing it, we've created space for others to live here too.

I keep saying "I wish I had a lover," yet today I'm realizing that I've never been as equal and as comfortable with anyone as I have with Susan. And that's at least as valuable as being lovers. Our love and trust are deep, in a strange quiet way that is easy to forget. She has taught me a lot about being generous and about compromise. When I don't feel it's my whole identity and independence that's at stake, it's easier to bend and give. I'm relearning to be in tune with someone else's needs, but for the first time without giving away myself. That self I carry with me into new encounters.

double teat may or may not have an extra orifice— if it does, this kid will be a problem when she grows up (her milk will squirt in two directions at once, which makes things very messy and difficult). The double teat may be corrected surgically, but you should probably consider raising this kid for meat or selling it to someone for that purpose rather than raising it as a breeding animal. A kid may have extra teats, which are actually a second, smaller set of teats located close to the larger, normal set. These extra teats may be clipped off the very young doe kid, and she will grow up to be a normal-appearing milker. Both of these problems are inherited, though, so it may be wiser to cull this little animal from your herd, and not to repeat the breeding that produced her. Bucks with either of these defects should definitely not be kept as breeding animals, as they will transmit this fault to their daughters and may produce a number of problem animals. Bucks should also be checked for two normally descended testicles. If one or both of the testicles are not properly descended, the buck has a condition called "cryptorchism" and may be partially or totally sterile. A buck with an over or undershot jaw, crooked face, or improper breed characteristics or coloring should not be kept as a breeding animal. Occasionally a kid will be born with external genitalia of both sexes—a true hermaphrodite—or have the outward appearance of one sex and the internal makeup of the other (pseudohermaphrodite). Hermaphrodites are supposed to occur more frequently in goats than in most other species of livestock, but I have never seen or heard of a real case. The one "hermaphrodite" I've encountered was a very blocky, heavy-set doe that a friend of mine purchased. This doe mounted her pen mates and generally behaved very aggressively. An old-time animal dealer offered my friend fifteen dollars for the animal, saying that it was most certainly a hermaphrodite. My friend sold it, thankful to be rid of the problem for even a low price—and six months later the "hermaphrodite" presented her first set of triplets! I've been wary ever since. The incidence of hermaphrodites born to hornless does mated with hornless bucks (this refers to the naturally hornless not the disbudded animal) is supposed to be quite high.

Kids begin to nibble at hay or grain or browse when they are only a few weeks old. They should have access to good-quality hay at this age and be given a tiny bit of grain daily. Feed a 14–16 per cent protein dairy or goat ration to your kids. This should include trace minerals. If it doesn't, provide these free choice to the kids. A kid must also have access to a salt lick and to fresh, clean water (it will learn to drink by itself as it gets a bit older). It is safe to give the kid all the hay it wants—for the first six months, it will eat a pound or less a day depending upon its size and the availability of other roughage, pasture, and so on. From six months on, the kid's needs will gradually increase until it is on a regular ration of three or more pounds of hay a day, or equivalent pasture or pellet/hay mix. The amount of grain fed to the kid should be gradually increased as it grows, until it is receiving about a pound a day as a six- or seven-month-old. This ration may be fed to the doeling right through until she is bred, providing she does not get too fat on it —or too thin. You will have to adjust your feeding program according to your individual animal and the type of feed you are offering. Our young goats do well on free-choice alfalfa hay, a pasture with browse, and a daily feeding of about a half pound of alfalfa pellets and a pound of grain per animal. This is all adjusted to the weight and growth and condition of the animals. Other breeders I know feed free-choice alfalfa pellets and browse, with or without grain; or alfalfa hay and grain; and so on. Your young goat should be well filled out but not overly fat. The best feed for developing a doe kid's body, her capacity to eat and later to produce milk is not a lot of milk or grain, but plenty of good-quality high-protein hay (or pellets and roughage). Remember that feed will greatly influence the size and ultimate productivity of the kid you are raising. A doeling with the very best inheritance for milk production will never do her best if she is not fed well throughout her formative years. Does and bucks can be seriously stunted in their first two years by improper feeding. Give your kids the very best start you can.

Within the first few weeks of its life, your kid should be disbudded if it is a naturally horned animal (see article on Disbudding), and bucks that are not to be kept as breeding animals should be castrated (see article). When the kid is a month old, it may begin to receive its regular immunization against such diseases as tetanus and enterotoxemia (see Appendix articles). While your kid is still young you will be teaching it (consciously or inadvertently) many habits that will stay with it as a mature animal. You will probably want to teach your kid to lead so that it will be easy to handle in this respect as an adult. If you want to show the kid, it will have to learn to lead while young, too. Use a soft collar and rope and be gentle. Usually a kid will trot along willingly with its person, so the collar and rope will be used in conjunction with this natural willingness to follow. Take your kid walking—goats are fine companions in the woods or fields, full of energy and interested in everything. If you plan to show this goat, teach it to walk and stand *before* you ever enter the show ring! Many a kid places low in its class because it balks or refuses to stand in place. Kids seem to enjoy the extra attention of being groomed and taught to show, but their independent nature can make this a lively experience for both of you! You should begin hoof trimming lessons early, too (see article for details on the actual trim). Your kid may not need its hooves trimmed until it is a few months old, but it will be easier to work with if the experience of having its feet picked up and handled is not a new one at that time. Goats love attention, and goat kids thrive on it. Once you've begun kid raising, you won't need any urging to spend a lot of time with your little one . . .

KIDS' HEALTH CARE

The best way to keep your kid well is to avoid the conditions that would make it ill. Kids are very susceptible to chills and dampness, certain bacterial infections that thrive under these conditions, and problems related to sudden changes in feed. Unclean stalls, dirty bottles or feed pans, or improper feeding are all likely to create health problems. If a kid isn't feeling well, it's usually very obvious. A kid that stands around with fluffed-up fur, won't take its bottle, or just looks depressed is cause for concern.

Scouring, the stockperson's term for diarrhea, is a fairly common and usually not too serious condition. A kid may scour from a chill, from a dirty bottle, or from a bite of something it shouldn't have eaten. If the kid appears well otherwise, make sure its next bottle is properly heated and clean, and feed it slightly less than normal. A pinch of powdered ginger in the milk will help clear up a mild case of scours. If the kid is old enough to be browsing, blackberry leaves (if you have them) are a good cure. Scouring that is really profuse, persistent and accompanied by other signs is more serious. If you know the kid has eaten something unfamiliar or spoiled, or has eaten too much grain, you should probably give it a dose of mineral or castor oil. Half an ounce will be plenty for a young kid. Mineral oil flushes the system of toxic material and soothes the lining of the intestines. In the case of a real poisoning, give the recommended antidote at once. A universal antidote (egg white, milk and/or inactivated charcoal) may have to do. A lump of washing soda will make the kid vomit. Sometimes black coffee helps the poisoned animal if an antidote is not available. Kaopectate (the veterinary type and the type sold for people are identical) may be given to a kid with persistent scours, though it is not wise to try and stop the symptom (scouring) before you understand why it is happening. Scouring maybe a symptom of a serious disease such as enterotoxemia or pneumonia; it may be an allergic reaction to some feed (I've seen our kids scour chronically on a particular brand of milk replacer, and become perfectly healthy when switched to goat milk or another brand of replacer). Scouring may also be a sign of serious worm infestation, but this is not likely to occur in a very young kid, and it will be accompanied by other signs (rough coat, pale inner eyelid, and so on). Most often, scouring is a temporary reaction to a temporary situation which you should be able to discover and eliminate.

Bloating is another reaction to spoiled feed, too much grain, or to sucking air when drinking a bottle. Bloat appears in the kid as a dramatically distended stomach, particularly large on the left side. The kid will be obviously uncomfortable, if not actually in pain. It will usually stand still, reluctant to move, and may cry. A kid bloats when the air expelling mechanism of the stomach doesn't function properly and gas builds up in the stomach (see Appendix article). Treatment should be immediate. Give the kid a few tablespoons of mineral or peanut oil, and follow this with a teaspoon of baking soda dissolved in water. Most kids will drink this willingly from their bottle. Then force the kid to walk around. Stop now and then to massage its stomach and let it rest. You can repeat the oil and soda/water every fifteen minutes or half hour until the kid begins to expel some of the gas built up in its stomach. It is essential to keep the kid moving until the gas is expelled, as what it wants to do is just lie down (which furthers the problem). When the kid has deflated to almost normal size and looks as though it is feeling better, you can relax your vigilance—but not until. Bloat can kill the kid if not treated thoroughly. Try to figure out what caused this problem and avoid a repetition. The most common cause of the bloat I've seen in kids has been sucking air while drinking a bottle. This may be caused by faulty feeding, but often it is the result of tiny holes that have worn in the rubber nipple. Check all nipples regularly for this problem. Once the kid is recovered, it should be kept on a light diet for the next few days, and watched carefully. The kid that has once bloated seems to be much more susceptible to this condition. The mineral oil may give the kid a mild case of scours, but these will clear up in a day or so.

Pneumonia is another problem I've seen in kids. The kid will refuse its food, and stand with its fur fluffed up. It looks miserable. In kids, pneumonia develops rapidly and is often fatal if not caught and treated in time. One definitive sign of pneumonia is a highly increased respiratory rate—the normal rate of twenty to twenty-four per minute may be tripled. Breathing is usually audible, raspy, and difficult as the disease progresses. The kid's temperature may be above normal in the beginning stages, or subnormal as the disease becomes more serious (a kid's normal body temperature will usually run slightly higher than the older goat's). The kid suffering with pneumonia must be kept warm in a heated box or stall—use hot water bottles if you don't have electricity, and blanket the kid snugly. Feeding should be frequent, with special care that the kid gets plenty of fluids in small amounts. Combiotic (an injectable penicillin/ streptomycin combination) should be given according to directions. With careful nursing, the kid may recover in four to six days.

Enterotoxemia in the kid is also called "pulpy kidney disease." It is best avoided by regular vaccination of pregnant does (see Appendix article). The sudden death of a healthy, fast-growing kid in an unvaccinated herd is often the first sign of this disease. A kid suffering with enterotoxemia has severe, bad smelling scours and may stagger or fall down. It will be very depressed and may cry with pain. In the final stages, it may have convulsions. A kid may have a mild or "subacute" case of enterotoxemia and linger for weeks with intermittent scouring and general weakness and loss of condition. It may die of this or recover after a long period. The kid with either a severe or subacute case may be treated with antitoxin and sulfa drugs and may recover totally.

Tetanus may be an indirect result of horn disbudding or castration of kids. The first sign is a stiffness of the jaws and neck. The kid will refuse its bottle, or try to nurse and be unable to. It may hold it ears or tail stiff, and look preoccupied. For more details on this disease, see the special article in the Appendix. Tetanus in kids is better prevented than treated. Protect the kid via immunization of its mother before she gives birth, and/or give the kid an injection of tetanus antitoxin at the time of disbuddng or castration.

Goat Pox or a similar condition seems to affect certain kids and can become quite a problem if not treated at once. Tiny blisters or pimples develop on the bare skin around the anus. These may break and develop into a raw, bloody area that is very sore. Treat the blisters as soon as they appear by applying tincture of iodine to each one daily until they dry up and disappear. Gentian violet may also be used. In severe cases, an antibacterial ointment called Panalog (brand name) is the best I've found. Repeat treatment with this ointment every few days until the condition clears up. Be sure the kid is being kept in a dry stall or box, as something called "urine scald" may occur (with similar symptoms) in kids kept on wet bedding.

Parasites, internal and external, may affect the health and growth of your kid if not attended to. See Sheep, "Keeping a Healthy Flock," for a discussion of parasites—most of which are similar in these two species of animals. Goat kids may be wormed regularly with commercial or herbal wormers, particularly if they are on pasture. Lice must be treated immediately if found, as they can make the kid anemic and stunted if infestation is severe.

disbudding a kid

The wild goat uses its horns for protection. For the domesticated goat, confined with other animals and handled by people, these same horns can be a dangerous adornment. Horned goats can unintentionally (or intentionally) hurt their companions or accidentally hurt the people who work (or play) with them. A horned goat also has an unfair advantage over hornless animals and will usually use her leverage to bully the others. Horns can hook in feeders and fences, leaving the animal helplessly caught. For all of these reasons, most goat kids are disbudded a few days after they are born. A mature goat may be dehorned (the horns cut off), but this painful and bloody operation is best avoided by early disbudding of any kids you have.

Disbudding is the process of burning down the horn buds of a young kid. It prevents the horns from growing and is, in the long run, the easiest, safest, and least painful way to dehorn a goat. I have always used an electric disbudding iron (available from a veterinary or stock supply house, and costing about $10 or $12). This method is simple and seems more humane than using caustic paste or allowing the horns to grow in some and using an elastrator (a rubber band that fits tightly around the base of the horn, cutting off circulation and causing it to atrophy). If you don't have electricity, you can make or buy an iron that may be heated in the fire and used in the same way as the electric iron.

The first step in disbudding should be making sure that the kid is protected against tetanus. This may mean giving it tetanus antitoxin, or simply having the foresight to have vaccinated the mother (see Appendix, Tetanus article). I like to disbud kids when they are from five days to a few weeks old. An older kid may be disbudded, but it is more traumatic for the kid and harder for you. When the kid is four to five days old, examine its head for horn buds. Tiny swirls of hair are present over each bud. Naturally hornless kids will not have these swirls—the hair will lie flat in place. Sometimes a kid will not develop its horns until it is almost a month old (this is rare, but I have seen two kids I thought were hornless begin to show the tiny horn nubs when they were this age). If your kid is horned and you can see the swirls and actually feel a tiny bump under each one, it is time to disbud.

Generally, buck kids develop their horns much more quickly than does. Some buck kids are born with tiny horns already showing. I usually wait on these until their litter-sister(s) is ready, and disbud all kids on the same day.

Plan to disbud the kid just before its normal feeding time. If the kid is nursing its mother, take it away and isolate it for a few hours prior to disbudding. The kid should have an empty stomach to insure it doesn't vomit and choke during this operation. No anesthetic is necessary for this operation—indeed, the anesthetic would be more dangerous to the kid than the disbudding. While there is some pain involved, it is passed quickly and the kid recovers immediately. While your iron is heating, take the kid and (using a pair of small scissors) clip the hair away from each horn bud in a circle about the size of a half dollar. You will be able to see the bud clearly then as a raised, tiny gray circle.

The kid must be held absolutely still while the buds are being burned. You may have a friend hold it or build a small confinement box. I've found that holding the kid myself is the safest—I don't have to worry about someone else letting go at the crucial moment, and can feel more in touch with the kid's responses. Try the following method of securing the kid: fold the kid's legs under it as though it were lying down and then kneel over the kid's body, holding it between your knees. If you're right-handed, take the kid's head with your left hand and press it firmly against your leg. Be sure that the kid can open its mouth to breathe, but press strongly enough so that it can't struggle loose. If you are disbudding a Nubian, be sure its ear is held out of the way. If you are working with an especially large or feisty kid, you may need a friend to hold its head still. Be prepared for some struggles and screams from the kid. The burning *does* hurt, and the kid will let you know it. The friend who taught me to disbud cautioned that a silent kid is the one to worry about, not the screamer. A suddenly quiet kid may have gone into mild shock. I have never had it happen, but if it did, getting the kid up on its feet and giving it a few brisk slaps should bring it out of it.

During the entire disbudding process, your iron should remain plugged in if electric. It is hot

enough to use when it will burn a perfect dark circle on a piece of wood (our porch is ringed with "tests," a record of many disbuddings). You must reheat the iron after each application. You may also want to use a stiff wire brush to clean it—the surface becomes incrusted with burned hair, and this makes it difficult for the iron to heat properly.

When you are certain that you have a good hold and your iron is hot, apply the iron directly over the horn bud. Hold it down firmly for at least six seconds. This will seem like an eternity the first few times—it helps to have the reality check of a friend with a watch! Remove the iron and look at your work. The circle you have burned should be copper in color. Parts may show the white of the skull. Parts that are black have not been burned deeply enough. If the circle is not evenly copper, reburn. You don't have to put much pressure on the iron when you are actually doing the burning—a firm, even, light pressure is what's needed. This is something you will develop a touch for. Older kids will take a bit more pressure and quite a bit more time

as their horns are already grown in and must be burned down. Next, let the iron reheat again and then hold it sideways and carefully use the rim of the iron to sear down the inner part of the bud. This should cause the kid no pain and can be done slowly. Never hold the iron on the kid's head for more than twelve seconds, though, as you would cause the whole area to heat up and could possibly damage the skull and brain. Many short burnings are better until you develop a sure touch. When you are searing the inner part of the bud, hold the body of the iron tilted up away from the kid's head to avoid scorching the areas around the horn bud. Let the iron reheat and do the other bud exactly the same. The middle part of each bud should be burned down fairly flat. If the kid is young, this is very simple and is done quickly. An older kid will take time. When both horn buds are done, let the kid up and apply a soothing antiseptic cream to each burn. The kid will probably be frisky, fine, and ready to take food in a matter of minutes.

As the horn buds heal, you should check every

few days to make sure they don't become infected. The scab that forms will sometimes hide an infection—so *smelling* each bud is a good idea! If a horn bud does get infected (oozing a bad-smelling, pus-filled liquid), you should clean it well with hydrogen peroxide or iodine (be careful that this doesn't run down into the kid's eyes—use a paper towel or cloth to protect them). This treatment should be repeated every few days until the infection clears up. One other problem I have seen in disbudding is that of bleeding. If you burn too deeply, the horn bud may begin to bleed. Or the kid might suddenly begin bleeding a day or even a week later—usually because it hits the bud area on the edge of a feed box or such. This bleeding can be a simple minor problem—or blood might come squirting out in what becomes a hemorrhage. The latter is something that happened to my favorite doe kid one year. I tried applying pressure to stop the bleeding, but each time I took my thumb away, blood would come gushing again. I tried applying ice, but the same thing happened. I was feeling pretty desperate when I remembered reading a paragraph in a magazine somewhere about applying cornstarch to this type of wound. A rush to the kitchen, a liberal sprinkling of cornstarch all over the bleeding horn bud—and in a matter of minutes the bleeding stopped. The cornstarch (or any fine powder—and in a real emergency, old spider webs!) makes a thick paste with the blood which gives the blood time to coagulate and block the wound. I just heard of a kid who bled to death from this sort of an injury—so the story (and wonders) of cornstarch seemed well worth repeating!

In a few weeks, the scabs will form over the horn buds and healing will be well underway. The scabs eventually fall off, and hair covers the scars, leaving a smooth, clean head. In cases where the buds have not been burned deeply enough, something called a "scur" may form. This is a partial section of horn, which will grow out as a misshapen hard piece that usually lies flat against the head. Scurs are very common in bucks. The scur will break off (usually when the animal is head-banging with a newcomer or such) and there may be some (severe) bleeding. It will then grow in again, break off again someday, and so on. If you find a scur developing in a young kid, you can reburn the bud and hope that you have eliminated the problem. Otherwise, you and your goat will have to live with it.

MILKING, UDDER CARE, AND HANDLING MILK

If you want good-tasting milk with a low bacteria count, you have to follow some basic principles and procedures in milking your goat and handling her milk. Goat's milk is very sensitive to odors, and all milking should be done in a clean area separate from the doe's sleeping shed and pen. A small, tight, cement-floor building would be ideal in that it could be regularly washed down and disinfected. Partitioning off a part of your hay barn or adding a small, separate room to your goat shed are two other alternatives. You will probably find milking most comfortable if you build a milking stand. A stand for a single goat should be roughly 3′ long, 2′–3′ wide, and 1½′–2′ off the ground. In front of this, position a stanchion for securing the doe's head with a box or bowl for feeding her grain while you are milking. Stanchions are designed in a multitude

of ways. A keyhole type may be cut out of boards. We have always used a simple arrangement of two small vertical posts with a rope or chain and clip that attaches to the goat's collar. For other ideas on stanchion design, visit friends and neighbors who keep goats and look through the various goatkeeping books and magazines. If you are handy with welding equipment, you could fashion an iron pipe stanchion that would be durable and could be disinfected.

Most goats learn very quickly to come into the milking room and jump up on the stand. Be very patient with your doe the first few times you introduce her to the milking routine—bad habits often result from impatience or rough, inconsiderate handling. Remember that the goat is wary in new surroundings. Her trust in you and her natural curiosity and intelligence will all help overcome fears and hesitations. A tasty bowl of grain also helps! If you buy a young doe, it is a good idea to introduce her to the milking room and stanchion well before she freshens. Bring her in for her daily grain ration and let her become comfortable with the routine. Once she is at ease with this, you can spend a few extra minutes sitting by her and stroking the area where her udder will be (or is beginning to form). By the time she actually freshens, your doeling will be perfectly happy and relaxed in her new role and place as a milker. If you have ever had to deal with a first freshener who is terrified of having her udder touched, afraid of the milking stand, and who kicks you and the milk pail around in her effort to communicate this fear, you've learned the value of this early introduction method!

The production of clean, good-flavored milk requires that the milk room, the animal, the udder, and all equipment used (as well as your hands) be as clean as possible. If your doe has long hair on her belly and flanks, she should be periodically clipped and brushed. Once the doe is in place and stanchioned, you should wash her udder with warm water (and a suitable disinfectant, if you wish) and dry it with a soft cloth. There are special dairy cloths and sponges made for this process, but cotton diapers are soft and absorbent and work just as well. Most does seem to enjoy having their udders washed and the warm water and handling helps stimulate the udder to release its milk. Another benefit is that you will be ready to milk with

clean, dry, and warm hands.

The very best milking pail I've found for goats is the four-quart stainless hooded pail illustrated. Although this pail is more expensive than aluminum, plastic, or enameled pails, it is well worth the initial investment. It will outlast a whole succession of those other pails (my actual experience). Because it is seamless and made of stainless steel, it can be easily and thoroughly cleaned and disinfected. The height and capacity of this pail are geared to the dairy goat. The half-moon hood keeps a lot of dirt and hair from falling into and contaminating the milk (even the most well-groomed, well-kept animal may shed hair and dirt particles into the unprotected pail). This pail is available from American Supply House (see Resources) and other dairy supply companies.

The first few streams of milk from each teat should be drawn into a strip cup as a check for mastitis (see Appendix article for an explanation of this cup) and then discarded. Even if you do not use a strip cup, it is a good idea to discard these first few streams: they usually have a high bacteria count that can adversely affect the quality and taste of the milk. A good milking technique is something you will develop with practice—don't be discouraged if it seems hard the first few times. Patience and gentleness are the primary requisites for this most soothing and pleasant "chore." The milk is produced and held in the spongy tissue of the udder. When the udder is stimulated, the goat's body secretes a substance which stimulates the "let down" or release of the milk. This stimulation may be mechanical—the washing of the udder and actual beginning of the milking (or the butting and sucking of a nursing kid). It may also be caused by the enactment of certain rituals which the doe associates with being milked (your arrival in the milking room, her coming into the stanchion, and so on). Whatever the stimulus, once the doe's body has begun the process of letting down her milk, she should be relieved of this milk as quickly and

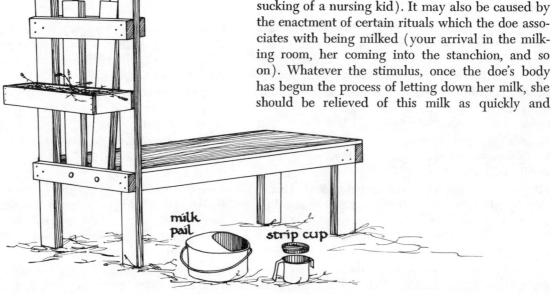

milk pail   strip cup

smoothly as possible. If the milking is delayed or interrupted, she may not release all of her milk and her production will be lowered. If a doe is subjected to irregular hours, interrupted milkings, and so on, her production may decrease over a long period of time and she may be wrongly judged to be a poorer milker than she actually could be. Extra attention to regular and considerate milking routines will help your doe produce to her best capacity. Actually, milking a goat is, I think, harder to describe than to do. The milk comes down from the cistern above into the teat. If you squeeze the teat, the milk is pushed back up into the cistern. When your release the pressure (stop squeezing), the milk flows down into the teat again. The milk must be blocked from going back up into the cistern if it is to be drawn out through the teat. You can do this by grasping the teat (as close to the top as possible) in the crotch of your hand between thumb and first finger—and then pressing your thumb and first finger together. The milk is now trapped in the teat. By pressing your second finger, third, and so on into your first, you can squeeze the milk out of the teat. Relax your hand, let more milk flow down into the teat, and repeat the whole procedure. When you feel confident that each of your hands has a feel for this technique, move on to milking with both hands at once and let a rhythm come to you. I like to milk by alternating pressure on one teat and then the other—so that the milk from one teat is hitting the pail while the other teat refills. Some people squeeze the milk from both teats simultaneously, rest and squeeze again. What feels comfortable for you will probably suit your goat too.

Each goat is slightly different to milk. Some have large holes and milk out easily; some have smaller holes, which makes the milk come out in tiny streams, taking longer. Teats may be large or small, elongated or bulbous. Some goats have very short teats that cannot be milked conventionally, but must be milked with two fingers. Hopefully this condition is found in a young doe who has just freshened. If so, her teats will probably lengthen out a little after she has been milked a few weeks. Teats this short in an older doe are a nuisance that can't be corrected (if you keep this doe, take particular care to breed her with a buck who sires daughters with good udders). After you've learned to milk and have milked a number of goats, you'll begin to appreciate the classic "perfect" udder with well-shaped, nicely sized teats and large holes.

When most of the milk seems to have been drawn out, massage the udder gently and then milk out the remainder. The last few squirts of milk are drawn in a slightly different manner, called "stripping." Use your thumb and first finger to close off the teat as before, then draw them (still closed fairly firmly) down the length of the teat. Repeat this a few times to remove the last traces of milk. You should then dip the end of each teat in an udder disinfectant or teat dip to prevent harmful bacteria from entering the teat canal (see Appendix, Mastitis article).

The milk should then be weighed and the weight recorded. You can use a regular kitchen scale or invest in a more accurate dairy scale (available from Sears, Roebuck or most animal supply houses) of the type illustrated. The twice-daily weighing and

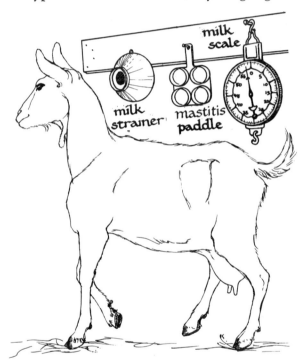

recording of milk from your doe will give you a way to evaluate her as a milker. It will also give you clues about her health, her heat periods, and so on. Most goats milk very consistently from day to day. Their gradual drop in production is barely noticeable—a few ounces over a period of days, for example. When a doe drops suddenly from her normal production, you should be alerted: is she not feeling well? She may be off her feed for any number of reasons—ranging from a simple problem of soiled hay or cutting teeth to serious illnesses such as pneumonia or ketosis (see Appendix articles). A drop in production may mean a beginning case of mastitis. It may also simply mean that your doe is in heat, or going into heat. Whatever the cause, the drop in production should spark an investigation and special concern. In the long run, carefully kept production records may help you sell your doe's kids, too. If you are able to show prospective buyers the kind of production your doe gives over long periods of time, they will have a better idea of the quality of kid(s) you are offering. Production records will also help you evaluate your bucks (when and if you get to this stage of goatkeeping) as you will be able to compare records of a buck's daughters with their dams and sisters. All in all, this is a habit that should become part of your milking routine and general practices.

Once the milk is weighed, it should be strained into a clean glass or stainless-steel container and set in ice water to chill. Unless you have a very well set up milking room, you will probably want to bring your milk back to the house for this handling. Strainers are inexpensive and may be purchased at rural grocery stores or from dairy or animal supply houses. Most of them make use of replaceable cotton filter pads and very fine mesh wire screens. You may also strain your milk through cheesecloth if you are careful to boil the cloth in between each use. Straining removes any particles of hair or dirt that may have found their way into the milk. Chilling the milk in ice water brings the temperature of the milk down much faster than simply placing the jars in a refrigerator or cold box. This is important in inhibiting the growth of bacteria that can ruin the flavor and healthfulness of your milk. Once the milk is chilled, it should be refrigerated. All jars, milking pails, and so on that you use for milk should be rinsed first with cold water and then washed with warm water and a suitable soap or detergent. Household soaps generally contain perfumes and/or coat utensils with a fine film of soap that can ruin the flavor of your milk. Use a dairy detergent or use washing soda to clean your milking utensils and jars. You may also want to use iodine of chlorine compounds (specially made for this use) to sanitize milking equipment just before using it. Whether you use the simpler washing soda method or the more commercialized special compounds, always rinse your pails and jars with scalding water and set them upside down in a rack to air dry. The more care you give to producing clean, good-tasting milk, the more you will enjoy having your goat!

Pasteurizing goat milk is, so far as I know, not a very common practice. Goats in this country are virtually tuberculosis free. Brucellosis is a fairly rare and regionalized problem in goat herds. Check with other goat owners and with your farm adviser or veterinarian about brucellosis in your area, and if there is any problem have your animal(s) tested. If your goat is otherwise healthy and you handle your milk carefully, your milk should be clean and wholesome. Pasteurization destroys nutrients and is probably an unnecessary addition to your milking procedure.

Not all goats will stand by patiently while you learn the gentle art of milking. Some goats will not stand for even the most experienced hand. These are usually does who have had some bad experience(s) on the milking stand, changed hands a lot, or have some sort of udder problem. If you have acquired (or raised) a doe who is extremely difficult about being milked, there are many alternatives to culling her outright. There is no animal I know as responsive to gentleness and *feelings* as the goat. First, be sure the goat is not acting up because *you* are in a bad mood, short-tempered, wishing you were doing something else. Goats are incredibly sensitive to our moods and sometimes will respond to nervousness with nervousness, or will kick the bucket over on an impatient milker. A new goat may just be frightened, uncertain of what you are doing, unsure of her surroundings. A few minutes of calm talking may soothe her. A new goat may also arrive with the habit of struggling with a former owner and be ready to take you on in the same spirit. Again, calmness and consideration may work wonders with this "difficult" goat. Once she realizes that you aren't going to hurt her or subject her to any rough treatment, she may settle down nicely. The language of the goat on the milk stanchion is something you should learn, too. Snorting and/or stamping usually means anger or fear. Ears held stiffly forward often indicate alarm. The goat that is too tense to eat her grain is also communicating her fear. An unfamiliar noise, a new person, even a familiar barn cat rubbing suddenly against a belly or leg can startle your goat. Don't make an assumption about her "ornery" nature without looking for what is disturbing her. There are other categories of disturbance, too. A tiny, almost imperceptible scratch on a teat may make your goat very reluctant to have her udder touched. I have found a tiny thorn embedded in the udder of a goat who had been nibbling rose bush prunings earlier in the day. The pain of a scratch can be eased by the pre-milking application of udder balm (special ointment made commercially for dairy cows—available at most feed stores) or any similar lubricating ointment. A larger rip or cut should be cleaned well and treated before and after each milking with some sort of antibiotic or healing ointment. Serious wounds may have to be sutured and the doe dried off for that lactation. If a doe is a heavy milker and has injured her udder, you may want to milk her four or more times a day to keep the pressure from building up in the udder and aggravating the condition further. Sometimes a newly freshened goat will have small, hard calcium "stones" which plug the ends of one or both teats. These are very painful and subtle; the milk may or may not be able to flow past them. If you suspect this problem, lightly roll the end of each teat between your fingers. You will be able to feel the stone—a tiny, hard object which can move freely in the teat unless it is caught firmly in the orifice. Usually soaking the teat in warm water and then working the stone gently will loosen it and allow it to pass out with the milk. Never force this, however. Try soaking again and massaging the teat. If this still doesn't work, you can make use of a teat dilator or substitute for this with a tube of mastitis ointment. The applicator this ointment comes in will serve as a substitute dilator that is sterile and of proper size. Wipe the end of the teat with alchohol and very gently insert the tube. This will usually break up or reposition the stone, allowing it to come out, or stretch the teat orifice enough to allow the stone to pass. Withdraw the tube and squeeze out milk and stone—this is a tremendous relief to the doe, who may then be normally milked out. Be absolutely careful when you do this not to introduce any dirt (with accompanying bacteria and fungi) into the udder. You may find this same doe with another stone the following day . . . treat

her the same way. This problem will not reoccur (in my experience) after the first few days of milking. A newly freshened goat may also have waxlike plugs in her teats. These are softer and come out easily. They often occur for a few days after freshening.

Chapped teats—not uncommon in cold weather —may also make for an unhappy goat. These may be treated with daily applications of udder balm or other ointment. Be sure you aren't causing this irritating condition by improperly drying the udder after washing or by keeping the doe in a poorly bedded house. The *Herbal Handbook for Farm and Stable* suggests bathing sore teats with a brew of dock leaves and elder blossom. The skin of the udder and teats is also susceptible to bacterial infections that cause small pus-filled eruptions or pimples. Left untreated, these may spread and break, creating a painful and messy situation. They should be painted with tincture of iodine or with a broad-spectrum antibacterial cream *before* they get serious. Very similar in appearance are the lesions of Goat Pox, a virus-caused disease that affects the udder, inner thighs, nose, and mouth of the animal. This is not a severe disease in goats, and the lesions can be treated in the way described above. Apparently goats are subject to two types of pox: the true Goat Pox, and a "false" or "springtime" pox. A case of true Pox will give the goat lifetime, naturally-acquired immunity once she recovers. The second type may reoccur in the same animal year after year. It is probably, in truth, a bacterial infection of one type or another that resembles true Pox in appearance. Again, treat these lesions with careful applications of iodine or antibacterial cream. If you are milking more than one goat, milk any doe

Work never stops in the country; it comes in cycles, varies in intensity with the seasons, but it never stops. I spent all afternoon splitting wood, came in and built a fire. Already I'm running out of wood again! Nothing is ever completely finished here. As soon as something is built, it becomes part of a new cycle. It will be patched, improved, repaired, expanded, but it is never just done. As soon as one garden is being harvested, the next is being planted; as soon as one season's kids are being sold, the next's are being conceived. The Andersons told me last week, "Take it easy, Jenny, and do it as you're able. If you live here eighty years, you won't get done all the things you can think of for that land."

with these udder eruptions last to avoid spreading the problem to your other doe(s).

A doe who is normally calm and manageable on the milk stand and suddenly becomes difficult (with no outer-appearing problems) may be suffering with mastitis. This means an infection of the udder tissue—ranging from a very minor problem to a devastating systemic infection. One of the early signs may be a sore, tender udder. There is a simple and fairly accurate test you can do yourself to determine whether this is your doe's problem. (For a fuller discussion of this problem, see the Mastitis article, in the Appendix.)

There are two more problems that affect the udders of newly freshened goats. The first makes a dramatic appearance in the milk itself: tiny capillaries within the udder burst and the milk has blood in it. There may be the faintest pinkish tinge to the milk, or there may be enough blood to turn the milk an unappetizing and frightening red. It is usually the unfamiliar distention caused by the newly produced milk that causes these capillaries to burst, and is a condition more commonly found in young does (particularly first fresheners). The doe may be milked out three or more times a day to minimize pressure on these capillaries until they are healed and strengthened. Usually the condition passes in a week or so. Sometimes it helps to feed the doe a slightly less rich diet during this period. The same problem may appear in any doe who is very roughly milked or milked by someone with very long fingernails. Here, the cause should be immediately apparent and easily corrected.

The second problem affecting the newly freshened goat is udder congestion, also known as caked udder. It usually occurs right after freshening and seems, again, to be more common in younger animals. The udder will appear full but little milk will come out. The udder tissue feels firm but not hot (in cases of mastitis, the udder will usually feel hot and the milk will appear abnormal). To further distinguish this condition from a possible case of mastitis, the animal does not seem to have a tender or sore udder as she would normally have with mastitis. The best remedy I know of for this condition is a combination of massage and frequent milkings. Application of towels soaked in warm water and Epsom salts may also help break down this congestion. Dry the udder thoroughly after these applications, then massage and milk out. Some goat breeders recommend the feeding of ascorbic acid to does with this condition. I've read, though, that ascorbic acid is destroyed in the rumen—so this treatment seems questionable. Also be sure that you aren't overfeeding grain to this doe. This is a common cause of caked udder in young does. Build her up to a proper grain level slowly, over a period of weeks. If the congested udder is treated carefully from the very beginning, the doe should come into her full milking with no damage to the udder tissue. It may take a week to reduce this congestion, but it should not reoccur during that lactation (so far as I know and have experienced).

If you diligently pursue and fail to find a reason for the bucket kicking, foot stamping doe's behavior, you still have some alternatives to simply selling her and passing her problems on to someone else. The bucket kicker can sometimes be cured by the determined application of the Quicker-Than-Thou bucket-moving method. This means simply that you use a bucket you can move out of the way with one hand and that you milk with one hand and keep a wary eye on the offending feet. The instant the doe makes a move to kick, slide the bucket out of the way. I have cured a couple of chronic kickers of their habit with this primitive but effective method. After a couple of weeks of kicking the empty air, these does seem to lose heart and settle down to their grain. You can then return to the speedier two-hand milking method with a certain smug feeling. . . . Don't get too relaxed, though. The very doe you thought you'd cured may send the bucket flying again some calm summer morning—just to keep you in shape. Some people resort to hobbling these does on the milk stand: not very sporting, but very effective. Permanent leather hobbles may be worn by the doe, with matching clips on the stand itself. These hobbles are not connected together, as those used to hobble horses. They are more like little leather collars, fastened between the coronet and pastern. More annoying and more difficult to deal with is the doe who lies down on you as soon as you touch her udder. There are two ways that I know of to deal with this sort of passive, non-violent refusal to be milked. The first is to be used spontaneously, when you are taken by surprise. It consists basically of coaxing the doe back to her feet, offering her a little more grain, and waiting until she seems calm again. You must then crouch on the milk stand and slide your knee under her belly. If you are lucky, she will not be able to lie down and you can proceed to milk her in this most awkward of positions. If you are even luckier, no casual visitor will stroll in and catch you in this compromising arrangement. . . . A more permanent and much more reasonable solution to this behavioral problem is the belly sling. This consists of a burlap sack nailed to the wall behind the milk stand that is run down under the doe's belly and fastened back up to the wall (you simply hook the material back up to the same nails that anchor the other end). This method is absolutely foolproof, hurts neither the doe nor you, leaves your hands free for milking, and after a period of time cures the doe of this unfortunate habit. As she finds herself suspended in the sling every time she tries to lie down, she grows accustomed to staying on her feet. After a while, you can stop using the sling. Don't get too sure of yourself and take the sling down, though. Like the chronic kicker, this doe will almost certainly surprise you someday with her same old routine.

Whether your milk comes from the most docile or difficult doe in the world you may have to deal with the problem of off-flavored milk. If your doe has a healthy udder and is well fed and in good general condition, and if you handle her milk with care, you may never have this problem. If you milk your doe in a clean area, use sanitary pails and strainers, keep your milk bottles shining but still have unpleasant-tasting milk, you may have some sort of feeding problem. Sometimes a mineral deficiency will cause a doe to give off-flavor milk— either temporarily or chronically. A free-choice trace mineral mixture may solve this problem. Sometimes a doe's milk will go off-flavor for a day or two then return to normal with no change in her care or feeding. Her body seems to be making some sort of demand and adjustment. Certain feeds and browse can affect the flavor of milk: garlic, turnips, cabbage, and any other strong-tasting feeds are obvious offenders. If your goat is grazing, she may periodically find something tasty that will make her milk less so. There are a few does who just give bad-tasting milk, regardless of feed or conditions. You may want to keep this doe anyway, and use her milk to raise calves or kids—but her doe kids may have the same characteristic (it is inheritable), so think hard about perpetuating this line.

Ropy or stringy milk may be caused by improper cleaning of milking utensils—or be a problem with the doe herself. Be sure that all utensils are absolutely clean and that the milk is cooled properly. If this doesn't help, isolate the milk of each doe and stand it in a warm place to see if it becomes ropy. The doe who is producing this kind of milk should be given plenty of green feed, particularly leaves and twigs. She should have access to trace minerals, too. The one case I've ever seen of ropy milk was due to poorly washed milk pails.

One other problem with milk that might be confused with the bloody milk discussed earlier is also related to the cleaning of pails and jars. Milk that turns pink after it has been sitting in the refrigerator a day or two may simply be dirty. There is a type of bacteria that grows in milk and causes a pinkish tinge. Extra attention to the handling of your milk should eliminate this.

A final, obvious, and very common problem with off-flavor milk is caused by the male of the species. Buck goats have a strong, pervasive odor during a good six months of the year (breeding season) and a lesser, but still pervasive, odor the remaining months. A doe kept anywhere in the vicinity of a buck may produce milk that has a distinct "goaty" flavor. Moving the buck to an area by himself or with other bucks will remove this problem from your milk. At chore time, bucks should be fed last or by someone who is not going to be in the milking room, as the bucks' odor can pass from clothes to milk with very subtle, unpleasant effects.

At the end of her lactation, the doe will have to be dried up. Normally, she will be dried up three months after being rebred (or two months before she is due to kid again). Some does will dry up (cease milking) gradually over a period of months. A good doe will milk persistently and present a bit more difficulty. I have found the best way to dry a doe up is to reduce her grain, then decide upon a day and cease milking her. She will be uncomfortable for a few days or longer (depending upon the amount of milk she had been producing) as her udder fills and the milk creates pressure. The best course is to simply remain firm and withhold grain. Gradually this doe will reabsorb this milk, cease producing more, and dry up. Now and then an unusually heavy milker may have to be milked out once or twice (twenty-fours hours apart) to relieve intense pressure. Some does are almost impossible to dry up and may have to be milked right through their pregnancy. Most, though, will dry up and will benefit greatly from this two-month dry period. They will be able to build up bodily reserves necessary to a healthy, long new lactation.

milk products

Milk is the most versatile food one can produce on the farm. Its three main components—fat, protein, and sugar (lactose)—can be separated and changed into totally different forms by churning, incubating, or boiling. All the same products can be made from cow's milk or goat's milk. The main differences between these two is the size of the fat globules. Those in goat's milk are smaller, harder to separate, and easier to digest. First I'll deal with whole milk and its possibilities. It should be strained and cooled (in ice water) immediately after milking to assure good flavor and low bacteria count.

YOGURT

Making yogurt is a means of preserving milk by changing the lactose to lactic acid, by the action of a pure culture bacteria. Milk should first be heat-shocked (warmed to 145° F for fifteen seconds). This kills pathogenic bacteria (staph, salmonella, etc.) but doesn't inactivate enzymes or affect nutrition. Then cool the milk to 110°. Add yogurt starter (this can be expensive Bulgarian powdered starter or plain commercial yogurt—1 tablespoon of yogurt to 1 pint of milk). Non-instant powdered milk or evaporated milk can be added to thicken the final product. Experiment with the amount. Stir well. Incubating at 110° is the most crucial step of yogurt making. A few degrees above or below will incubate other bacteria and impair flavor and consistency. The streptococcus thermophilic bacteria is responsible for this and will be lost at a lower temperature. Lactobacillus bulgaricus will grow from 100° to 115°, so within that range you'll get yogurt. Here are some ways of holding the temperature: quart jars set in 110° water inside a canner, styrofoam ice case, or insulated box with a low-watt light bulb or an electric heating pad set on "low" inside; the pilot light on a gas stove; or, most exact, an electric yogurt maker. The mother culture (starter) should be changed every two weeks. We've found that goat's milk yogurt becomes thicker after you've used your own starter for about three consecutive batches.

CHEESE

Cheese is made by separating the protein (curds) from the milk sugar (whey). If it is made with whole milk, rather than skim, the fat adds a creamy consistency to the cheese. There are many ways to effect the separation of curds and whey: letting the milk sour slowly, adding rennet (an enzyme from a calf or kid's stomach), or adding cultured buttermilk, lemon juice, or vinegar. How you form the curd and how you handle it afterward determines which of the hundreds of varieties of cheese you'll have. The next chapter describes cheesemaking in detail.

SEPARATING

The next things to do with milk all involve separating the cream out. If you have cow's milk this can be scooped off the top of milk that has set overnight. Goat's milk cream rises after several days and should sit in the refrigerator, covered, in a large, flat pan to get the most cream. Of course, the most efficient method is the cream separator, a centrifugal machine (either electric or manual) that spins out every bit of cream. From our separator we get one quart of cream per three gallons of milk. If you're planning to make cream products, be sure and choose your breed of animal accordingly. Jersey or Guernsey cows and Nubian or La Mancha goats have the highest percentage of butterfat in their milk.

BUTTERMAKING

Butter can be made from whole milk, but this method incorporates too much liquid in the butter and it won't keep as long. Separating the cream and using it is best. Churning can be done in a number of ways. Starting with the smallest amount of cream (one pint), you can shake it vigorously in a canning jar for about twenty minutes. Or, if you have a quart or two of cream, a glass jar with a beater attached is best. Sears sells electric models, and hand-turned ones are available at junk-antique stores for seven to fifteen dollars. Or rig up your own with an old egg beater. And finally, if you're into serious buttermaking, the big crock with wooden dasher is still available. A rule of thumb is that for proper concussion of the cream your container should be one-third full.

Let your cream warm to room temperature before you begin churning. In our glass-jar churner the fat gathers into lumps after about twenty or thirty minutes of churning (make sure the cream is really splashing around when you're churning). The butter granules should be large enough that you can pour the buttermilk off easily and then rinse the granules in cold water two or three times until all the milk is gone. Put the mass of lumpy butter on a cold, wet board and flatten it with a cold, wet wooden paddle into a one-inch-thick slab. Salt can be lightly sprinkled over it at this point if you wish. Now work the butter, piling it up and spreading it out, until all the milky water is out and the consistency is proper. "Proper" is smooth and easily spread. Next put the butter into a mold or crock and refrigerate. Yield is about one pound of butter per quart of thick cream.

SOUR CREAM

Stir together 3 cups whole milk, 2 cups thick cream, and 1 cup cultured buttermilk. (To make buttermilk, add ¼–½ cup of commercial buttermilk to 1 quart whole milk. Let stand overnight and refrigerate.) Warm the sour cream mixture to 70°. Pour into 3 warm pint jars and incubate at 68°–70° for 12–24 hours. You sacrifice taste for thickness in leaving it longer.

ICE CREAM

For 2 quarts of ice cream, mix 3 beaten eggs, ½ cup of warmed honey, 1 teaspoon vanilla, pinch of salt, 1 cup cream (or evaporated milk), and 1 quart whole, rich milk. Chill this mixture before putting it in the ice-cream freezer and cranking. This is a basic vanilla recipe and it can be elaborated on with 2 cups of honeyed fruit, chocolate chips, peppermint extract or candy, a cup of cold coffee, etc.

Pour the chilled mixture into the inner bucket of an ice-cream freezer. Fit in the dasher and put on the lid. Between this and the outer wooden bucket pack alternate layers of cracked ice and rock salt. The best-textured ice cream comes from steady churning (with the crank that fits over the freezer and attaches to the dasher). You don't need to be speedy. Using plenty of salt and ice is important. The salt lowers the melting temperature of the ice. The ice, needing heat to melt it, draws the heat out of the ice cream, thickening it.

HOW TO OPERATE AND CARE FOR
A HAND-CRANKED CREAM SEPARATOR

SETTING UP

Secure gear box to a level, steady platform with screws. A vibrating separator will damage the parts as they revolve at high speeds.

Fill the oil cup with a light motor oil (preferably special separator oil available from DeLaval Pacific Company. Parts for DeLaval Separators are still available from them.

Assemble bowl parts in order shown in diagram with all marks and notches lined up and facing front. There are as many as nine bowl discs fitting one on top of the other. The bottom one is sometimes specially caulked on both sides. Secure bowl nut tightly on top of the bowl body with a special spanner. These pins ⌐───────⌐ fit into holes on bowl nut.

Set bowl with all notches facing front on worm spindle with its notch also facing front. Bowl should settle down on spindle. Put on skim-milk cover (usually the shorter spout) then cream cover, then regulating cover, then float, finally supply can.

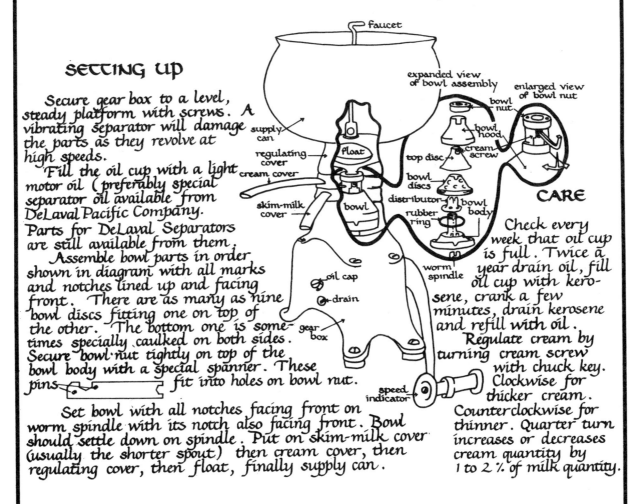

faucet

expanded view of bowl assembly

enlarged view of bowl nut

supply can

regulating cover

cream cover

float

skim-milk cover

bowl

oil cap

drain

gear box

bowl nut

bowl hood

cream screw

top disc

bowl discs

distributor

rubber ring

bowl body

worm spindle

speed indicator

CARE

Check every week that oil cup is full. Twice a year drain oil, fill oil cup with kerosene, crank a few minutes, drain kerosene and refill with oil.

Regulate cream by turning cream screw with chuck key. Clockwise for thicker cream. Counterclockwise for thinner. Quarter turn increases or decreases cream quantity by 1 to 2 % of milk quantity.

OPERATION

Milk must be warm (85-95°). Pour into supply can with faucet closed. Commence cranking and increase speed until the bell (speed indicator) stops ringing (on our model this happens at 60 rpm). Too fast will result in less and thicker cream. Too slow - thin, milky cream. Open faucet. Be sure you have a container under cream spout and skim-milk spout. When all milk has been separated, close faucet, pour a gallon of hot water in supply can and run this through to clean left-over cream off parts. Then let the spinning bowl run down by itself before dismantling. All parts that have had contact with milk should be washed immediately in hot water and washing soda and air-dryed.

cheese making

I have never yet made a cheese that was really like what I buy in the store, but I have made some delicious, if unnamed cheeses in my time from very simple recipes. Cheese is basically just curdled milk —the butterfat and milk solids form the cheese; the whey is drained off. Milk for cheese can be curdled naturally or by the addition of a starter, usually buttermilk, lemon juice, rennet, vinegar, or a cultured starter. Hard cheeses, like cheddar, use rennet and are more difficult to make than soft cheeses. Recipes for hard cheeses are included in the Resources section. I usually make soft cheeses from lemon juice and vinegar instead.

A dairy thermometer calibrated from 75° to the boiling point is necessary for most rennet cheeses and is helpful for making other kinds. They cost just a few dollars. The only other equipment you need is a large enamel pan (one gallon of milk makes about one pound of cheese; aged cheeses should be made from three to four gallons of milk) and a cheese press. A simple one can be made by cutting both ends out of a large tin can, filling it with cheese and weighting the top. For larger cheeses, I use two boards, pegged at each corner, with bricks on top. You also need cheesecloth for draining the curd.

The very simplest cheese is made by just letting whole milk go sour. Friends of mine regularly make cheese this way—setting gallon jars of milk in the sun for a day. The curd is then drained into a cheesecloth, rinsed with boiling water, and pressed into a bowl. This cheese is rich (creamy) but is sour to my taste; many people like it, however.

Basic Soft Cheese can be used as cottage cheese, pressed for slicing or pressed and aged for sharper flavor. I make it from goat's milk, but am sure cow's would do as well. Heat the milk (it can be sour or fresh) to the boiling point. Then add about ¼ cup lemon juice or vinegar for each gallon of milk. Lemon juice seems to give a better texture (less rubbery). To save money, I usually add 2 tablespoons lemon juice first, then however much vinegar is needed. Stir the boiling mixture, and it should curdle pretty rapidly. If it doesn't, add more lemon juice or vinegar. Experiment to get just the right amount of acid—if you use too little, the curd won't separate completely, the whey will remain milky, and you'll get a fine curd; if you use too much the curd will be rubbery. Just right is a clean, solid curd with a clear whey. When the whey is pretty clear, drain the curds into a cheesecloth and hang it up to drain. When the curd has cooled and drained for ten to fifteen minutes, add 1 tablespoon salt per gallon of milk used.

This is the basic cheese. It can be used as cottage cheese and is especially delicious in lasagne. The flavor may be changed by adding caraway or cumin seeds, garlic powder, or sautéed onions or sesame seeds. The onions and sesame seeds improve the texture of aged cheeses, probably because of the oil used in sautéeing. The basic cheese itself can be sautéed in oil or butter for a different texture. For regular use, we press it by simply placing it in a bowl with a plate and a weight on top. After two days, it is a mild-flavored cheese suitable for slicing or grating. It will keep about two weeks in the refrigerator.

Aged Soft Cheese: I have gotten all kinds of results from aged cheeses—some spectacularly good, a very few hardly edible. Aging cheese is fun and exciting, the only problem being its unpredictability. Once I cut open a cheese and found a fine "bleu" cheese when I thought I would find my usual sharpish goat cheese—with no idea how the mold got there or how to do it again.

Once the curd has been formed, drained in cheesecloth, and salted, the aged cheese needs to be pressed for one to three days. If you are using a press made of wooden boards, the cheese needs to be shaped into a round (about 3″ or 4″ high and as big in diameter as you have cheese—you really need to use at least four gallons of milk for a good aged cheese). The round should be shaped by hand to be as smooth and free from cracks (which can mold) as possible. The round of cheese gets a cheesecloth bandage wrapped tightly around its circumference and pinned in place. This bandage keeps the cheese from spreading and flattening as it is pressed. The cheese is then pressed with heavy weights on top. The pressing drains the remaining whey and firms up the cheese.

After the cheese is pressed it can either be dipped in paraffin or kept oiled during the aging. I have had trouble heating paraffin to the proper temperature so it won't crack off, so now I always oil my cheeses. To do this, I let the cheese dry on a rack until it forms a rind (one to two days). Then I rub it well with salad oil; I turn it and oil it every four days after that. The cheese should age from three weeks to three months, becoming sharper and harder as time goes on. A three-month goat cheese is very similar to parmesan. If the cheese gets mold on it during aging, it should be wiped with vinegar to kill the mold and then reoiled. Once the aging is finished, cheeses should be wrapped tightly and stored in a cold place.

Whey Cheese: The whey from basic soft cheese can be fed to chickens, dogs, and cats. It can also be made into a sweet-tasting whey cheese. This is done by boiling the whey until only the milk sugar is left in the pan (up to eight hours cooking). The milk sugar will look like thin brownish mush. Add a little raw or brown sugar. Spoon this paste into a mold (which can be made from tin foil) and let it cool. This cheese should be aged several days before eating. It has a distinctive sweetish taste, vaguely like maple sugar, which people seem to really like or dislike.

Cabin Cheese is similar to basic soft cheese but is made with rennet. Heat 1 gallon of whole milk to 90° and add 1 tablet of junket rennet or ¼ tablet of cheese rennet (either one dissolved in ¼ cup of cold water). Let the milk and rennet sit undisturbed in a cool place for twenty-four hours. Cut the curd into large chunks and pour it into a double thickness of cheesecloth in a colander or strainer. Save the whey, which is used to make ricotta. Stir the curd to release the whey. Tie the cheesecloth corners together and hang it up to drip for several hours, stirring occasionally. Salt the cheese to taste and use it as cottage cheese or press it in a bowl for slicing. This is a rich, delicious cheese.

Ricotta Cheese is made from the whey of rennet-curdled cheeses. Heat the whey from 1 gallon of milk. When you see a slight creamy film on top, add 1 cup whole milk. Then heat it slowly, without stirring, until it is nearly boiling and curds form on the top. At this point either skim off the curds and drain in cheesecloth or add 2 or 3 tablespoons of vinegar, remove from heat, stir, cool, and strain. Add salt. The whey from ricotta is good food for chickens, very high in riboflavin.

Herbed Cream Cheese is delicious on crackers or bagels and disappears as fast as I can make it. To make it you need: ½ gallon of whole milk, 1–2 pints of cream depending on its thickness (if bought this should be non-sterilized), ½–1½ cups powdered milk, and 2 tablespoons buttermilk. The consistency of the cheese is better if you do not use absolutely fresh milk; I usually let it keep in the refrigerator for a week before using. Combine everything except the buttermilk in a pot and heat to 90°. Pour into a glass or crockery bowl, add the buttermilk and cover. Let sit in a warm place eighteen to twenty-four hours, until it clabbers. This part is like making yogurt, and the clabbered milk has the consistency of thick yogurt.

Line two colanders with two layers of cheesecloth and set in the sink. Pour half the cheese into each and let it drain ten or fifteen minutes. Then set each colander in a large bowl or pot with some drainage room. Lap the cheesecloth over the cheese (this prevents a "skin" from forming). Cover the top of each with plastic wrap and refrigerate for twenty-four hours. Be sure to pour off the whey if it begins to cover the cheese.

Turn the cheese into a bowl and season to taste. If you add about 1 tablespoon salt you will have a very rich cream cheese. I like to add 3 tablespoons chopped chives, a good handful of minced fresh parsley, and 2 cloves of garlic, pressed. Put this mixture back into fresh cheesecloth in one colander, lap the cheesecloth over, return colander to bowl, cover with plastic and refrigerate for another twenty-four hours. Store the cheese in jars or crocks in the refrigerator. It's good right away or you can store it for at least a month.

The amount and kind of milk one uses is very flexible; sometimes I use more cream, sometimes less; once I used a whole quart of buttermilk—it all works.

raising sheep

Our sheep raising began as most of our learning to farm: a vague notion that "it might be nice to" coupled with the sudden opportunity to buy—in this case, our choice of a small backyard flock that was being dispersed. We went to look at the sheep for sale with absolutely no idea of what was "desirable" in sheep, or what we would want them for. Somewhere in the future was perhaps a spinning wheel—and lambs—and looms. The sheep were an outgrown 4-H project, bottle-fed and hand-raised. They were newly shorn, clean and sleek, came at a run to their people's call. As we talked they milled around us. Words and advice passed lightly over our heads—what to feed them, when to breed, what was good wool or a good fleece, weight, crossbreeding . . . the sheep were there, nibbling at the grass, nosing against us. Their shed had a fine smell and look to it. The wooden feed racks were worn to a smooth oily shine. It all felt *good* . . . we bought the two lambs at market price— $.32 a pound—and the older ewes for $15 to $20 each. It was a fortunate beginning—four fine wool-

type ewes, a purebred ram, all of them good animals and very tame.

A good beginning, but wrong way around. With the sheep confined to a tiny pen, we began to fence them a pasture. There was an open, grassy hillside that had once been horse pasture and had most of its barbed wire fence still intact. We reasoned that adding four or more strands to the existing fence would make it sheep tight. So we began. We tore ourselves, our clothes, our boots. We stretched, nailed, and cut—learned about the springy, ornery, dangerous nature of barbed wire. When it was completed, we didn't stop to admire our first fencing job; in went the sheep. They did a little experimental turn through their new place, tested the grass, sniffed at the fence, then one after the other popped out between the lower strands. If a head went through, the rest of the sheep would follow. A lot of wool was left on the barbs. We rounded them up (a bucket of grain and familiar "come, sheep" and they would arrive at a gallop and follow you anywhere) and put them back in through the gate. In a few minutes they were out. They were free, and quite content to stay around awhile, and we had our first sheep-training lesson . . .

Starting a Small Flock: More than twenty breeds of sheep are raised in this country. Crossbreedings produce countless more types. Characteristics that vary from breed to breed are body size and conformation, breeding season, number of lambs per season, type of lamb and its growth rate. Wool varies tremendously—in density, staple length, average fleece weight, grease content, softness, and fineness. "White" wool will vary from snow white to cream or yellow. Fleeces weigh from six to twenty pounds, depending on breed, age, and size of sheep. If you plan to use your own wool, try different types to see how it spins and works up. Decide what is most important in your sheep flock and look for a breed or blend of breeds that will best suit your needs and your land.

There are fine-wool sheep (Merino, Rambouillet), mutton-type sheep (Hampshire, Southdown, Shropshire, Suffolk), medium-wool- or crossbred-wool-type sheep (Corriedale, Columbia). Dorset sheep (medium wool, light fleece) will breed any time of year and produce off-season lambs (high market value). Coarse, carpet-type wool with an incredibly long staple comes from large sheep, such as the Lincoln. Dual-purpose breeds (the Targhee) have been developed to give good wool and good meat-type lambs. A new breed to this country, the Finnish Landrace, characteristically throws multiple lambs (five per ewe per lambing is not unusual, triplets are common). One outstanding characteristic in a breed may be at the sacrifice of something else you want in your sheep. The Merino, for example, has a beautiful, fine fleece, but short in staple. Merino lambs would not finish out (a stock-raising term referring to the type of carcass of an animal) as well nor grow as fast as lambs from the mutton-type breeds. The prolific Finnsheep grows a mediocre fleece and is small in size. Careful and intelligent crossbreeding can balance and enhance desirable characteristics. A flock of fine-wool ewes, for example, bred to a large mutton ram, will produce excellent fleeces and good, fast-growing meat lambs.

Choose a type of sheep that will do well in your area and under your conditions. Some sheep are bred to withstand cold; others do not do well in wet climates. Cheviots are exceptionally hardy; Rambouillets are not supposed to take rain because their fleeces dry out too slowly. Some sheep do well

under range conditions, and others are good in small farm flocks. Talk with anyone in your area who raises sheep to find out what breeds you should consider. Availability of breeding stock and the type of market available (if your sheep will be a commercial venture) should be explored. Then within the breeds that will do well on your land and under your conditions, choose the type that most appeals to you. An important part of sheep-raising is liking the animals you work with!

The best times to buy sheep are spring (early or late—whenever weaned lambs go to market in your area) and in early fall (when flocks are culled). A good beginning flock would be a number of crossbred ewes and a purebred ram. The number will depend upon how much land and capital you have available and what you plan to do with your sheep. *Understock* your pastures. Begin with a small flock and build it up as you learn. Even if you want only half a dozen ewes, you'd do well to start with two or three the first year. For a flock of thirty or fewer ewes, you'll need only one ram. He should be the very best animal you can find. Extra care in choosing and an extra investment for your ram will pay off in the lambs he sires. Using a purebred ram on a flock of crossbred ewes will produce lambs that are consistently better in every respect than their mothers (a breeding process called upgrading). The initial investment for crossbred ewes will be small. Prices for sheep vary from year to year, area to area, season to season. You can probably buy a young crossbred ewe for $20 to $25. Old ewes often sell for as little as $5.00. Wethers (castrated rams), ewe lambs, and ram lambs sold at market price, by the pound, may run $15 to $25 and up. A purebred ewe or ram will cost more. If you buy your animals at auction, you may get them considerably cheaper—but beware of the

292

"bargain" that may turn out to be an infertile ram, a barren ewe, or otherwise defective animal. If you live in sheep country, you may be able to start a small flock with some orphaned lambs—called "bummers." These are lambs that have been deserted by ewes that have hard lambings, whose mothers have died, or they may be one of a set of twins or triplets that an old or underfed ewe can't provide with milk. Some shepherds give these lambs away rather than take the time and expense to bottle-raise them. Make arrangements before lambing begins to take these lambs if you can. Try to avoid ewe lambs from mothers who didn't have enough milk (for one or two lambs under normal feed conditions) or who deserted the lamb for no apparent reason—these traits are inherited. Rams or wethers wouldn't present this problem—and a ewe lamb whose mother refuses her after a difficult lambing, or whose mother died or was killed, could be expected to take *her* lambs normally.

Choosing the Individual Animal: In choosing a sheep, look for a large, well-developed animal with strong legs and a good straight back. Size is important: a large ewe will produce more and faster-growing lambs and will also produce more wool. An open-faced ewe (with very little or no wool on her face) will be most productive. If you want twin lambs, buy twins. Twinning is an inherited characteristic. Fleece should be dense (especially on the rump—where any lack shows most clearly), of uniform length, and with good crimp (the finer wool is tightly crimped, or waved), and greasiness. Look for breed characteristics in coloring, body type and size, and wool. Don't buy an animal with an over- or undershot jaw (upper or lower jaw protrudes noticeably). Check to see that ewes' udders are free of lumps and that teats haven't been clipped off in shearing. Look at the animal's teeth. A full mouth of fairly good teeth means years of grazing and health regardless of the animal's age. A

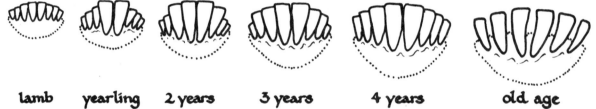

lamb   yearling   2 years   3 years   4 years   old age

When you are buying sheep, you should know that the mouth of the individual sheep yields information about its age. In the front of its mouth, a sheep has only one row of teeth. These are located on the bottom jaw. With these teeth, it nips off grass and forage. A set of back teeth on the upper and lower jaws function as grinders. The front teeth can tell you the approximate age of the animal you are looking at. A lamb under a year of age will have a set of small, even, off-white teeth. When the lamb approaches its first birthday, a pair of large teeth will push their way up in the center of the tooth row, displacing the smaller baby teeth. The yearling carries this set of two large, six small teeth until it is a two-year-old. Then another pair of adult teeth will erupt, one on either side of the "yearling" teeth. The two-year-old sheep then has a set of four large and four small teeth. As a three-year-old, this sheep will replace two more of its baby teeth with adult teeth. The mouth of the four-year-old sheep will carry a full set of eight large teeth. After age four, a sheep's mouth is less revealing. The teeth gradually wear down and also spread apart. Teeth will wear out more quickly if the sheep is grazing in a sandy or rugged area, so it may be difficult to tell if tooth wear is a sign of age or environment. The spreading of the teeth is a more accurate indication of aging; the sheep at seven or eight years of age will have definite spaces between the teeth. A sheep with one or more teeth missing is said to have a "broken mouth." The old sheep with no teeth at all is called a "gummer." A gummer or a sheep with badly worn teeth may require special feed such as chopped hay and/or grain supplements.

The first lamb was born today. Premature and dead. Olivia, the mother, seems to be all right though. I had a dream a few weeks ago that the lambs were born tiny (like mice) and pink. And that I struggled to save them, but they were too small to feed. This lamb today was small and pink, its fleece plastered against its body, thin and sparse. For a moment it was nightmarishly like my dream. Judith came to help. She calmed my fears; I was afraid Olivia would die too. I realize I know much less than I thought I did about animal care—I didn't know what to do. This is my first animal death. The beginning of a long cycle. It seems even harder to have death come before life, than to have an old one die giving birth. Hopes for the future stillborn.

young sheep that's been grazing over sandy soil or on hard foraging may have a really "old" mouth—and a shortened productive life. If you have a chance to buy some older ewes that look healthy and have fair mouths, consider that they will generally be easier to manage—they've been through years of shearing, lambing, dipping, handling. Their extra experience at lambing time will be invaluable to you. Even an old "gummer" (a sheep with no teeth at all) will produce lambs and fleece for you if you feed and treat her well (she'll need extra feed—and probably chopped feed because she doesn't digest it efficiently). Look at the animal's feet to be sure they're in fair shape. If the hooves are terribly crooked or obviously overgrown to the point of being deformed, don't buy the animal. A hoof that is very soft and/or oozes a dark gummy material may indicate foot rot. This condition, if not too advanced, can be corrected (treatments in any sheep and most livestock books are simple and usually work). Don't buy an animal with abscesses (lumps or swellings) on the face or under the jaw. These are usually caused by an infectious disease, although they may be from wood slivers or minor wounds. An abscess may be lanced and treated—but you're taking the chance of infecting other animals or dealing with what might be a serious infection. In general, choose animals that seem alert and healthy. An animal that is losing clumps of wool may be suffering from internal or external parasites or have severe nutritional deficiencies. Also try to buy animals that you like on sight.

Black Sheep: If you want black or colored fleeces,

you should know that a dark sheep lightens as it ages. A coal-black lamb will often fade to steel-gray or even off-white. This fading takes three or four years, and you will get a series of subtly colored browns and grays—but a "black" sheep that really holds its color is rare. One breeder we know of feeds her sheep blackstrap molasses and says it keeps the fleeces dark. This is probably because of lack of copper and possibly other trace minerals in the diet that cause loss of pigmentation. Wethers (castrated rams) seem to hold their dark color best. They also produce exceptionally heavy, good-quality fleeces—so if you want sheep only for fleece and don't want to bother with breeding and lambing, you may do well to keep a few wethers. Black sheep also tend to bleach out in the sun—so look at the undercolor of the fleece rather than the outer appearance of the sheep.

Most black sheep are not of a particular breed. They appear in purebred flocks or in crossbreeding. Some are from Karakul stock bred with good wool ewes such as Merino and Corriedale in an upgrading process. Karakul sheep are characteristically dark in color but have coarse wool that does not spin up well. They are raised chiefly for the pelts of their lambs (which are marketed as Persian lamb coats and hats) and are hard to find—at least in our area. We bought a couple of "Karakul-Suffolk" cross sheep with fleece that looked more like Merino . . . so look carefully at the animal you're buying before you take the seller's word on breeding. These sheep both had very dark fleeces when we bought them, but one faded to a beautiful silver-gray and the other to a deep brown-gray. The ewe throws coal-black lambs that hold their

294

color to about age three. We breed the ram with our other ewes and get black or spotted lambs from the dark ewes and white lambs (sometimes with light brown spots if we breed the ram's white daughter back to him) from the white ewes. A woman who has been breeding black sheep for twelve years wrote that she still gets one white lamb out of every eight when crossing black/black, but that "that white lamb bred to a black gives about half black lambs." In large-scale sheep raising, black sheep are frowned upon. One black fleece in a bag (100 pounds of wool sacked to go to commercial market) will spoil the bag. Even dark fibers in a fleece are undesirable. Some flocks include a black ewe or wether with every hundred sheep as a counting device. Unpopular as they are, dark sheep are difficult to find and usually expensive to buy. Black ewe lambs—if you can talk the breeder into parting with one—are $35 and up (again, in our area). Rams are cheaper and easier to find. There are some breeders around who have developed true-throwing black sheep with nice fleeces—but you'll have to *search* for them.

Pasture, Feeds, and Feeding: Choice pasture for your sheep is open land with good drainage. Low, boggy areas encourage parasites (liver flukes) and foot rot (a common disease). A few trees will provide necessary shade in the hot months. Thick tangles of vines should be removed—we know of a ewe lamb who caught herself up in blackberry vines and strangled. If a lot of bracken fern is in the pasture, you should clip it down (repeat until it dies away). Bracken fern is cumulatively poisonous to most livestock. Find out what other plants on your land are poisonous to sheep (consult your farm adviser, the *Merck Veterinary Manual,* or livestock books) and remove them.

You may choose to graze your sheep on the native grasses. At certain times of year, when grass is dry and sparse or very green and watery, you may have to supplement this feed with hay or concentrates. Generally only a fourth to one-half pound of protein supplement or a pound or two of grain is all that is necessary. For extra protein there are "range blocs" you can set out. The abundance and quality of feed will depend on your soil and location. Planting annual crops such as rye, wheat, oats, and/or Canadian field peas will provide good temporary pasture. Legume-grass mixtures such as clover and rye make excellent feed. The use of natural fertilizers will increase the productivity and nutritional balance of your pasture. Irrigating will

also increase its yield.

Keeping a few sheep in a small pen and hay-feeding them is feasible but not very economical. In areas with severe winters you may *have* to hay-feed your flock. Each sheep will need about three to four pounds of hay a day. A mixture of ⅔ alfalfa and ⅓ oat hay is good. Other legume hays with some roughage may be fed, or a mix of these and corn or corn silage. If you keep goats, you can feed their leftover stalks (still plenty of nutritive value) to your sheep. Sheep also like apples, Swiss chard, comfrey, cornstalks, turnips, carrots, and lettuce.

Grain is fed to sheep during breeding season to increase their fertility. This is called "flushing." Ewes should get a half pound daily the two months before lambing. This may be continued while they are nursing their lambs, though good-quality hay or pasture should suffice. Lambs may be fed a little grain from the age of two weeks on and will eat grass and hay *very* early. "Creep-feeding" is the practice of feeding lambs in an area the ewes can't get to. Grains suggested for sheep are sorghums, corn, barley (mix with corn or wheat), and oats. We've fed our sheep a barley-molasses mix and also a sixteen per cent protein dairy chow. Alfalfa pellets and silage may also be fed. Always provide your sheep with salt, trace minerals (may be combined with salt), and plenty of clean, fresh water. Keep your sheep well fed but not overfed: they should be solid, not fat. Good nutrition will give them good heavy fleeces, large strong lambs, and natural resistance to disease.

The amount and type of land you fence will greatly influence the economics of your sheep keeping. If the soil is poor to average and you want to graze your sheep, you'll need one acre per animal. This ratio changes according to the quality of your pasture—one acre of exceptional soil, well-fertilized and planted as part of a rotating system can support up to eight sheep.

Fencing: In small-scale sheep raising, one of the biggest investments of time and money is fencing. A good quality, medium heavy, woven wire fence is best (close-spaced, smooth boards would do nicely if you have a lumber windfall). Height should be at least 36". Three to 6 strands of barbed wire above that will discourage dogs from jumping over, and one farmers' bulletin suggests "an apron of woven wire 18 inches wide along the ground will prevent predatory animals from burrowing under." Field fencing with wire spaced 12" apart is used around large pastures because young lambs

sheep shed.

designed for small homestead flock of 6-8 sheep

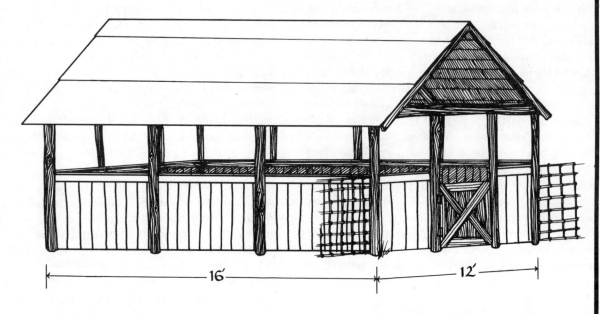

16'   12'

This building is designed to be simple pole framing with siding going up only about 3 or 4' remaining open above. In cold climates some sides may have to be closed. Leave a good overhang on roof to keep rain from blowing back in. With sides open, feeders may be filled from outside the building. Two 3' x 5' lambing stalls may be converted to creep feed lambs.

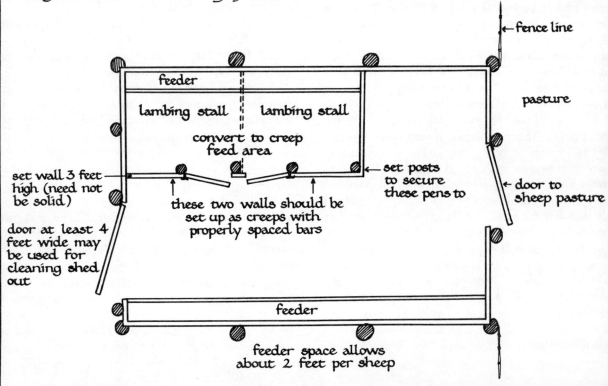

→ fence line

feeder

lambing stall   lambing stall

convert to creep
feed area

pasture

set wall 3 feet
high (need not
be solid)

set posts
to secure
these pens to

door to
sheep pasture

these two walls should be
set up as creeps with
properly spaced bars

door at least 4
feet wide may
be used for
cleaning shed
out

feeder

feeder space allows
about 2 feet per sheep

can go back and forth through the spaces. If spaces are closer, both lambs and older animals will stick their heads through and get caught. The lambs who go through the fence won't wander far from their mothers. If your pasture is small and close to your house (so you can hear or see a caught animal), you can use wire with a 6″ spacing. This wire has a great advantage in being dog-proof. Wire comes in four weights. Medium or heavy will do for sheep. Extra-galvanized will last longest. Avoid using low-strung barbed wire or rough boards—they damage the fleece.

Divide your pasture and cross-fence so that you can rotate the flock and avoid overgrazing the land.

Buildings: Besides fenced areas for your sheep to graze and possibly a creep-feeding arrangement for your lambs, you will need some minimal shelter for your flock. A low, two- or three-sided shed will protect them during heavy rains; a more elaborate building may be necessary in snow. Building some movable partitions will give you a way to isolate ewes and their new lambs when necessary. Or you can make an area of the building adaptable to use either as lambing stalls or creep-feed area (as shown). A feeding rack, manger, or trough should be part of your housing. Plan your shed for easy cleaning, dry floors, no drafts, and protection from the worst weather. Build with smooth lumber to avoid snagging wool. If you have a really large flock, you will want a clean area for shearing—possibly a whole room with a wooden floor. With a small flock, you can do your shearing outdoors on a tarp, or in any area you can sweep fairly clean beforehand.

Routine Care: Once you have your pastures, sheds, and sheep together, what does raising a small flock of sheep involve? In terms of time and energy, not a tremendous amount. Lambing season will demand a lot, and shearing takes some concentrated effort, but generally your sheep won't take much of your day. Periodically they will need their hooves trimmed (twice a year usually keeps them fairly well if they get normal wear). They'll need periodic worming, tagging, and so on (see "Keeping a Healthy Flock").

Contrary to all myths and sayings about sheep, you'll find that your sheep have distinctive personalities. There are sheep who will bound to greet you, let you rub their ears, nuzzle you in return—and those that run at the sight of you even after months of daily care and feeding. There are aggressive sheep and timid ones, sheep who form fast friendships with cats and horses, and others who will butt any other creature entering their pasture. Each sheep will be part of the flock, but each has its own being, unique and personable. It's one of the surprises and pleasures of raising sheep to discover these differences!

Basic to sheep raising is a sense of closeness to the seasons, to the earth. A sense of order and calm. Your sheep will give you lambs and wool, will eat up your goats' leftovers, trim your grass, fertilize your pasture—and grow on you. There is something infinitely comforting, quieting, and satisfying about sheep—a peacefulness they teach you, or return you to . . .

BREEDING

With the exception of a few types that will breed and bear lambs at any time of year, most sheep have a regular breeding season. This season usually begins when the weather begins to cool in late summer and extends through most of the winter. The breed of sheep you raise will determine your breeding season: Rambouillets and Merinos, for instance, come in heat much earlier in the summer than most other breeds. The weather in your particular area will also influence breeding season. Extended hot periods may make your rams temporarily sterile; early cool weather may bring the ewes in heat sooner than normal. You can learn more about your particular flock by reading up on your breed's habits and by talking with other local sheep people. In a year or two, experience with your own flock will give you this particular knowledge.

Within the available breeding season, you will want to fix your breeding months and schedules according to the way you want your lambs to come. If you plan to sell the lambs commercially, try for the very earliest breeding possible. These early lambs will usually bring a good price on the market. Try to breed all of your ewes at about the same time so that you will have lambs of a fairly uniform size. These will sell better as a lot and will grow out better as the lambs will be about the

same size and can compete equally for their food. Try and plan your lambings for a time when you can give most of your attention to your flock. Finally, take into consideration the condition of your pastures when the lambs will be coming, and the availability of other feed. You will want to be raising your lambs when feed is plentiful and of the best quality in your area.

A few weeks before you plan to turn your ram(s) in with your ewes for breeding, you should flush the ewe flock. This means putting the ewes on a higher standard of nutrition, either by moving them to an exceptionally fine pasture or by giving them supplemental feed. Flushing results in higher fertility, and also tends to bring most of the ewes into heat at about the same time. You may flush the ewes by giving them a supplement feeding of legume hay or by giving about a half pound a day of corn, oats, mixed grains, and so on per ewe.

Ewes come in heat about every sixteen days (with a possible variation of thirteen to nineteen days). The *average* heat period is about thirty hours, but this may vary from three to seventy-three hours. During this period, the ewe will stand for and receive the ram. You will probably want to run the ram and ewes together for a month and a half or longer, to be sure each ewe is bred and set-

Ewe lamb born tonight, snow white and wonderful. There is nothing so beautiful as a new lamb. First born sheep for Black Sheep Farm as white as can be. Much as I love the ones I have, this one·feels different. The first born. One to love unreservedly and for years. The experience has changed me, made me more emotional and tender. I wonder now what Mary Blair thinks when she sells her lambs to me—an intimate tie with herself, given to others.

The ten-acre fence is now long (three weeks) finished and the young blacks are out there. Its accomplishment has left me somewhat depressed and lost—I have no major direction to my life now, "only" a consolidation and completing of what is already here. It comes at a time of so much turmoil anyway about what I'm doing with my life. Will this be a homestead or a farm? Will it be either? Can I do well what I have started, will fear or energy or money stop me? I can't afford to gentlewoman farm; I must expand to make a profit, which means going more into debt, or cut back. The state of the economy worries me. I fear losing my job and realize how little security I have. Finishing the fence should have left me filled with confidence and energy; instead I'm immersed in doubts about whether I can really make this place a ranch.

But, oh, that lamb. She is so beautiful and I don't question any future that contains more and more. I love the sheep, they each have personalities and voices to me. Each is special and known.

tled. A mature ram can service (breed) about thirty ewes if he is in good condition and not too old. A ram is in his prime up to age five and can be used many years beyond that age. A young ram (seven to ten months old) or a yearling should never be used to service more than ten or twelve ewes. Before turning the ram in with the ewes, you should shear him and trim his hooves. Shearing will lower his body temperature and make him more fertile and more vigorous. The ewes, if they have not been very recently shorn, should be tagged, and their hooves trimmed as well. At this point you will also want to separate out your ewe lambs—they should not be bred until they are at least a year old and preferably older, and well grown. Breeding a ewe too young will stunt her growth, and she will produce small, unthrifty lambs.

If you have a very small flock and keep a close eye on them, you will probably see many of your ewes bred. In larger flocks, two methods are used to keep breeding date records and to make sure that the rams are actually servicing the ewes. One method is the use of breast paint on the ram. A thick paste of raw linseed oil and lampblack or a similar commercial paint (available at feed stores in sheep country) is applied to the ram's chest weekly. When the ram mounts a ewe to breed her, some of this paint will rub off on her rump and mark her. The second method is similar, but instead of the paint a special "marking harness" is used. This harness, which holds a marking crayon, is worn by the ram and similarly marks the bred ewes. The use of paint or harness will also give you an idea of the fertility of your ram, if he is servicing the ewes over and over again and they are not settling, he may be past his prime as a breeder or have some temporary or serious breeding problem. Whether you can watch for and catch your breedings or make use of marking paint or harness, you will have a record of your expected lambs and can plan flock management around these dates.

The gestation period of the ewe varies from 144–152 days. Fine-wool breeds tend to have longer gestations, medium-wool breeds the shorter gestations, and long-wool breeds average in between. During their gestation, ewes should be kept in good condition, well but not overfed. It is a good idea to have the ewe flock on a gradually increasing standard of nutrition during the last month of pregnancy, when the fetuses are doing most of their growing and the ewe is storing reserves for the demands of lactation. This means giving a grain supplement and giving supplementary hay if the pasture is not in excellent growth. It is also important to keep the ewes exercised. If they are grazing a large pasture, this will happen normally. If, however, they are confined to a small lot or pen, they should be encouraged to move about. Scattering their hay around the lot will give them some exercise as they move about to feed. Underexercised ewes tend to have more problems lambing and are more susceptible to pregnancy-related diseases such as ketosis (see Appendix article). It is equally important to keep the ewe flock calm and untroubled by dogs, other animals, and the rams. Cattle, horses, and hogs should not be run with the pregnant ewes. Once the ewes are settled, the ram(s) should be removed to a separate pasture. These ewes should not be moved to unfamiliar quarters or switched to new feeds during the last month of their pregnancies. All radical changes in management can lead to problems such as abortion or ketosis. Generally, you should keep your flock quiet and stable during this period. Prepare yourself for lambing and enjoy this period of rest . . .

Coming back from Sonoma, I multiplied heads of sheep by price of fleece by cost of fence for mile after mile, driving up the coast. I have decided to do it, to borrow the money and build the fences and make this place a farm. But making the decision has been agony. Partly it's the money, the going into debt. My whole life I've dealt with money in small, paid-in-cash sums. I spend $600 to fence ten acres and get a small sheep flock that will continue to eat more than it pays for, instead of spending $3,000 to get a flock that will support itself and possibly me too. I am untrained to deal with big money, to trust myself enough to take risks that depend on me, to create a way to support myself instead of supporting others. The very borrowing of $3,000 (think how little $3,000 is in this society) feels like an act of courageous effrontery.

And then besides the money there's the commitment, the years it will take to make the farm really work. And even though I'm there already—planning and dreaming years ahead of myself, so involved in today and tomorrow that I can't ever find time when I *want* to visit the city—I'm still very scared by this settling down, this end of my youth. For so long I've slid along from college to job to job, from East Coast to West, into and through a marriage, to this place, without ever once saying, "well, this is it, this is my life." Well, here I am. This is my life. I am settled here, and I will be tending lambs and patching fences and paying debts five years from now. I know this is right for me and for this time in my life and for this time in America. The committment has not come easily. Working at it has made it deep and real, but the conscious, public statement has been hard.

LAMBING

Lambing season on the Mendocino coast is wet and chilly. It begins in late November or December and may continue on through March. Drenching rains, high winds, and three-day storms are all common fare. In between are days of cold, clear brilliance, and nights of frost—and sometimes snow. Strange weather for lambs? Other places they're born to deep snow and sub-zero temperatures or blizzards. Our winter-green fields suddenly look milder . . .

Lambing season actually begins a month earlier, when range ewes are brought in from their hillside grazing close to the sheds and pens. On our small farm, we don't move the sheep anywhere. We begin to notice that the ewes are "showing"—their sides bulge, their bellies hang lower and wider. Their udders begin to fill. For the two months before lambing, the ewes must be in their best condition. This means extra feeding of about three quar-

ters of a pound daily per ewe of mixed grains added to their normal ration of grazing and/or hay. If the ewes are being hay fed, the hay is scattered around their field each feeding to encourage them to exercise. As always, the ewes need plenty of fresh water and access to salt. Some shepherds advise worming the ewes two weeks prior to lambing to protect them and the newborn lambs from excessive loads of internal parasites.

When the ewes' udders are well filled out, we move the ram(s) to another pen (they will sometimes bother a ewe who is about to lamb). We tag the ewes (clip the wool away from their rear quarters). Usually the ewe will pull the wool away from her udder—if not, we clip that too. Any feed boxes that the ewes might get their front feet up on are removed from the pasture. One year we had a ewe heavy with lamb who stood with her front feet up on a box part of almost every day. The pull of gravity on the unborn lamb caused her vagina to slide out. This is a fairly common problem with sheep who lamb in hilly country and lie with their rear quarters downhill too often. It is called a "prolapsed vagina" and may result in a much more serious condition, a prolapsed uterus. We've learned, too, to have a place ready to isolate reluctant mothers with their lambs. And we've learned to keep close to home, with an eye on the ewes—special attention at lambing time is often needed.

The few days before a ewe is about to lamb, changes are usually noticeable. Her udder gets very tight, swollen, pink, and shiny. The unborn lamb changes position, causing a hollow to appear just below and in front of the ewe's hip bones. When she is close to lambing, a ewe will go off by herself. If she doesn't come in to feed with the flock, you can usually expect lambs within a few hours. Often a ewe will wait to lamb at night, or when you aren't around. Some prefer stormy nights or the earliest of morning hours. If you think a ewe is about to lamb—and especially if her lambing is her first—it is a good idea to confine her in a small pen or area of her own. This pen should be dry, as draft-free as possible, and deeply bedded with straw. A pen located in the shed where she can see and hear the rest of the flock will relieve anxieties about being isolated.

Just before beginning her labor, the ewe will circle, paw at the ground, and turn to look back toward her tail. She may have a thick, yellowish mucous discharge and her vulva will probably be unusually swollen and pink. The ewe will appear very restless, circling and pawing, changing her mind and moving to another part of the pen, and so on. Finally she will lie down and begin to strain. Her labor may last from a few minutes to a few hours and vary in intensity. If her water breaks and the lamb doesn't come fairly soon thereafter, you might suspect trouble. Normally, a lamb is born with its front feet extended and its nose resting

(pointed forward) on its front legs. If one hoof appears, or the tail first, or if the ewe strains heavily and nothing appears, the lamb is probably positioned wrong. A leg bent back or a head turned improperly may catch the lamb in the birth canal. Twins may get tangled and jammed in the canal, particularly if the first of the two has come with a leg or its head bent awkwardly. Single lambs are sometimes so large that they are difficult to deliver. Always give the ewe sufficient time to deliver her lamb(s) herself. Her straining and contractions may seem interminable your first few lambings, but have patience. If you determine finally that you have to help her, proceed with gentleness and determination. It will help to have a friend there to hold the ewe if she struggles (as well as for moral support). First wash your hands well with a mild soap and warm water. Wash the ewe's vulva and surrounding bare skin as well. Most of the procedure for delivering a difficult lamb is identical to that used in delivering a kid, so you should refer to the "Kidding" section of this book for more details. The ewe's vulva will stretch to receive your hand much as it stretches to pass the lamb. If the lamb is, indeed, stuck in the birth canal, you will feel it almost immediately. Try to visualize what you are feeling. Close your eyes and concentrate: if you feel a hoof, a leg, a nose, try to follow the connecting bone or neck to see how the lamb is positioned. Legs that are folded back may be gently pushed and pulled straight. If the head is bent back against the body, push the entire lamb back a little in the passage and try to straighten the head. A lamb may be delivered back legs first—this is a common position in the second of twins—as long as the legs are straight back and not bent in some strange way. This lamb must never be left with half of its body protruding from the ewe and the other half inside, as the pressure on the umbilical cord may cut off its oxygen supply and suffocate it. A lamb coming normally, though, may be sus-

pended for quite a few nerve-wracking minutes with only its front half out and not be endangered. This lamb is receiving its oxygen from its mother—but if that supply is cut off by pressure, it may still open its mouth (you may have to clear the nostrils of mucous) and breathe. Once you have a problem lamb readjusted to what should be its normal birth position, you can usually relax and let the ewe deliver the lamb. If she is very weak, though, or her contractions become weak, you may have to deliver the lamb. Get a good firm hold on both front feet if possible and pull very, very gently *with* the ewe's contractions. Pull down toward her hocks. The lamb will probably come sliding out in a fluid, easy motion once the head and shoulders clear. In the case of twin lambs which have become simultaneously stuck in the birth canal, you will have to push one lamb (after untangling it) back up toward the womb and deliver the first in line. The ewe will then usually deliver the second lamb normally. Any time you have to help deliver a lamb, there is danger of infection for the ewe. Farm supply houses and most pharmacies sell special medicated boluses which are inserted into the ewe's vagina (as far up toward the womb as possible) to disinfect and protect it. Watch this ewe carefully for the next week or so for any bad-smelling, unusual-looking discharge that would indicate infection. If this happens, you will want to treat the ewe systemically with a broad-spectrum antibiotic until the symptoms disappear. In the case of a really serious infection, you may want to call in a veterinarian to flush the ewe out with a suitable antiseptic/medication.

Once the lamb is delivered, clear its nostrils and mouth of mucous. Normally the ewe will do this, but if she has had a difficult labor, she may not. If the sack the lamb is encased in doesn't break as soon as the lamb is born, you may have to tear it loose. Put the lamb close to the ewe's head so that she can lick it clean if she doesn't move to do this.

If she's exhausted and won't or can't tend to this, you may have to take over. Rub the lamb briskly but gently with an old towel, a handful of straw, or a clean burlap sack. All new lambs should have their navel cords dipped in tincture of iodine to prevent some very serious diseases and infections. This is absolutely imperative, whether the lamb is delivered normally by a doting mother or helped into the world by you. The newborn lamb is covered with a sticky, orange-colored mucous. An old-time livestock book I read suggested this was to make the lamb the most obvious thing on the landscape, so that the mother could easily find it. A normal lamb is up and about in five or ten minutes, a little unsteady but eager and lively. It will try to nurse its mother's legs, stomach, or udder looking for the teat. You may have to help it. At this point, and again once or twice within the next few days, you should check the ewe's teats to be sure that the milk can flow normally. Squeeze each teat gently—the milk should come in an even stream at once. Sometimes a waxlike plug will prevent the milk from flowing and, consequently, the lamb from nursing. This plug can be worked loose by a little gentle massaging and squeezed out with the first streams of milk. The ewe's udder should also be checked for unusual lumps, swelling, or hotness. Any of these might indicate a mild or beginning case of mastitis (see "Keeping a Healthy Flock"; also Appendix article). If a ewe refuses to let her lamb nurse after accepting it for a number of days, you should likewise check her udder for signs of mastitis.

If you find a newborn lamb that is extremely weak and with no energy (it can't or won't stand, has no interest in nursing), it has probably been chilled. This happens to lambs born in very cold weather, particularly if the mother doesn't clean the lamb thoroughly (the licking and nudging involved in cleaning the newborn lamb not only removes the mucous and afterbirth, but stimulates

the lamb's circulation). The chilled lamb must be warmed as quickly as possible. Bring it into the house and put it close to a warm stove, under a heat lamp, or in a very warm room. Wrap and rub it with warm towels. Some sheep books suggest immersing the lamb's body in warm water to raise the body temperature immediately (dry it well after this dunking and keep it near the stove or heater). When the lamb is warmer, it should look a little more energetic and alert. If the lamb is strong enough to stand, you may wrap it in a blanket (to minimize the shock of changing temperatures and chilling again) and take it back out to the barn or shed to its mother. Stay by the two to see that the mother accepts her lamb back and that the lamb is actively nursing. If the lamb is very weak, you may want to milk the ewe (only take a few ounces) and bottle- or tube-feed the lamb (see section on Bummers) the first few times.

If a ewe has a difficult delivery or has twins and will accept only one of them, she should be isolated in a pen or stall with her lamb(s). Smearing a handful of afterbirth on the lamb and encouraging the ewe to lick it first from your fingers and then from the lamb may result in her acceptance of the lamb. For the first few days, a ewe knows her lamb only by its smell (later she knows its voice) and this trick makes the lamb more positively hers. You may try this same approach when trying to convince a ewe to accept an orphaned lamb if her own has died or been killed (see "Raising a Bummer Lamb," following). If one twin is favored to the exclusion of the other, you may have to take both lambs away for a little while and then give the ewe the "unfavored" one first. Once she has accepted this one, the other may be put with her too.

Shortly after lambing, the ewe should be given some warm water and molasses (about a half cup to a quart of water) to drink. Some ewes will refuse this treat, but others accept it as a ready source of tasty nourishment. The ewe should pass her afterbirth within a few hours and be up to graze or nibble hay. She will be very concerned with her new lamb, calling to it and nudging and licking it lovingly. In the rare cases of retained afterbirth (wait at least twenty-four hours, and be sure the afterbirth hasn't been pawed into bedding or dropped in a different part of the pasture; some ewes will eat part or all of the afterbirth, but in most cases you will see the afterbirth at least beginning to come out), the ewe must be treated by you. You can either use a special bolus (available from most farm supply stores and country veterinarians)

which is inserted into the ewe's vagina to stimulate passage of the afterbirth, or use an herbal drench. The *Herbal Handbook for Farm and Stable* sugggests a brew of linseed, raspberry leaf, and molasses. Another old-time treatment is to give the ewe some of her own milk to drink—something in the colostrum apparently helps to stimulate the passing of the afterbirth.

For the first few weeks after lambing, the ewe should be fed lightly. Build her up to plentiful good-quality hay and some grain (oats especially) for a heavy milk flow. Good pasture is ideal feed for the newly lactating mother. Root crops (carrots, turnips, etc.) fed with hay are also good. The lambs will begin to nibble at grass or hay when they are very young. When they're a few weeks old, you may start feeding them a little grain.

A new lamb is infinitely curious—it will come to your outstretched hand to see what you have (or are) with very little fear. During the first few weeks, you should give each lamb special attention if you have the time. Handle and pet and talk to it to help it get used to trusting you and willing to come to you. When it is a little older, spending the few minutes a day hand-feeding it a bit of grain or special hay will make all the difference in how it relates to people. A lamb that is calm and trusting around people will be a pleasure to shear, move from pasture to pasture, or handle for such routine care as hoof trimming and tagging. Lambs are also endlessly playful—they will bounce at chickens and cats that happen through their pasture, invent games of tag and running amongst themselves, and spring about with sheer uninhibited joy. In between their playing they do a lot of napping and eating.

When a lamb is a couple of days to a few weeks old, its tail should be docked (see "Docking"). Ram lambs that are not of exceptional quality (good enough for breeding stock) should be castrated at or near to the same time (see "Castrating"). When the lambs are four to six weeks old, they should be immunized against tetanus, enterotoxemia, and so on. All lambs should be fed well and consistently. Don't switch feeds suddenly or overfeed. Watch to see that the lambs are nursing normally. A lamb's tail should wag freely (as long as it has one!)—if it doesn't, check to see that it isn't pinned down with an accumulation of sticky feces. This is one problem I've seen many times in our lambs. A lamb with a pinned tail should be caught, and its rearquarters washed or soaked with warm water until the tail can can be pulled loose.

The area under and around the tail should be cleaned as well as possible and then dried. Lamb's feces are normally an orange color for the first few days, probably a result of the rich first milk. They may be sticky and rather loose. If an older lamb begins scouring (having diarrhea), watch it closely and treat it if the scouring continues for more than twenty-four hours. Sometimes a little powdered ginger mixed in some warm water and given as a drench will suffice. Or the lamb may be given three or four tablespoons of mineral oil, which soothes the intestinal lining and moves out any irritating material (this oil will accentuate the scouring at first but may relieve it in the long run). If the lamb continues to scour, or shows other symptoms such as weakness, indifference to food, and so on, you may want to treat it for possible bacterial infections. One good product I've used for lambs with digestive problems that have seemed serious is Sulmet—an orally given sulfa drug (follow directions on the bottle and be sure to continue treatment a day or two after the symptoms disappear). Bad hay (moldy, dusty, or containing strange bits of vegetation), spoiled grain (wet, moldy, etc.), and too much grain can all cause scouring. Any lamb that is scouring badly should be isolated with its mother in a dry, draft-free pen while it is being treated. If your flock isn't protected against this disease, you might also suspect enterotoxemia in a lamb that begins to scour very suddenly (see Appendix article). Lambs are also subject to a supposedly common (I've never seen this condition but have read of it many times) problem called inverted lashes. One or more of the lamb's eyelashes will turn in under the lid and irritate and damage the eye. These lashes should be removed. The eyelid may then be nicked lightly to cause scar tissue to form, holding that part of the lid in place and preventing more problems. Lambs should also be protected against severe infestations of internal and external parasites (see "Keeping a Healthy Flock").

Lambs are generally fast-growing, sturdy, healthy, and inquisitive animals. They need space to exercise, good feed, and a little daily care. Lambs leap, spin, frolic, and gambol—there aren't enough verbs for them. Often their sideways dances across the pasture will catch an old ewe unaware: she'll go springing with them, careless and abandoned, then stop short with a most amazed look on her face . . . and go back to grazing.

A season of lambing—with all that it involves, teaches you and stirs in you—will probably give your sheep flock a very special, enduring place in your heart and your life.

RAISING A BUMMER LAMB

Now and then you will be faced with or offered an orphaned lamb or one whose mother simply won't or can't take it. The orphaned lamb may be weeks or months old; the unaccepted lamb will usually be a few hours old. These lambs are called "bummers" regardless of age. Each presents certain problems having to do with its age. Each can be reared through a healthy, lively lambhood if you're willing to put in some extra time and energy. There is no animal I have ever known more trusting, sweet, and totally endearing than a bummer lamb. You will probably find your effort not a task but a delight!

The newborn lamb, deserted or orphaned, is in a most serious plight: without the colostrum from its mother, it has no antibodies with which to ward off disease, no source of initial strength, no intake of warm nourishment to offset the sudden change of temperature from inside the womb to the outside environment. It is susceptible to chills and pneu-

monia or scouring. It is also vulnerable to attacks from predators and unfriendly butting from other sheep. Left to itself, it would most likely die within twenty-four hours. In my experience, the weaker or secondborn of a set of twins or triplets is likely to be deserted—particularly if the ewe is lambing for the first time and is not an experienced mother. Sometimes a ewe will leave her lamb if she is harassed during lambing, or if she has a very difficult delivery. Now and then a ewe will appear who is simply an unwilling mother, and cannot be convinced by the most elegant trickery or extended solitary confinement to take her lamb. You may also find yourself with a lamb whose mother has been killed by dogs or has died during lambing.

If the lamb's mother is alive and you try isolating her with the lamb and she still refuses it, you must see that the lamb gets its colostrum. Sometimes you can simply corner and hold the ewe and guide the lamb to nurse. Some ewes will accept this treatment (it may even lead them to accepting the lamb); others will become furious (or frightened) and stamp and kick. This latter situation is dangerous for the lamb—and short of hobbling the mother, you may have no choice but to milk the ewe and bottle-feed the lamb this first milk. A ewe is not too much harder to milk than a goat. Her teats are tiny, of course, and you will have to milk with two fingers rather than your whole hand. Milk out only a few ounces, duplicating the lamb's normal appetite at first. If you can't milk the ewe or have been brought the lamb from a dead or unknown mother, you will have to make some substitute for this colostrum. If you've had the foresight (and good fortune) to keep frozen colostrum from another sheep, a goat, or a cow on hand, you can thaw and use this. Or you can make up a formula. The best I've seen is this one, taken from *The Shepherd* magazine: mix one and a half pints warm milk with one beaten egg, one teaspoon sugar or corn syrup, and one half teaspoon cod liver oil. One final word on the reluctant mother before you go off to bottle-feed your lamb: never *assume* that because you have isolated a lamb with its mother and restrained her and let the lamb nurse that the ewe will now take the lamb. It is very likely that the ewe will *still* refuse to mother her lamb. You should spend enough time with the two of them to see for certain what is happening. Once you actually see the lamb nurse again, on its own, you can go off with a clear mind. If it tries to nurse and is once more rejected—you can try restraining the ewe again, but chances are you have a bummer on your hands. It is dangerous, too, to leave an unwanted lamb with a ewe who might stamp on or butt the "stranger" sharing her pen. Ewes are not as docile about readopting a rejected lamb as some sheep books would lead you to believe! All of this applies too to "grafting" a lamb onto another ewe who has a single lamb or has just lost her lamb. You may come up with a ewe willing to mother this lamb. She must have a lamb of almost exactly the same age—never graft a lamb to a ewe whose own lamb is older, as the older lamb will take all of the milk and starve the newcomer. Techniques for disguising the lamb you hope to graft onto the new mother include smearing this lamb with afterbirth from her own lamb (if she has just had one), or covering the lamb with her lamb's skinned pelt (if her lamb has died). The intended mother and the orphan must be kept in a solitary pen and watched carefully. If you have a small flock of sheep,

chances of having a willing and able mother coincidentally with your bummer lamb are small. Back to the bottle . . .

Feeding a baby lamb is much like feeding a baby goat (see "Raising a Kid"). Gentleness, patience, regularity, cleanliness are the key words. For the first few days, the lamb should be fed as often as you can manage (five or six times a day—or more—and at least two or three times in the night). You can use a lamb nipple (available at most feed stores) and a small bottle of any type. The colostrum must be heated carefully as it tends to scorch. When it feels hot on your wrist (not burning hot, but much warmer than blood temperature), it is ready. A few ounces per feeding is adequate. The lamb should always be hungry *after* it eats. Watch a ewe nurse her lamb someday and you will see that she allows it only a few minutes to suck and then walks away. The lamb's digestive system can only handle a little milk at a time. Overfeeding can result in enterotoxemia, bloat, scouring, and other problems—may literally kill the lamb. Duplicating the lamb's normal nursing habits, you must feed many times a day.

Colostrum should be given to the lamb for the first few days. After that, you have to choose the best substitute for ewe's milk. Cow's milk is *highly unsuitable* for a baby lamb. It is difficult for the lamb to digest and is too low in fat content. Ewe's milk contains almost twice as much fat as cow's milk, as well as twice as much digestible protein, almost double calcium, and half again as much phosphorus. Only the sugar content is about the same. I have had really good success raising bummer lambs on fresh goat milk. This milk is easier for the lamb to digest and a little closer in composition to ewe's milk, though it still lacks the fat content (even from a herd of high butterfat-producing Nubians and La Manchas). Lambs raised on goat milk seem to thrive. Another excellent substitute is the special high-fat milk replacer made for lambs. This milk replacer is powdered and must be mixed with water. It is available at most feed stores or can be special-ordered through them. The very best brand in my experience is Land O' Lakes. I've mixed this half and half with goat milk or used it straight when we've been short on milk. The lambs do very well on it and seem to like it. Be sure you specify *lamb* milk replacer when you go to buy or order at the feed store—most companies also make a milk replacer for calves, which is unsuitable for your lamb. If you feed any kind of powdered milk other than the replacer, you may have trouble with your lamb. I've seen one case of mild rickets in a lamb fed powdered milk from the local grocery store. The vitamin D content was too low. This lamb recovered completely and rapidly when it was switched to goat's milk. Some cases of poorly growing, sickly lambs were traced directly to the government-surplus powdered milk they were being fed. Chronic scouring, slow growth, and general ill health seem to be the results of feeding lambs nutritionally insufficient milk.

For the first few weeks, your lamb should be fed three or four times a day—whichever your schedule allows. The amount of milk per feeding should be gradually increased until the lamb is getting about a quart a day all together. This amount can be gradually increased, too, until the month-old lamb is drinking a quart and a half per day. It will probably be nibbling at pasture or hay by now, too. You can cut down to three feedings a day if you haven't already. Another period of slowly increasing the amount of milk per feeding should put your lamb on a regular half gallon a day ration. This may be given in two feedings, but is better for the lamb to keep it on three feedings if you can. It should stay on this amount and schedule almost until it's weaned. A lamb can be weaned anywhere from six weeks of age to four or more months. Most of the bummers I've raised have more or less weaned themselves at three and a half months— some will stay on the bottle longer, few any shorter (given a choice). About two weeks before you plan to wean your lamb, cut it down to two feedings a day (less than a quart per feeding). After a week, cut down to one feeding. A week later, your lamb can be weaned. It should be grazing well by now, and be introduced to a small amount of grain each day.

If your lamb comes to you newborn, it will need some careful handling for a few weeks. Without a mother to cuddle with, it may have to be kept in the house if the weather is chilly. Or it can live in the barn with a heat lamp at night. It definitely needs a snug, well-bedded shelter—a box in the house or barn, or a little stall of its own. I've always raised these lambs with our goat kids. They don't seem to snuggle up at night the way the kids do, but they enjoy the companionship and warmth of other young animals. The only problem with this arrangement is a long-term one: the lambs grow up thinking they're goats, and it's very hard to persuade them to join the sheep flock! At any rate—the lamb can be out in the day if the weather isn't terrible, and put in at night. As its coat of wool grows in, it can take more exposure. Even then, it will need the protection of a shelter and a bed of deep straw.

Your lamb will also need another type of protection—from dogs and other animals. Don't be fooled by an overly affectionate dog who engages in "playing" with the lamb. This play can lead to a serious game in which the lamb is likely to be mauled. Have firm rules about the lamb's "playmates" and stick by them—even with your own dog(s). A lamb is one of the most vulnerable and totally defenseless creatures there is. A well-fenced, dog-proof yard should be available for the lamb when you aren't there with it. You should also be aware that other animals—larger goats, pigs, sheep, horses, and so on—all pose a potential (if often quite unintentional) threat. The first lamb I ever raised was accidentally killed by our goats. It was roaming loose around the farm and somehow squeezed through a pole fence into their pasture. They apparently began butting at this strange new creature as goats will do amongst themselves—and broke its neck. It was a very sad lesson in protective/preventive fencing. Pigs have been known to kill and eat baby lambs that wander into their pens, and sheep and horses might easily hurt a small lamb. All possibilities should be considered even if they seem unlikely. It seems better to be overly cautious than regretful later.

If the bummer you find yourself raising comes as an older lamb—a week or two, or even a month— you will have some other problems. These lambs usually come from ewes who have been killed (often by dogs) and they have probably suffered from exposure and hunger, if not directly from an attack. They are often not brought in from fields or range until they are too weak to run away (a vigorous, newly orphaned lamb is almost impossible to catch). You should consider giving this lamb an immediate, protective/preventive series of penicillin shots against an almost certain case of exposure-caused pneumonia. You can use Combiotic (a brand name penicillin-streptomycin combination) or Bicillin (this drug was suggested in a sheep raising magazine because a single shot gives a week's protection). I would not advocate loading a lamb —or any animal—up on antibiotics for no good reason, but in this case the lamb is very susceptible to pneumonia, and pneumonia is better prevented than treated. The older orphaned lamb will take some very special feeding and care. If it has been nursing its mother and pretty much running wild, it

I carried the fence posts down into the canyon today for the seventy-acre fence, setting them in the holes I dug yesterday. I ferried them down on my back, one or two at a time according to the weight, then went back up for more, up a hill-side that would satisfy a Sierra Club hiker seeking a challenge. I'm tired, tired out, tired through, too tired to feel proud or happy, only aching muscles. It's the kind of tired the fantasies leave out.

Setting fence posts is a way of getting to know the land intimately, every bump and hollow and gully and rise of it. I look at a section of last summer's fence and my feet remember the feel of the ground in that spot. It's an all-body memory; I know this land with my legs and my back and my arms and my feet—bone-deep knowledge.

will probably be very frightened of people and *very* unwilling to take a bottle. You will have to be as patient and gentle and persistent as you can. Keep the lamb in a small confined area so there will be a minimum chase involved in catching it. This will make things easier on both you and the lamb. Spend as much time as you can there, until it becomes accustomed to your presence. The first few bottles will probably be traumatic since you will have to patiently teach the lamb that the bottle is a source of food, not something strange and terrifying. Never force a lamb to drink, as you can cause inhalation pneumonia if milk goes into the lungs rather than down into the stomach. Usually a few dribbles of warm milk around the lips will encourage the hungry lamb to try sucking. It may have a hard time adjusting to the nipple, but eventually it will learn, and the hard part will be over. If the lamb is in poor shape, you might add a little honey to the first few bottles as a tonic and a source of easily digested immediate nutrition. A very weak lamb will have to have its head supported and milk dripped into its mouth a bit at a time. This lamb may also be tube-fed, but this is a process that should be done by a skilled veterinarian or experi-

enced person. In tube-feeding, liquid is run through a tube directly into the lamb's stomach. There is a serious danger of liquid being run into the lungs if the tube is improperly placed. The very weak lamb should be kept warm and quiet and tended rather constantly. I've seen nearly dead lambs spark back to life wrapped in a towel, kept near the wood stove, and lovingly, persistently fed. If you have the time and devotion to give to this little animal, you will be well rewarded when that soft woolly head gives its first shake of returning energy and a tiny baaa greets you . . .

Whether your lamb comes a rejected newborn or an older orphan, be sure you take care of routine lamb matters: tail docking, immunization, hoof trimming, and castrating of rams. If you aren't keeping an objective distance and raising this animal for meat, you might find yourself involved in a very special relationship with a lamb who follows you about, comes to a call, and has a growly little voice reserved especially for you. On the practical side, bummers grow up to be wonderfully tame, easy to manage, and personable too. A better leader for a small farm flock I can't imagine.

docking

Lambs are born with long woolly tails that bounce when they bounce and wag when they nurse. Tails are docked (cut off), for two main reasons. First, for cleanliness in the grown animal—a long tail will catch urine and feces and become very soiled. A short tail means a cleaner fleece. In ewes, a short tail makes lambing more sanitary and thus safer (less chance of infection). Long-tailed ewes are also supposedly harder to get bred in the first place. The second reason for docking has to do with the first. There are two types of flies—blowfly and screwworm fly—that will lay their eggs in soiled, matted fleece such as that found under and around a long tail. When the eggs hatch, the maggots burrow down into the flesh of the animal. This condition, called blown, can cause serious infection if not tended, and could lead to the death of the animal. Sheep that are put out to pasture would be particularly susceptible if they had long tails. Even a farm or homestead sheep kept as part of a small, well-tended flock would be susceptible to such an attack. Docking seems a simple precaution you should take for the long-term health of your sheep.

Docking is usually done when a lamb is three to fourteen days old. There are a number of methods used. You will usually need a friend to hold the lamb or a specially made box or chute to confine it. The tail stub you leave should be from an inch to two inches long. In large flocks, rams and wethers are sometimes left with longer tails (up to four inches) for easy identification. If you look at the wooless underside of the tail, you will notice two folds or ridges of skin where the tail connects to the body. I usually cut or band the tail just beyond these connecting pieces. This leaves the proper length. Before you cut or band, be sure the area is fairly clean—warm water and soap or alcohol will remove dirt that might cause later infection.

Very young lambs may be docked with a sharp knife. I have used a small pair of garden pruning shears in much the same way. Cut the tail and then press or pinch to stop the stub from bleeding too much. A sprinkling of blood-stop powder (available from vet supply houses—*very* inexpensive) will check bleeding and disinfect the wound as well. Another method used to stop bleeding is touching the end of the severed artery briefly with a hot iron. This cauterizes, sterilizes, and seals the wound. Cutting with a cold knife or shears is not recommended for lambs over ten days old as they might bleed to death. The open wound caused by this method (unless cauterized) may be a problem if lambs are confined to small pens (too unsanitary) or if it is fly season. You should definitely swab the cut tail with iodine or similar disinfectant once it stops bleeding. When we first got our sheep, I docked our lambs this way with no problems other than my own initial squeamishness. It isn't easy to clip or shear off a wiggly, bony piece of your first lambs!

A second method of cutting is with a hot iron. Docking irons can be bought or made—a blunt chisel might work fairly well. Sheep books advise "a black heat, just turning red, rather than a red heat." The wound made by this method should be reasonably sterile and not bleed—but it will take longer to heal than the cold knife or shear wound.

Tissue-crushing instruments made for castrating may be used for docking. Either the scissors-type emasculator or Burdizzo clamp will squeeze the large blood vessels and actually sever the tails of small lambs. Larger lambs may need to be cut after the clamp has crushed some of the tissue. This method should be fairly bloodless.

The rubber-band method of docking is not advised in most sheep books, but I have used it for years with no problems. This method requires an instrument called an elastrator, which is not expensive and may also be used for castration. The elas-

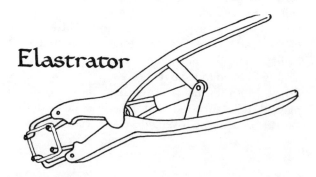

Elastrator

trator merely stretches a small, thick rubber band which then may be fitted over and snapped around the tail. It cuts off circulation, causing the tail to eventually atrophy and drop off. I like this method because neither you nor the lamb has to suffer the trauma of cutting. It is bloodless and quick. The lamb will often circle around uncomfortably for a while, but the pain or annoyance soon stops. If you do your lambs extra young—three days old, if they are strong and healthy—the tails drop off sooner. It usually takes five to ten days for the tail to drop off. We live in a cool coastal area so have few fly problems—one of the disadvantages of this method in other areas. Our farm adviser said we would have problems with tetanus if we used this method, but in five years we didn't lose any lambs to tetanus and now we've begun routinely vaccinating them against it. We have either been lucky or the problem really isn't related—but be prepared for it, just in case . . .

CASTRATION OF SHEEP AND GOATS

All male animals that are not to be kept for breeding should be castrated or gelded at a fairly young age. This operation involves removing the testicles or destroying their function. Most farm animals may be castrated at home in a simple operation. Some—horses and mules, for example—should be castrated by a veterinarian or experienced person. Castration will affect the temperament and growth of the animal. If you are raising animals for meat, this operation will have definite effect upon growth rate and type of carcass the animal produces.

Sheep and Goats: Wethered kids and lambs are considered to have better quality meat, with a larger percentage of their weight more usable. If you plan to keep a male kid as a pet or companion for another goat or a horse, he will be more tractable (and probably happier) as a wether. Wethered sheep have the finest of fleeces, as all of their energy beyond body maintenance and growth goes into their fleeces.

There are three basic methods for castrating either a lamb or a kid: cutting with a knife, using an elastrator, or crushing the cords. Regardless of method, the animal may be castrated as soon as the testicles have descended. The younger the animal is, the easier this operation is and the sooner the animal recovers. Most people castrate lambs when they dock their tails (four to fourteen days of age). Kids may be castrated at three days old. Most livestock books say that castrating should be done on a clear, warm day. Probably the mild weather makes it easier for the animal to recover from the mild shock of this operation. As with tail docking and other processes, it will help to have a friend available to hold the animal. The lamb or kid should be positioned on its back with all four feet drawn together and held, or in a sitting position on its rump with feet held.

The most common method of castrating is cutting with a knife. A small, sharp pocketknife will do. The lower one third of the scrotum is slit open or cut off. Squeeze the base of the scrotum (where it connects to the body) to cause the testicles to descend. Each testicle should in turn be grasped and pulled out slowly and firmly until the cord

breaks. There should be very little blood. It is important to pull the cord rather than sever it with a knife—a sharp cut could cause serious bleeding. You may, if you want, scrape the cord with a dull knife until it breaks rather than simply pull. This will cut the cord, but it will be a ragged break which will not cause hemorrhaging. This method of castrating leaves an open wound which can drain freely. You should apply an antiseptic to the wound to prevent infection. If it is fly season, watch this wound for fly strike (see "Keeping a Healthy Flock").

The second method of castrating employs an instrument called a Burdizzo or emasculator. This is a bloodless operation in which the cord that leads to the testicles is crushed. The testicles gradually dry up and wither away. There is no blood, no danger of infection from an open wound, and very little pain for the animal. It seems to be, however, the most risky of the three methods. If the cord isn't entirely crushed, the animal is not fully castrated. This happens pretty commonly. This (plus the expense of the instrument used) makes the Burdizzo method one you will probably consider least.

After watching a friend castrate our goat kids by cutting, I decided to try the elastrator which worked so well for tail docking. This instrument makes castration bloodless and almost painless. It may be done on a very young kid or an older animal. I usually castrate our kids and lambs at about five days of age. Have a friend hold the kid as shown. Stretch the rubber band on the instrument

312

and slip it over the scrotum, still holding the band stretched. You should then check to see that both testicles are descended. If not, pressure at the base of the scrotum should make them descend. Let the elastrator close and roll the band off the instrument into place. The circulation to the testicles is cut off by the tight band, and the scrotum and testicles will atrophy, dry up and drop off within a few weeks. There is little danger of infection with this method, but you should check periodically to make sure that none is beginning. The kid or lamb may be very uncomfortable at first. Some will circle, cry, and fall on their sides. This can be very alarming if you aren't prepared for it! Within a few hours, the animal is usually feeling better, though it may be uninterested in nursing for a day or so. Fortunately, most animals don't react this dramatically. In my experience, using the elastrator is a simple, safe, and effective method of castration.

If you are castrating animals in an area where horses or mules have been or are kept, you should protect your lambs and kids with tetanus antitoxin given at the time of castrating. Young animals may be protected if their mothers have been actively vaccinated against tetanus (see Immunology and Tetanus sections). Whether you are using the cutting or elastration method, you should protect these animals. Those castrated with the Burdizzo or emasculator probably won't need protection for this operation.

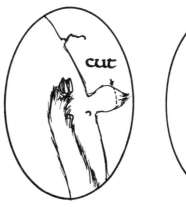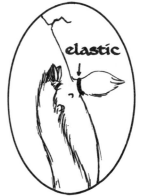

Women's night at the coffeehouse tonight—the first time I've gone in months. I felt self-conscious and alien at first, trying to socialize and wishing no one would notice me. Someone started playing records and I began to dance, shaking tensions loose. I looked up to see Margaret beside me, eyes laughing, and we began to dance together as we used to in the group. Helen, Judith, Tory, and more, joined us until almost everyone was dancing in easy comaraderie. We finished hours later with a joyous boot stomping jig to Katie's fiddle and Sally's piano.

The warmth of those women surrounds me tonight, pressed tighter than my quilts. I can feel their presence scattered across on ridge after ridge. There is a deep unspoken bond between us of growing shared, of years of knowing each other. It's so easy to forget how deep that bond is. I doubt I'd be raising sheep, or Katie performing her music, or Tory building her house, or Sarah running her school if it hadn't been for the support of all of us giving each other courage. It *does* make a difference to live in a world full of sisters!

hoof trimming of sheep and goats

The hooves of a wild animal are worn down by constant travel over rock and hard-packed earth. Because they aren't exposed to this natural wear, the hooves of your domestic animal need regular attention. You can provide large rocks or concrete areas to help wear down your animal's hooves, but usually a monthly or bi-monthly trimming is necessary. Neglecting hooves can lead to badly deformed feet, lameness, and hoof disease.

Structurally the hoof is simple.

From the side and bottom it looks like this:

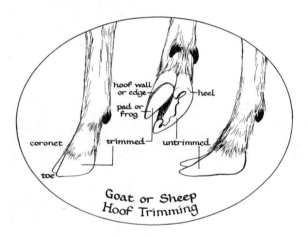

Goat or Sheep
Hoof Trimming

The hoof walls grow much like our fingernails. If left to grow too long, they begin to fold over and cover the frog (the soft inner part of the hoof). Manure and dirt trapped by this overgrowth of hoof can cause bacterial infection or decay of the inner hoof. The toe of the hoof will grow long, sometimes curving or folding in and throwing off the animal's natural walking balance. From the side, the hoof will then look as it does in the drawing. The heel will also grow down and throw the foot off balance. In trimming, you should try to restore the hoof to as near-perfect a state as possible. If the goat's hooves have been allowed to become badly overgrown, it may take two or three monthly trimmings before they can be totally corrected. Sometimes a hoof has been so neglected the bone structure of the foot has been permanently deformed. In that case, trim as carefully as you can to give the foot maximum natural balance.

A small pocketknife, with blade well sharpened, is a fine hoof-trimming tool; one that locks open is ideal. You can also use a pair of small pruning shears or buy a pair of special hoof-trimming shears. A furrier's knife with a fairly broad blade and a curving end may also be used. In addition to your trimming tool, have a bottle of iodine or similar antiseptic handy.

Most goats will tolerate a hoof trimming well if you go at it gently and quickly. Generally, goats don't like to be held still—so if the trimming goes on too long, your goat may begin to kick and carry on. Try to be efficient and patient! Usually you can tie your goat by her collar or secure her head in a stanchion. Some intractables must be thrown down on the ground or otherwise confined. Sheep should be thrown down and held on their backs—a position from which they can't get up. Facing the opposite direction from the goat, pick her foot up between your legs. You'll have good control this way and also be able to make knife strokes *away* from your body and the goat's. If your goat tries to kick, don't let go of her hoof (if you do, you are teaching her that she can make you stop). Using the tip of your knife or shears, dig all manure and dirt out from the hoof so you can plainly see the level of the frog and what must be trimmed away. You will be trimming the hoof walls to be even with or only *slightly* longer than the frog surface. Trim the overgrown sides first. Hold your knife parallel to the frog surface (if you cut at an angle, you'll be throwing the hoof off balance) and use short, light strokes. Normally the outer edge of each toe needs more trimming than the inner edge. Sometimes (especially in wet weather) the wall of the hoof will separate from the inner hoof, leaving a space that gets packed with dirt and may become infected. Clean this out as well as you can and drench with iodine. You may have to repeat this cleaning and drenching three or four times until the hoof begins to heal. Next, trim back the point of the toe if it is overgrown. If it is badly overgrown, you may have to cut it back quite bluntly. Once the walls and toes are trimmed, you should slice the heel down to the same level. The heel is softer and can be pared in thin slices. When you are done with this, the entire bottom surface of the hoof should be level. Lay your knife blade across the surface of both parts of the hoof to check this. Both halves of the foot must be trimmed equally so that the animal's weight is evenly distributed.

If at any point in your trimming you begin to see tiny pink dots or reddish areas, be cautious. These indicate that you are nearing the blood supply of the hoof. On some goats the blood supply is quite near the surface. If by accident you cut too deeply and the hoof begins to bleed, apply pressure with your thumb until the bleeding stops. A few dabs of iodine will protect and seal the cut. You may also find areas of hoof rot or decay if an animal's feet have been neglected. These will vary in size and seriousness. They are usually pockets of evil-smelling liquidy material. Clean them out as well as you can and pour iodine or a similar antiseptic into the cavity. Repeat in a few days. There are specific remedies sold by country pharmacies and feed stores for serious hoof rot.

Make hoof trimming a routine part of your animal's care to keep her walking on healthy, normal hooves. Don't forget that rams, bucks, lambs, and kids also need this regular attention.

Went out this evening after chores to check the spring box in the sheep pasture. There was a great blue heron standing in the marshy part of the spring when I got there—a powerful, primitive bird, now rare around here. As I sat down beside the spring I saw the first wild iris of this season opening just where the first iris of last year did. There is such magic on this hillside where springtime first comes—the water rushes downward from the spring toward the creek, toward the sea; the perfect whole dome of the sky encircles me, the light sliding from evening into darkness. There is no tension or pain this meadow can't erase; I could live here fifty years and still not fully comprehend its power.

SHEARING

Shearing sheep, like vaccinating your animals or repairing your pump, is what I call a "real" country skill. Not that these skills are any more difficult or necessary than any of the other tasks I do regularly, but they are the ones I never imagined I might someday be doing.

Shearing is, in fact, not difficult at all, though it is something that you do better with practice. The more familiar you become with the animals' bodies, the easier it will be to do a smooth clean job. I have been pleased to discover in this, my second season, that my hand seems to instinctively remember the best positions, a body memory that is very difficult to translate into words. Now, twelve sheep later, I'll try to do so . . .

The only equipment you need for shearing is shears. For small flocks, hand shears are fine and relatively inexpensive. They cost from $6 to $10 and look like strange large scissors. The best ones are made of English Sheffield steel, and you will want the best as they will hold their edge for years.

Shears need to be kept sharp (you can feel them dragging when they're dull); this is a little tricky to do as the baffled edge must be kept exact. I take mine to the local saw sharpener (who once raised sheep) each time they get dull—every three to five fleeces, depending on the amount of dirt in the fleece. The shears should also be kept oiled and free from rust. For larger flocks (twenty or more sheep) you will probably want to get electric clippers. These cost at least $60 and come in two types—those that run on a gasoline generator (in the field) or directly on AC current. These directions are for hand shears; the motions are somewhat different with the clippers, and diagrams for them can be found in sheep books and magazines.

Large flocks are usually shorn in a shearing room or pen, with a clean wooden floor. I have found a tarp or bedspread (double-bed size) on the ground works just as well to keep the fleece clean. Last year, I just let my sheep outside the pen and sheared them on a bedspread there. That kept the other sheep (still in the pen) from butting me or bothering the one being shorn. This works fine if you have friendly, tame sheep or if you are careful to keep control of the animal when you are done (there's not much to hold onto on a shorn sheep!). In other words, you need to be sure you can get the sheep back in the pen. This year, with my flock running fairly wild on a large pasture, I have made a shearing pen—a small fenced pen inside which they are regularly fed. The call of "Sheep! Sheep!" and a little banging on the grain bucket will bring them every time. I then corner the one I want to shear, get her down on her back and let the others loose. Sheep seem to have short memories or greedy stomachs, for they will fall for the same trick day after day.

When to shear depends on your location, climate, and the variables of any particular year. Most sheep are shorn once a year in the spring or early summer. A few long-wool breeds, with very coarse fleece are shorn both spring and fall. You should wait to shear until the weather is really warm with no chilling nights—so there is no danger of the animal catching cold or even pneumonia.

One of the most important aspects of a fleece is its greasiness—the amount of natural lanolin it contains. This makes spinning easier and provides water resistance in unwashed yarn. In order to bring out the most grease, you want to wait to shear until you've had two or three sunny days in a row and then do it on a warm, sunny day. If you live in an area with a lot of summer rain or fog, you may have to settle for the first sunny day. If you have to shear on a cloudy one, make sure the weather is quite warm and there is no chance of rain.

What time of day to shear is somewhat flexible. I prefer late morning or early afternoon, once the sun is really hot and the fleece greasy but there is still enough time before sundown so that the animal won't feel chilled. The one thing to be careful of is to hold off shearing for several hours after you have fed your sheep since they can't regurgitate and chew their cud while on their backs. For the same reason, you should never shear sheep that are on really wet, green pasture. The green grass causes lots of gas in the stomach which the sheep is unable to expel while on its back. Whole flocks have been found dead twelve hours after shearing because of bloat.

Whether or not to shear any lambs you are keeping (not butchering) seems to be an open question. It depends on the length of their fleece and the weather. Can you get a fleece with a long enough staple to be worth spinning and still leave enough time between shearing and cold weather for the lamb to grow back a protective coat? I sheared two of my December lambs in late August last year and got average fleeces (my sheep have exceptionally long staple). My third lamb was a triplet, slow to grow and small to start with, so I let her go unshorn all winter and did her first this spring. Her fleece was long and thick then, but it was also badly matted with mud and hay from the winter. I will probably go back to shearing lambs in August (our winters are mild and don't begin until late November). An average fleece is better than a badly matted one. The only problem I see with this system is that the following year, those yearlings must be shorn relatively late (July) in order to have a full year's fleece—which spreads my shearing out more than I'd like.

My first sheep took me 3½ very nervous hours to shear; now with sharp shears and a docile animal, I can do one in less than an hour. But I was a bit dismayed when I first heard that professional shearers can do "eight an hour." (It helps a bit to know they use electric clippers.) Two sheep a day is about my limit, after that my muscles are too tired for me to feel in control; I actually prefer to do only one a day. Shearing is an all-body task—legs, back, as well as arms—and that first sheep left me so sore I hobbled like a hunchback for several days. Fortunately, the limbering process seems cumulative, for the first one this year left me only very slightly stiff. I prefer to shear with an assistant who will help hold the sheep. This is mostly because I have a bad back which gets quite cramped from stooping. It is perfectly possible to shear alone. Being an assistant for someone else, however, is a good way to learn.

I learned to shear from a friend, and her teaching technique is the best I know. We each took a sheep and worked side by side, she instructing me as we went. The next day, while it was all still fresh, I sheared one of my own ewes by myself. Since then I have learned more with each successive sheep as the body contours have become more familiar to me. My shearing style now is an amalgamation of two other shearers' techniques and my own experience. It seems to work. If there is no one to learn from, you can do so from these instructions. Shearing is easy and the technique comes very naturally once you understand the principles.

A well-shorn fleece will come off pretty much in one piece—a wonderful, warm, velvety mat. What

place your shears with confidence and will have few, if any, second cuts. If you do cut the sheep, check to see if you have just nicked the skin or if it is a real wound. Treat any cuts with iodine or gentian violet after you have finished shearing and removed the fleece (they stain). Check wounds for several days to make sure they're healing properly, without abscesses or infection. A few skin nicks are normal and nothing to worry about. Sheep have few nerve endings near the skin and are not very sensitive.

What you try to do when shearing is to keep the animal pretty much immobilized. This is fairly easy to do with sheep, which are ungainly and awkward anyway. In theory, a sheep can't get up off its back; they have been known to get trapped on their backs and die that way. I happen to own a couple of notable exceptions to this rule, but in general it is true. So, you begin by shearing the belly and legs with the sheep on its back. To get it there, reach under the belly and catch the two legs on the far side from you. Pull on these and the animal will come down; then roll it on its back. Do this as gently (but firmly) as you can. Sheep have some difficulty breathing on their back, so if yours seems

you want to do is to cut as closely to the skin as you can, without cutting the skin, and to do so evenly all over the body. A second cut is when your shears cut too high, leaving a ½″ or more of fleece still on the body that must be cut again. Second cuts are undesirable, since the staple of the first cut is shortened and the second is unspinnable. It's easy to make a lot of second cuts on your first sheep, being overcautious about the skin. As you get more experienced, you will learn where to

I've been thinking about the changes I've been going through, the lessening of my anger and my rhetoric and the growth of a greater openness to people, to the reality of who they are. I watch the women who never left men, who've remained in stable couples, some of whom never shared in woman consciousness. Now I am friends with some of these women again after silence and alienation and anger. Watching Winona kid this time, watching that birth that did not come without pain or struggle, I knew that I have not returned full circle to where I was but have spiraled to a profoundly different place. That from the anger and pain and struggle has come a deep trust of myself and a freedom that defies expression: I have become myself. This confidence is not superficial or egotistical, it was birthed slowly from real experience; a continual testing and stretching of limits, confrontations with fear. For some other women confidence may have been nurtured in childhood; for a rare few it may come from personality; many may never choose or be forced into this struggle—security does not particularly foster growth. But for me, confidence has come from anger and isolation and work well done.

Daily I discover that the effect of confidence is not a static state but a daring to try more. Where I would have hesitated or hidden in fear before, now I act. And from those actions will probably come more doubt and struggle, and more trust. I hope I never rest too long in safety; I prefer the intensity of growth.

318

restless or begins to struggle, try altering the position of the head. They will also struggle if you nick them or if you are shearing some particularly sensitive spot (teats, tail, backbone, and testicles are ticklish on some of my sheep). Just keep the animal firmly in position (watch out for back legs) and wait till it calms down. I've found talking or singing helps.

First position:

Now to start: the first shearing position is with you standing, knees bent, and the animal's back and shoulders resting on your knees in a half-sitting position. You then bend from the waist to shear. This gives you the most reach and gets the

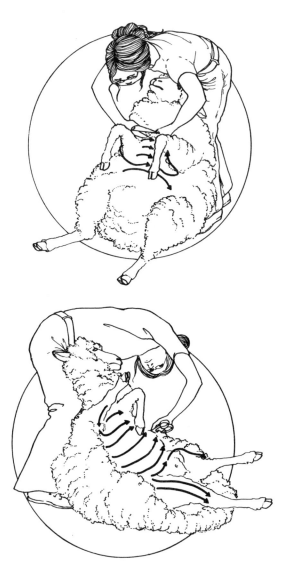

best stretch on the animal's body. My back won't always take the strain of this position (much of the weight is carried on your legs), and I sometimes

shear in a kneeling position with the animal's head on my knees. This also works, though it is difficult to reach all the way down the body. I begin shearing on the chest, just above the front legs. There may be a natural opening there or you may have to part the fleece. When you look at the fleece, you will see it grows in straight lines, almost in rows. You want to cut along these lines, but not very deeply into the fleece (¼″ at most). Slide the point of your shears into the fleece from your right-hand side (all these directions presume you are right-handed). Begin clipping until you reach the left-hand side of the belly. Then come back and start the next row. This will go easier once you have a clear space to work in. Good shearing should be with long, smooth strokes, which follow the curves of the animal's body. Trying to cut too much fleece at a time (going too deeply) makes for short, choppy strokes. You should shear across the belly, perpendicular to an imaginary line down the center of the body. When I reach the front legs, I shear them, too, from underneath. I stretch each leg and trim off all the hair and matted fleece. Be careful here as the skin on the legs is loose and easy to cut. The hair on the end of each leg is just that, hair not wool, and should be discarded. With the leg on your right stretched out from the body (hold this with your elbow) you'll find you can shear a straight line down the right leg, across the belly, and out the left front leg. When the legs are mostly clear continue down the belly. The fleece will be coming off in a solid mat, and you should keep rolling it back from where you're cutting. Be careful to only shear to the edge of the stomach (approximately a line between the front and hind legs on each side). The back begins to curve downward there, and it is easy to make second cuts if you go too far to either side. Watch out for the penis or the teats as you shear as it is possible to cut them. The hind legs are done basically the same way as the front ones, as much as possible from this first position. On a ram, the wool on the testicles will also have to be cut. This is of poor quality and can be discarded. On a ewe, there will be a bare space where the udder is (rams have teats too, remember). Then on both there is more fleece up to the anus and the base of the tail. Shear clear up to the tail while the sheep is on its back.

Second Position:

Next you want to shear around the head and neck, down to the outside of the front legs. This is done with the sheep still on its back. You can remain in the same half-crouch position or you can

319

kneel with one knee on the ground and one raised, the sheep's head on the raised knee.

I begin by carefully clipping the wool around the ears and eyes (the ear is easy to cut, so be careful). Then begin shearing across the top of the head, down the right side of the face, across the

neck (beneath the chin, the skin is loose, so, again, be careful), and then up the left side of the face to where you began. You will have to shift the animal's head to be able to cut all the way around. The idea is that the fleece should roll back from the head and neck in a large ruff. I usually make several cuts down one side of the face to the center of the neck before I shift the head and do the other side. But the basic motion is circular—around and around the head. When you get down to the front legs you should pull them through the gap between the body and the cut fleece. Then trim any remaining wool off the legs. Before beginning to shear down the back, lift the cut fleece on the breast over the front legs and over the sheep's head so that it lies on the animal's back.

Third Position:

This is the home stretch—the back is the easiest part of the sheep to shear and has the best and most beautiful fleece. This is where you will really get to see the quality of your wool. To shear the back you need to roll the animal onto its side. Start with the animal's right side on the ground and its head toward you. You will sit or kneel with the sheep's head and forelegs thrown over your knee. This is how you keep control of the animal—by preventing it from getting its front legs or head on the ground (so it can't get up). If I'm working with an assistant, she takes my place under the front legs when I'm halfway done and I sit farther

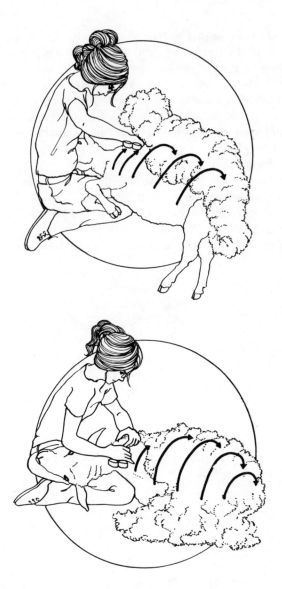

down the body. This lessens the pull on my back.

The reason for having the animal on its right side is so that you can begin shearing from a clear edge. You shear from the edge of the fleece (where you stopped cutting from underneath) to the animal's backbone. This can be hard to find on a well-fed sheep but try to be sure you're cutting to the middle of the back. If you don't, you'll have a hard time doing the other side without second cuts. Use the same long, smooth strokes with the shears that you used before, cutting one row at a time and not too deeply. Keep rolling the fleece back down the back as you go. Be sure your line of cutting remains perpendicular to the backbone as you go. If you begin to cut at an angle, the fleece will not come off in straight rows for spinning. Continue all the way down the back on this side until you reach the tail.

Then roll the animal over so you can shear the other side. Do this by rolling it across its back, so its legs never touch the ground. Working down the right side, you begin shearing at the backbone and cut to where the fleece ends at the stomach. Continue the same as before until the whole fleece comes off at the tail. It's a good idea to take this time too to trim the sheep's hooves before it's up and about again. Once that's done, you can roll it off the fleece and onto its feet and let it go. (I always find my first glimpse of the fleeceless sheep a bit of a shock—that small thin body emerging butterfly-like from its roly-poly cocoon.) And it's fun to watch the surprised reactions of her flock mates and lambs; they call and sniff and skeptically survey the newcomer, looking for their friend or mother.

Once my sheep is back with the flock, running free, I spread the fleece out on the bedspread and begin to clean it. I pull all the tags (dirty, matted parts), second cuts (the short, unspinnable parts), and the worst of the leg hair. Then I fold the fleece to form a neat square, roll it up, and carefully stuff it into a grain sack for storage. Keeping records year by year of the weight and quality of each fleece has helped me make decisions about feed and shearing dates.

SHEARING ANGORAS

Shearing angora goats is more like clipping a dog than shearing a sheep. The goats have hair which is best removed with scissors while they remain standing. We used to shear our angoras once a year, taking off a fleece with 10" to 12" staple, a good third of which was terribly matted. We were barely able to work with these fleeces, combing them first on carders and making large rolags. Then a friend from Missouri wrote that they shear twice a year there, once in early spring and again in early fall. This year we took the fleece off a four-month-old kid. The staple was about 6" and a lot more workable. We also bathed her first, and there were no mats and the fleece was silky clean.

The method of shearing is just to begin along the top of the back and clip down the sides. The fleece doesn't hold together like wool so it doesn't much matter what comes off first. If the goat is securely tied, it will not fuss and you can clip it pretty quickly.

SPINNING

Spinning is one of my favorite skills. It is perhaps one of the few things I will ever do that is a whole process—I raise and shear the sheep from which I spin the yarn from which I weave the cloth. And though I would never want to return to the days when women spun all that everyone wore, I do love participating in every part of the process: from the birth of the lamb to the tying off from the loom. That's a rare kind of pleasure these days.

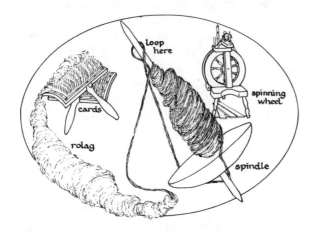

As you become more proficient, you may want a spinning wheel that produces more yarn with less effort, but you can spin beautiful fine yarn on a drop spindle too. Spinning is easy to learn and almost immediately you can produce thick and (interestingly) bumpy yarn. With more practice, you will be able to spin as fine as you want.

The most important thing you need to begin spinning is good fleece. Fleeces vary a lot in quality and it's especially important to have a good one when you're learning. Using a bad fleece the first few times I tried to spin, I became really frustrated as my yarn continued to break, dropping the spindle to the floor. I was about to quit when I tried a friend's spindle and a new bag of wool, and discovered that I was not at fault. There is still an awkwardness to learning how to spin and it takes a lot of practice before yarn spins easily through your fingers. But a good fleece will hasten the process.

The first thing to look for in a fleece is the "crimp"—the number of waves in each inch of fiber. Fine-wool sheep have a very tight crimp (many waves)—these are Merinos and Rambouillets, among others. The mutton or long-wool

sheep have a very loose crimp (only two or three waves to the inch). Fine-wool fleeces spin really easily on a spinning wheel into very strong, thin yarn. For hand spinning, I prefer the medium-wool fleeces (Corriedale, Columbia, Targhee). These have slightly less crimp than the fine wools, but have a longer staple and spin more easily on a drop spindle. Staple is the length of each fiber; long staple is highly desirable (if it's not at the expense of crimp) since it cards more easily and makes a stronger yarn. The other things to look for in a fleece are greasiness and color. Rub the wool with your hands and see if your fingers feel slick. This grease will make the fibers cling better when spinning and will make a water-resistant yarn for knitting or weaving, as long as it's left unwashed. White fleeces will often look yellowish or beige in

their natural state because of the grease; they will usually scour to snowy white—though there is a surprising variation in "white." If your "black" fleece looks brown, part the fleece and look at the base of the fibers. The sun will bleach out the ends of the fibers to light brown. Silver or gray running through an older ewe's black fleece makes a beautiful naturally variegated yarn, so reconsider if you have been wanting only pure black. Once you've found a fleece that looks good, spread it out and look more closely. Make sure you've gotten a whole fleece (good back wool as well as the less desirable belly and leg wool) and that it's fairly clean ($2.00 a pound is steep for manure). When our weaving co-operative first started, my friend Susan arrived one day glowing from her wonderful find—four fleeces of a beautiful red-brown color bought from an old woman known as Crazy Mary. In the bright light of day, we discovered that Not-So-Crazy Mary in her dimly lit barn had sold Susan four bags of leg wool and manure tags! Don't be discouraged if the price of fleece seems high to you. A really good fleece at $1.50 to $2.00 a pound will give you far more yarn for less effort than a poor one will at $.75 a pound.

Besides fleece, you need cards and a spindle. Cards, which resemble curry combs or dog brushes, comb and straighten the wool fibers to ready them for spinning. Cards have curved or flat wooden backs, wood handles, and metal teeth set in leather on the wood backs. They vary in quality and cost from about $7.00 to $15, some with very stiff teeth and some with fairly loose ones. The stiff ones wear better and card more wool at a time; the loose ones card less wool and sometimes lose teeth. Yet some spinners clearly prefer one kind and some the other—what you learn on determines a lot what you like to use. Spindles are basically just a notched dowel through a round wooden disc. They vary in size and weight. I prefer heavier spindles with a broad base. Heavy spindles spin around longer and the broad base lets much more yarn pile up on the shaft before it has to be skeined.

Mark your new cards "left" and "right" and always hold them in the same hand. The teeth wear in one direction, so you want to continue carding that way. A rubber band around one card handle saves constant checking for which is which. Sit holding the left card in your left hand, handle pointing upward and teeth side face up pointing toward the center of your lap. Now pull off a section of fleece. (If it is sheared well you will see sections of fairly straight fiber almost in rows. Try to

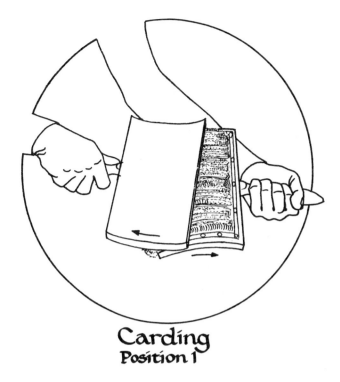

Carding
Position 1

keep this intact as you pull it off the fleece.) Look at the wool and make sure the loose end points down and that the uncut part (that which was the farthest from the sheep's body) is at the top of the card. Straighten the bits of fleece slightly and stick them on the top row of teeth that runs the width of the card. The card should be covered with a fairly even, not too thick layer. Now take the right card in your right hand, teeth facing the teeth of the left card. Hold the left card fairly stationary and brush the right card down over it. Some wool will come off onto the right card. Continue carding like this until the wool is straight and smooth. Matted or twisted places should be removed and carding continued. A lot of dirt and grease get carded too— onto your lap—but I prefer unwashed fleece since it spins more easily. Beginners seem to have trouble with matted spirals forming as they card. If this happens, just remove them and go on. As proficiency increases, spirals will happen less and less. Keep a bag near you into which to put these cast-off "nits"—they are a nuisance if they get mixed back into your good fleece, but they make good pillow or quilt stuffing if saved.

When the wool on both cards is well combed, straight, and unmatted, reverse the right card so its handle also points upward but with the teeth still facing the teeth of the left card. You will have to shift your hand position to make this comfortable.

Now, with bottom edge of the right card touching the top edge of the left, pull down making sure teeth touch all the way down (almost roll the card so the teeth on the top edge also come in contact and stay in contact until they leave the bottom of the left card). All the wool should now be on the

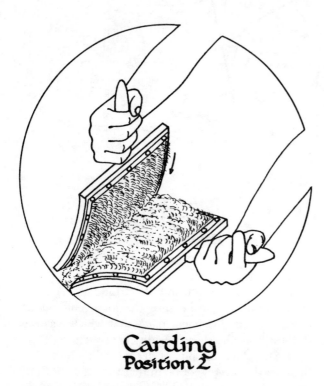

Carding
Position 2

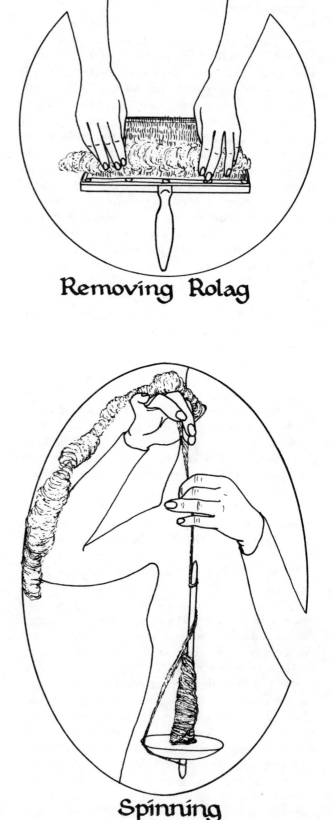

Removing Rolag

right card. Reverse the cards, holding the right card in your left hand exactly as you held the left card and vice versa. Repeat the combing. All the wool will now be on the left card (in your right hand). Reverse the cards and repeat one more time. Now all the wool should be on the right card in your right hand, in one loosely attached sheet. Set the left card down, put the right card in your lap and gently roll up the wool into a fairly tight cylinder. This is a rolag. Rolags spin best the day they are made. If you have to store them, stack them gently. Do not pack or squash them. Rolags may also be laid directly in woven pieces, unspun, for a loose tufted effect.

To begin spinning, tie a short piece of hand-spun wool to your spindle just above the base. Then loop the wool under the base around the dowel and bring it up to the top of the dowel with a twist into the notch. A short piece should extend from the notch. Twist the yarn to see which way tightens it, which way loosens it. You must continue spinning in the direction that tightens. Most spinners

Spinning

Driving home today, I suddenly realized that this really *is* going to be a sheep ranch, that I have done, and am doing, and will do it. That I'm making my livelihood from the land. The canyon is fenced now. There are sheep out there on pastures that were open hillsides two years ago.

The very act of building this place, the simple actions of tamping dirt, stretching wire, dumping hay in feeders, has profoundly changed my sense of self. I'm doing things I never dreamed I could do, and I'm doing them easily without even considering whether I really can. Last night I was talking with Susan about fencing the front meadow for feeder calves, and I realized that I could say that realistically, no fantasizing, no bragging; I can fence the front meadow as soon as I get done with the hay barn and get a little more money.

Like almost every other farmer in America today, I'm in debt and hoping for a good season. I'm only at the beginning now, and I know there are many struggles to come and overcome and come again: Someday I too, like my neighbors, will be counting carcasses killed by a marauding dog or watching the spring oats wash away in an "unheard of" late storm. No matter how prepared I am, there is always that vulnerability—to the weather, other animals, disease—that seems to strike when things are finally going smoothly. But inside me there also is this incredible joy: This life is real and good, and it has made me strong and real and good too.

quickly come to prefer spinning in one direction, clockwise or counterclockwise, and use starts that go in that direction. Loosen the wool at the end of your starting piece till the fibers (for an inch or two) look like carded wool. Pull out the end of your rolag until it is fairly thin, then mix its fibers with those of the starting piece, giving them a slight twist. These two will catch, mingle, and become one piece of yarn as you spin. From then on all your yarn will come from your rolags. Yarn comes from twisting together wool fibers that are stretched thin. The twist comes from the spin you give the spindle. The stretching comes from the gravitational pull of the drop spindle and from your pulling out the rolag as you feed it in. Hold the spindle in your left hand, the newly joined start and rolag in your right, with the rolag looped over your right arm. It must be kept out of the way or it will catch on the spindle and tangle. Give the spindle a swift twist with your left hand in the direction that tightens the wool (double-check that this is happening), let go of the spindle and let it spin on its own. Slide your left hand up to help twist and pull the fibers. It especially needs help catching here. As the yarn begins to spin, move your left hand up steadily, always at the place where spun and unspun meet. Use your left hand to aid the twisting and your right hand to pull out the fibers in the rolag. As the spindle slows down, set it spinning again. Continue making yarn until the spindle almost touches the ground. Then keep tension on your new-spun yarn by catching it on your right elbow or under your chin or somewhere (or it will start to unspin) and loosen the yarn from the spindle top and bottom. Wind the new yarn onto the spindle shaft above the base until there's just enough left to catch at the bottom and at the notch. Then start again, reattaching your rolag if it has come off. Sometimes at this point I do some preliminary pulling out of the rolag so it will be easier as I go along. The more you spin, the easier it is to gauge how much wool to feed into the spindle. If the spindle begins to turn backward whenever you stop, it is "overspun"—turning faster than you're feeding wool. Hold your yarn just above the notch on the shaft with your left hand, hold the rolag end with your right, and let the spindle spin backward until it stops. This way, tension is released without unspinning the yarn. If your yarn twists and tangles on itself, it is overspun, twisted too much, so you should pull out the rolag and move along faster.

As your skill increases, both with carding and with spinning, one good turn of the spindle will let you spin fine, even yarn clear to the ground. When you want bumpy or thick yarn for special effects, just don't pull the yarn out so fast as you spin. For variegated yarn it is best to mix two colors of fleece (brown and white, silver and black, whatever you want) as you card. You can also spin rolags of alternately colored wool for a similar but more regular variegation.

After the spindle is full, the yarn should be skeined for washing or balled for weaving. I often wash the whole woven piece when it is done rather than each batch of yarn separately. To wash the yarn, soak it for an hour or more in warm water and a mild soap. Then rinse it several times in warm and then cold water and hang it to dry. For dyeing, the wool needs to be washed even more thoroughly than this, as grease prevents the dye from setting. Hand-spun yarn in natural or naturally dyed colors makes beautiful hand-woven or crocheted goods. For knitting, however, it must be plied first (two strands twisted together for greater strength). This requires some spinning skill, to obtain yarn fine enough to ply without excessive bulk.

Learning spinning and carding is not hard! Find someone who spins and watch the process closely or reread this chapter, look at the pictures, and begin to try. In a few hours you will find yarn beginning to flow from your fingers. Those first few skeins will feel almost like gold as dirty twisted fleece, almost miraculously becomes fine smooth yarn.

keepiNG A HEALThy Flock

Diseases and problems with sheep can be divided into four major types: contagious diseases, non-contagious diseases and problems, internal parasites, and external parasites. Within these categories there are some overlaps—mastitis, for example, can be of a non-contagious *or* a highly infectious type. Ketosis is not a contagious disease per se but may occur in an epidemic-like outbreak under certain conditions. My own experience with sheep is limited to our geographical area and to the keeping of a small farm flock of largely fine- and medium-wool sheep. Other conditions and different areas may present quite different problems and emphasis. Besides familiarizing yourself with the following, you should probably read as much sheep-related information as you can (see Resources for some recommendations). Spend time talking with other sheep raisers, your farm adviser, and so on to learn more about sheep in your area. Most importantly, you will learn from the animals themselves.

The first step in keeping a healthy flock is to buy healthy animals, and to keep them in clean pastures and sheds. A good feeding and management program will be your most valuable asset in keeping your flock well. When you buy a new animal, isolate it from the rest of your flock for a month or

longer to be sure it is not going to introduce parasites or disease into the flock. Work out a regular schedule of vaccinating your animals against diseases in your area *before* they strike. Worm your flock routinely and keep all animals tagged and their hooves trimmed. With all of these precautions and a basic knowledge of problems to watch for, you should not have to consult this section too often!

CONTAGIOUS DISEASES

Abscesses: An infection of the lymph system of sheep, which manifests itself in abscesses or boils both externally or internally. The lymph nodes themselves become abscessed—filled with a thick, puslike material that is odorless and greenish-yellow in color. Externally, these abscesses appear on the shoulder or thigh. This disease, called caseous lymphadenitis, is found in sheep in the western U.S. It is most common in sheep from two to four years old. Like other abscesses (see Appendix article), these should be allowed to develop until they are ready to be lanced Treat with tincture of iodine, antibacterial cream, or a strong antiseptic. If the disease has developed to include internal abscesses, the animal may become progressively weaker and thinner and eventually may die. There is little, so far as I know, that can be done with these cases. This disease passes from an infected animal to an uninfected one through wounds—particularly at shearing time. Material from a ruptured or lanced abscess may contaminate pens and equipment. Your safest course is to cull out any infected animals, or at least to shear them last and treat abscesses carefully. All wounds on sheep should be cleaned with an antiseptic and protected with iodine, antibacterial topical ointment, and so on.

Anthrax: Anthrax is a very serious, highly contagious disease that affects almost all types of animals, including people. It is spread in a variety of ways: by the discharges of infected animals, through infected objects such as hides, wool, and meat scraps, by surface draining of the soil, and so on. It is characterized by the sudden death of the animal, often with bloody discharges from all the body openings and swelling of body parts. If you live in an area where anthrax is known to appear, be sure your sheep are vaccinated annually. If you don't live in an anthrax area, do *not* vaccinate, as the spore vaccines are dangerous as contaminants. If an animal dies of what you suspect is anthrax, call in a veterinarian and notify your farm adviser.

Black Disease: Another disease that is found only in certain areas, black disease is also highly infectious and fatal. It affects only animals badly infested with liver flukes, and is seasonal, occurring in late summer and fall. The animal will die rather quickly, with few signs. If you live in an area where this disease occurs, vaccinate your sheep.

Foot Rot: This is an infectious disease of the feet that results in lameness to a mild or serious degree. Only one foot might be involved, or all four. The animal will begin to limp and if not treated, this limp becomes more pronounced and more painful. You should catch the animal at once and look carefully at each foot. In hoof rot, the horn or outer hard part of the hoof separates from the soft inner part. There is usually a little puslike liquid with a very bad odor which may be scraped from the cavity between horn and soft part of hoof. Each infected hoof should be scraped as clean as possible and flushed with or soaked in a strong antiseptic or specific foot rot solution (available at farm supply

stores). The animal should be kept on clean, dry ground after this treatment. Repeat the treatment in a week. I have used a good strong tincture of iodine on some foot rot in our sheep and angora goats and had successful cures. More commonly, a solution of copper sulfate (bluestone) or a 2 per cent formalin solution is used. If you have an outbreak of foot rot that affects a lot of animals, you may want to use a foot bath treatment. A long, narrow trough is filled about two inches deep with the sulfate or formalin solution. Each sheep must have its feet trimmed and scraped and must be made to walk slowly through the solution (if you've ever worked with sheep, you'll know this sounds a great deal easier than it is). The whole procedure must be repeated weekly until the flock is cured.

Foot rot is aggravated by damp pastures and muddy areas. It is known to spread more rapidly in Merino and Merino-cross sheep. Keep your sheep on dry ground and check regularly for this condition during damp weather. Whenever you introduce a new animal to the flock, check its feet carefully. It might be a wise precaution to dip its feet before allowing it to join your flock.

Listeriosis (*Listerellosis*): This disease, more commonly known as "circling disease," is an unsolved problem (thus far) in sheep. In older livestock books, it is noted as a little-understood, contagious disease that has no cure. The name suggests the symptoms: the affected animal staggers in circles, always in one direction; it has a high temperature; it is depressed and has stringy salivation, a nasal discharge, and unco-ordinated jaw, eye, and ear muscles. This is a bacterial infection affecting the brain of the animal and causing encephalitis. It is usually fatal within a day or two. According to recent articles in *The Shepherd* magazine (see Resources), this disease is connected with feeding silage. Outbreaks of the disease seem to occur more frequently when animals are fed silage from a building that is in poor condition and has allowed rain or snow leakage and consequent spoiling of the feed. If you feed silage, be certain it is not spoiled. As there is no known cure for this disease, affected animals should be immediately isolated from the flock and the rest watched carefully. Feed should be checked for spoilage. If you think you have an animal that has died of this disease, a competently done autopsy can verify your suspicions. (Enterotoxemia and pregnancy toxemia are two diseases that might be confused with listeriosis.)

Mastitis: Although mastitis is listed in most sheep books as a contagious disease, it usually affects a single animal and is most often mechanically caused. Mastitis is an infection of the udder. It destroys the productive capacity of the udder gland and can make the ewe very sick. Any ewe who refuses to let her lamb nurse should be caught and checked for this condition. If her udder is hot, lumpy, and/or congested, she is probably suffering with this infection. Mastitis can be cured if treated promptly (see Appendix article). If it is neglected, it can develop into the gangrenous infection known in sheep as "Blue Bag." The udder literally turns blue as the infection becomes severe. The ewe may die from an untreated case of Blue Bag, or she may recover but lose one or both halves or her udder (the half affected with gangrene may atrophy and slough off). This is one disease you must watch for when your ewes are lactating—it can be caused by the smallest injury, and it can affect the healthiest, most well cared for animal.

Pink Eye: Pink eye is a very contagious disease. Though it is not serious in sheep, it may occur sporadically in a flock. It seems to be aggravated by dusty, dry conditions, the presence of flies, and by feeding in tall grass. The eyes will become red and watery at first, but as the disease develops, the discharge becomes thicker. The animal will try to stay in the shade and may become partially, temporarily, blind. Lactating animals may drop in their milk production, and some animals may run a slight fever.

Treatment should be specific and prompt. Isolate infected animals and treat with a commercial preparation for pink eye (or an alternative). A similar condition in one of our goats responded well to the *Herbal Handbook* treatment: very diluted fresh lemon juice squeezed into the eyes three or four times daily. I was skeptical of this treatment in the beginning, but it cured the condition completely in about three days. Pink eye may be passed from goat to sheep and vice versa. It may not, however, be passed from cattle to sheep or between goats and cattle. People are susceptible to this infection.

Pneumonia: Pneumonia is an inflammation of the lungs that can be caused by bacteria or viruses or mechanical means (very commonly, this type of pneumonia is caused by improper drenching). Signs of pneumonia include an initial high temperature, increased respiratory and pulse rate, congestion, and a dry, hacking cough. The animal should be isolated in a well-bedded draft-free stall and treated with broad-spectrum antibiotics. Sheep which are badly infested with lungworms are particularly susceptible to pneumonia, as are bummer lambs that have suffered exposure and hunger. Sheep taken to shows are also especially vulnerable —they may go through drastic temperature changes in shipping (if precautions aren't taken) or become overheated while in the show ring.

Progressive Pneumonia: This is a slow developing chronic respiratory disease of sheep usually found in the western range states. Affected animals have characteristically difficult breathing and a "double expiratory effort" (a kind of gasping wheeze follows each exhalation when the animal breathes). These sheep lose condition over a period of months. They may exhibit few signs—no cough (or a very slight one), no bronchial discharge, and so on. The victims of this subtle but devastating disease are called "lungers." They are usually culled from the flock when discovered because their generally poor condition makes them unable to raise a healthy lamb and they rarely regain their own health.

Shipping Fever: This is actually another form of pneumonia that affects sheep when they are moved from one area to another, or one farm to another. Changes in temperature and humidity, loss of condition due to feed changes, and the stress of the move combine to lower the animal's resistance. Signs of the disease are similar to those described for pneumonia—a rise in temperature, cough, rapid pulse and respiration, loss of appetite, and discharge from the eyes and nose. Shipping fever can be fatal if not handled immediately. Treatment is similar to that for other forms of pneumonia. This disease can be prevented by careful handling during shipping—make sure the animals are not overcrowded, not subjected to drastic temperature changes, and that they are fed regularly. It is a good idea not to feed grain to an animal being shipped (even one being moved in your truck or car a fairly short distance). With a grain-filled stomach, the animal is less likely to eat the hay it needs to maintain its normal body heat (ruminants convert roughage to digestible nutrients and in doing so, create body heat). It will then be doubly susceptible to a chill. Another good precaution is to give the animal(s) broad-spectrum antibiotics before, during, and after shipping. I have never dealt with this disease in our sheep but have had a few

329

cases in goats that were shipped from other farms and to shows. In all cases, the goats developed classic pneumonia symptoms about three days after they arrived home. I treated them with Combiotic (a penicillin-streptomycin combination), kept them stalled in warm, dry isolation, and gave a lot of extra care to them (including special feed, warm water with molasses to drink, and so on). All recovered in a week or so. After this experience, I think that prophylactic use of antibiotics might be easier on the animals and their owners! Shipping fever is not a disease limited to the large commercial flock being shipped to market or moved to new pasture—it can affect the two or three sheep you purchase and haul home in your pickup truck. Watch these animals carefully for the first week or so and treat any beginning cases before they become serious.

Sore Mouth: Highly contagious and transmittable to humans, sore mouth is an annoying, if not really serious, disease. It usually appears following some contact with an infected flock or carrier animal. Sore mouth is a virus-caused disease that can be transmitted on inanimate objects (particularly dung and bedding from infected farms) as well as on the animals themselves. It can exist a long time in the dormant state and then erupt in your flock with particular virulence. You may not know an animal has sore mouth until the disease has progressed through its initial stages. Initially, small sores appear on the lips, nose, and tongue of the animal. These become enlarged and filled with pus, and may rupture and crust over. Eventually these sores will form scabs and drop off, leaving no scar. The worst effect of this disease is a general loss of condition as the animal has difficulty eating. It is the reluctance to eat and a generally depressed appearance that may be your first clue to an outbreak of the disease. Sore mouth will run its course in a few weeks, and the animal will have built up an active immunity to it.

If sore mouth is known to be in your area, or if you take your animals to shows or to other farms to breed, you should vaccinate your flock against this disease. The vaccine is very inexpensive and easy to give. If some of your animals come down with sore mouth, you should vaccinate the whole flock at once. Affected animals should be isolated if possible. The sores themselves may be treated with iodine or a 2 per cent solution of potassium permanganate. Whenever you handle animals with sore mouth, wash your hands well with a good antiseptic. All equipment, sheds, and so on should be disinfected to keep this disease from spreading. Visitors to your farm should be cautioned that sore mouth is a soil-borne virus and might spread from area to area on their boots or clothing.

Vibrionic Abortion: Abortion during the last six weeks of pregnancy may be caused by an organism known as vibrio fetus. This is an infectious disease of sheep most prevalent in the Rocky Mountain area. The causative organism is ingested by the grazing ewe. Ten to twenty-one days later, she will abort. She may carry her lamb full term and deliver it very weak or dead. The ewe will usually recover and may have normal lambings in subsequent years, though she may remain a carrier of the disease. There is no treatment for this disease that I have read about. Aborting ewes should be isolated and the flock moved from the area if possible. There is a bacterin available which, if given annually, can protect against future outbreaks of this disease.

NON-CONTAGIOUS DISEASES AND PROBLEMS

Dogs and Other Predators: Probably the greatest losses of lambs and grown sheep are suffered because of predator and dog problems. Both the small farm flock and the free-ranging large commercial flock are subject to attack. Whether you live in an area where wild animals prey on sheep, or in an area where a neighbor's dog is the one in question, you will have to find a solution to this country-wide problem. Sheep have very few defenses against dogs and similar predators. Part of the care of your flock will be to protect them against being torn, ripped and mauled, injured or killed by "playful" or serious predators. Use dog-proof fencing if you can afford it. If you live in an area where there are running dogs or wild animals, you may have to buy a gun and learn how to use it. This sounds severe, but if you have ever seen a lamb or sheep torn apart by a dog, you will probably feel quite differently about this. A dog or a pair of dogs can maim or kill twenty or thirty sheep in a *very* short time. Often they will just tear at the animals indiscriminately as the sheep try to run. Pregnant ewes or ewes with young lambs are especially vulnerable. Post your land with signs that state clearly that loose dogs will be shot. Take the time to talk with all of your neighbors about this problem. Explain your position and ask them to keep their dogs and visitor's dogs confined. No breed of dog is above sheep killing—I've heard of everything from beagles to Great Danes running down sheep! The most well-trained, gentle dog can slip suddenly into a villain's role, so don't take chances. Be firm with your friends, too, when they visit with their dog(s). Better yet, ask them to leave the dog at home. Prevention is by far the best cure for this problem.

Many preventives are discussed, tried, passed around, praised, or criticized. Some people have had success by putting bells on their sheep. This alarm system can alert you if your sheep are being chased. It may discourage a wild animal or even a dog from chasing. Another possibility is running a full-grown, horned buck goat or wethered goat with your sheep flock. Goats will sometimes take on a dog rather than run from it—and a goat with a heavy set of horns and a determination to fight can make a formidable guard. This is a traditional practice in other sheep-raising countries. If you run a buck, though, be prepared for your ewes to present you with "geeps" at lambing time. Supposed crossings of sheep and goats are rumored to happen now and then. Some experiments are being done with providing ewes with scented collars (the scent is actually a dog and predator repellent) and with ultra-sonic devices in their collars—but these methods are probably too impractical to really suggest seriously at this time. The best alternative I know of is to run a guard dog with your flock. This dog will have to be a very special one with an exceptional disposition and a devotion to the sheep. Of all of the dogs I have heard of, one most suited to herding sheep is the Great Pyrenees. This is a large, mild, calm dog that has been found to be gentle and protective of flocks it is used with. I have had experience with other types of dogs: the kelpy (Australian sheep dog), a German shepherd, an English setter, and a border collie/Australian shepherd cross. Of these dogs, the kelpy and the collie/shepherd cross were suited to the sheep and the farm in general. The kelpy is a small, active, very bright dog that seems to have a natural bent for herding. My collie/shepherd cross has proven to be a wonderful guard dog and is completely gentle with all of the animals ranging from new lambs to grown horses to tiny chicks. The setter proved to be too much into bird-chasing and too nervous to be around the sheep. The German shepherd related very nicely to the sheep until she unexpectedly chewed up her favorite lamb one morning! My advice is to start with a dog of a breed known to work well with sheep and other livestock. Buy a trained dog or work with a puppy yourself. A good dog that will stay with and protect the flock is probably your very best help in keeping your sheep safe.

Abscesses: Besides abscesses that are caused by infectious disease, sheep will get abscesses from splinters, feed, and so on. These mechanically caused abscesses are commonly found on the feet and on the neck and face, though they might be on any part of the body. Foot abscesses are very common. An injury to the feet will occur when the sheep are grazing and even the tiniest puncture or scrape might seal up and become infected. When we were running our sheep in a recently cleared pasture, we had a lot of trouble with small abscesses that occurred between the toes (or claws) of the sheeps' hooves. The sheep were puncturing their feet on fallen branches, and these puncture wounds caused swellings and sometimes a mild limp. Often the abscess was so small it would not affect the animal's walk at all, and I would find it only when I went to trim the hooves. The treatment I used was simple: lance and drain the abscess and flush it with iodine. These abscesses healed very quickly, usually with just one treatment. Abscesses also occur around the coronet (top line of the hoof) or at the pastern. They should be treated similarly.

Face and neck abscesses may occur if the animals are using feeders made of splintery wood. In our flock, feeders made of old redwood were the problem. When these feeders were sanded (or worn) smooth, the abscesses ceased. Fences made of splintery boards will also cause abscesses as the sheep reach through the boards to the greener grass outside their pasture. Hay or pasture that contains foxtail or any other barbed grasses or grains may give your sheep facial abscesses. All of these mechanically caused abscesses may be differentiated from more serious bacterially caused ones by the presence of a puncture or a splinter or piece of barbed grain. Remove the cause (if it is still embedded) and lance and drain the abscess. Again, tincture of iodine seems to be an excellent flushing medium.

Bloat: Sheep run on rich legume pasture or fed concentrates, or flocks put out on very green grass are the animals most likely to bloat. A bloated sheep will stand with its head down, obviously in pain or at least unwilling to move or eat. The sides of the animal—and particularly the left side—will be swollen or distended. For a discussion of how to diagnose and treat this problem see the Appendix, article on bloat. Immediate treatment is imperative if the animal is to survive a serious case of bloat.

Enterotoxemia: This is a fatal disease that is more common in lambs but might strike an animal of any age. The sudden death of a healthy, well-fed animal (or animals) is the first sign of enterotoxemia in many instances. This problem is not a contagious one, but it may strike many animals under similar conditions so that it will appear to be going from animal to animal. Enterotoxemia is covered in the Appendix. You will probably want to vaccinate your flock against this disease.

Lamb Diseases: The diseases below are specific to lambs.

Navel Ill or *Joint Ill:* This is a condition that develops when the lamb is a couple of weeks old. It is an infection of the umbilicus which spreads throughout the body and affects the joints. The lamb will have swollen joints (particularly the knees) and may stagger about or just lie down and thrash. It may have congested eyes and be generally ill and weak. This condition may be treated with antibiotics (penicillin-dihydrostreptomycin injections), sulfa drugs, or a combination of antibiotics, vitamins, and electrolytes called Bovi-form. Antihemorrhagic septicemia serum may also be given to prevent or treat this condition (consult a veterinarian or order from an animal supply house). Routinely dipping each newborn lamb's navel cord in iodine will help to prevent this disease. The herbal preventive treatment is to dip the navel in a mixture of equal parts garlic and sage, brewed in one and a half pints of water, with one tablespoon of witch hazel added (*Herbal Handbook for Farm and Stable*).

White Muscle Disease: This disease is apparently caused by a deficiency of vitamin E, and is common in certain areas where pasture and feed is lacking in this vitamin. Affected lambs become stiff in their hind legs and unable to get around. They may get pneumonia and die of this secondary problem. Or they might literally starve to death. The disease may be treated with daily doses of cold-pressed wheat germ oil (10 cc. per lamb). Selenium-tocopherol injections (given intramuscularly or subcutaneously) will act more quickly than

the wheat germ oil if the lamb is in very poor condition. Feeding pregnant ewes with selenium-vitamin E supplements may help avert this problem in areas where it is known to occur, but the placental-transfer is poor and lambs may still be affected.

General stiffness in a lamb may be caused by a sudden change in feed—as when it is separated from its mother or moved to a new pasture. This is a transitory stiffness that will probably pass before becoming more severe. It can be prevented by making feed changes gradual rather than abrupt whenever you can.

Rickets: Lambs may develop rickets if they are kept in pastures that are heavily wooded and dark, or if they are kept indoors for long periods of time. They may also develop rickets if they are fed milk deficient in vitamin D or are kept too long on a strictly milk diet. The lamb becomes stiff in the joints, particularly in the front legs, which may become bowed and very weak. The condition may be reversed by feeding the lamb irradiated yeast or vitamin D oil. The lamb should also be put out in the sunshine if this is possible. I have seen one case of rickets in a baby lamb that had been fed a powdered milk replacer that was inadequate in its vitamin D supply. Switching this lamb to a diet of fresh goat milk, which was rich in the vitamin needed, resulted in a dramatic recovery. The lamb began growing again normally, and within a week was frisking about on strong, normal legs. If rickets are allowed to develop over a long period of time, the young animal's bones may become soft and spongy and permanently deformed.

Wool Balls: This is a condition you may not discover until one of your lambs dies and is autopsied. The stomach will be found packed with clumps of wool swallowed by the lamb. These clumps block the normal passage of food from stomach to intestine and kill the lamb. Clipping the ewes' udders before they lamb and keeping them in good condition while they are nursing (ewes in poor condition will shed their wool) will help prevent this condition. Lambs must also be fed well.

Worms: Lambs are subject to heavy infestation of worms as soon as they begin grazing, particularly if they are kept under intensive conditions and if their mothers have not been regularly wormed. These lambs will become thin, weak, and will grow poorly. All lambs should be given the benefits of clean pastures, good food, and regular worming. Use either commercial wormers or herbal treatments (see later section on "Internal Parasites").

Poisoning: Poisonous plants and moldy hay are two sources of potential poisoning in your flock. You should familiarize yourself with plants in your area that are specifically poisonous to sheep and be sure that they are not present in your pastures. The *Merck Veterinary Manual* has a fairly extensive list, by geographical area, of plants poisonous to different types of livestock under various conditions. Textbooks on sheep or on general livestock care usually include lists, too. Or you may consult your farm adviser and other local farmers. Be careful not to feed moldy or spoiled hay, and check all bales for odd-looking plants. Symptoms of poisoning vary greatly, from trembling and depression to violent death.

Pregnancy-related Diseases: Ewes are subject to ketosis, mastitis, milk fever, and prolapses of the vagina and uterus. These problem are discussed at length in the Appendix. Briefly, though:

Ketosis usually affects a ewe a few weeks before she is due to lamb. Symptoms include a general dullness and disinterest in grazing, and the ewe may be found standing with her head pressed against something or lying on the ground. This disease particularly affects ewes carrying twin lambs or a single very large lamb. Treatment is simple and usually very effective.

Mastitis is an infection of the udder, and may lead to a severe systemic involvement with high fever and so on.

Milk Fever will affect a ewe who has lambed very recently, particularly if she is a very heavy milker. The characteristic pose of an animal suffering with this condition is lying down with her head turned back against her side. She will appear stupefied and weak. Treatment is very specific and unless it is done soon, the ewe will probably die.

Prolapse of Vagina and/or Uterus occurs in an animal in late pregnancy. The first sign is a mass of tissue protruding from the vulva which may vary in size from a half inch to two or three inches in length. Prolapses must be treated as soon as they are discovered or they may develop into a very serious problem.

Tetanus: Docking and castrating of lambs, parturition in ewes, and shearing of any animals all create possible vulnerability to tetanus. Read the special section for symptoms and treatment and prevention of this disease. Sheep kept on land where horses or mules have grazed are particularly vulnerable to tetanus and must be vaccinated. Classic signs of a developing case of tetanus include disinterest in or inability to eat, stiffening of the muscles of jaw and neck, and extreme reactions to sound and movement.

Vitamin D Deficiency: A long period of little sunlight can lead to vitamin D deficiency in young and old animals. Black sheep and unshorn sheep with very heavy fleeces are particularly susceptible. I have seen two mild cases of this disease—both of which occurred in our flock during a long, rainy, sunless spell. Both animals were black ewes with dense, long fleeces. The first symptom is a mild limp. The animal acts as though it has foot rot or an injury, but signs of neither can be found. It took me a lot of reading and cross-referencing to diagnose this problem, but both cases responded immediately to proper treatment, so I knew I'd found the cause. Injections of vitamins A and D (this combined form was the only D available) brought about a complete and dramatic recovery.

Severe vitamin D deficiency over an extended period of time can lead to serious symptoms and even death. The animal's joints become stiff and swollen—in particular, it will drag its back feet. The spinal column may become humped, the animal's breathing labored, and so on. Rickets, as discussed in "Lamb Diseases," above, is similar to this problem. Vitamin D deficiency is fairly rare, but if you raise black sheep or keep your flock indoors a lot, you should know about it.

Wool Blindness: Quite literally, this condition happens when a sheep's wool grows too thickly around its eyes, covering them and causing blindness. All breeds of sheep with wool on their faces should be periodically clipped for this reason. A young, purebred Rambouillet ewe we had needed monthly clippings because her face wool grew in so quickly and threatened her sight!

INTERNAL PARASITES

Sheep infested with internal parasites show similar symptoms regardless of the type of parasite involved: listlessness, loss of weight and condition, and weakness. Another sign of heavy infestation is a swelling under the chin known as "bottle-jaw." If the animal becomes anemic as a result of its burden of parasites, the inner eyelids will become pale and whitish-gray rather than their normal red to pink color. Young animals may grow poorly. An animal may become so weakened that it will eventually die—either as a direct result of its worms, or because it becomes vulnerable to a secondary disease, such as pneumonia. All in all, internal parasites may be the most serious problem you face in your flock.

Sheep are subject to infestations by stomach worms, liver flukes, tapeworms, lungworms, sheep bots, and coccidiosis. The life cycle of most of these parasites involves a period outside the body of the sheep. The eggs are shed in the feces of the sheep, develop in the soil or in an intermediate host, and are picked up again by the grazing sheep. Eggs of the broad tapeworm, for example, are shed in the feces of an infected animal and eaten by a type of grass mite which, in turn, is eaten along with grass blades by the sheep. Pastures become contaminated with eggs and larvae of parasites as the sheep graze. The longer the sheep are confined to these areas, the more heavily infested the areas become. The sheep, in turn, become more and more seriously infested. Treating the sheep with commercial or herbal wormers may have an immediate effect in freeing them of most of their parasites for a short time. If they are left in contaminated pastures, though, they soon become reinfested. Pasture rotation is one way to minimize this reinfestation.

Larvae and eggs of some parasites can survive on the ground for eight weeks or even longer, particularly if the pasture is damp or boggy. To break the cycle of reinfestation, you will have to rotate your sheep from pasture to pasture, leaving the contaminated ground to rest for as long a period as necessary. Dry land exposed to bright sunlight will be free of parasite eggs much sooner than damp or shady pastures. You may hasten the cleansing of a pasture by spreading lime, by plowing and planting the land, and so on. Or you may use the pasture for grazing other livestock if the grass and other vegetation is still plentiful. Traditionally, cattle are grazed during this rest period because they are not vulnerable to sheep parasites and vice versa. If you have limited space and are hay-feeding your flock, try to keep feed off the ground as much as possible. Following a system of pen rotation in this case might help. Keeping areas around

feeders and waterers dry also helps. In general, keep all pens, sheds, and feeding equipment as clean as possible. A healthy, well-fed animal is less vulnerable to heavy infestations than a poorly fed one, so keeping your flock in top condition feed-wise will also reduce worm problems.

If your sheep are showing signs of worm infestation, or if you decide to start a regular preventive program of worming, you have two basic approaches in deciding what preparations to give them. The first is more expensive and not usually used unless really necessary. This involves taking fecal samples to a veterinarian or laboratory for examination. The samples must be taken very fresh and lifted cleanly (without dirt or bedding material) with a spatula or knife, put in a clean glass jar, and brought in immediately. With a properly taken sample, your veterinarian should be able to isolate and identify eggs of parasites and tell you what you are dealing with in your sheep. You can then choose a worming preparation that is specific for that species (or combination of species) or parasite. More commonly, sheep are wormed with a combination of drugs that remove most frequently found species. This second alternative is much cheaper and easier. Some of the preparations available are:

Phenothiazine—this is a traditional compound used against stomach and intestinal worms. It is given as a drench, in capsules, or fed at a continuous low level in a free-choice salt mixture. Because some species of parasites can develop a resistance to this drug, it is being less commonly used now than in the past.

Thiabendazole—another drug used to control stomach and intestinal worms. It is given as a drench or in capsules or boluses. Thibenzole is one brand name for this drug. It seems to be a good, effective control for certain types of parasites but does not control tapeworms or lungworms.

Tramisol—this worming medicine is a little broader in effectiveness. It controls lungworms, stomach, and intestinal worms of most types. It is available as a drench or in oblets.

Hexachloroethane—a specific for treating liver flukes. Available as a drench.

If you don't want to use these chemical wormers on your sheep, you may choose to use herbs. One well-known natural vermifuge is garlic—it may be grated or pressed and mixed with bran and molasses (or molasses and other grains) to make it palatable to the sheep. The *Herbal Handbook for Farm and Stable* discusses treatment of serious cases of worms in fair detail. There are also botanical worming compounds available for goats that should work well with sheep. One of these compounds is available from American Supply House (see Resources)—I've used it on our goat herd for years and have literally no problem with heavy worm infestations.

Regardless of the preparations—herbal or chemical—you decide to use on your sheep, set up a regular program of worming and follow it conscientiously. To begin, worm all of your animals and move them to a clean area/pasture/pen. In three months or so, dose the entire flock and move them again. You will have to judge by the animals' appearance and condition how often they should be wormed and moved. Any new animals that are bought should be isolated for a few weeks and wormed before they are put in with the flock. This will minimize the introduction of new parasites. It is a good idea to worm your ewes shortly before they lamb and move them to a clean pasture. Lambs do little grazing for their first six or eight weeks, but after this they are susceptible to eggs and larvae in the pasture. They are especially vulnerable to worm infestation when they are weaned, as their standard of nutrition and therefore their general condition drops at this time. It would be wise to worm your lambs at or just prior to weaning and move them to a clean pasture.

If you are short on pasture or cross-fencing, you can still keep your flock healthy and fairly parasite free. We worm our small farm flock three to four times a year with a broad-spectrum wormer. They graze freely in the day and are locked in a smaller pen with a shelter at night. The hay and grain they are fed is placed in feeders. Although they are grazing the same land and keep to the same general areas, they rarely have problems with parasites —at least to the point of showing any symptoms. They stay in good condition year round.

Besides the parasites discussed above, sheep may be affected by sheep bots and coccidiosis. The sheep botfly lays its eggs around the nostrils of the sheep. The larvae—called bots—crawl up the nasal passages into the animal's sinuses. The sheep will have a runny nose, sneeze or snort, and shake its head. The larvae may stay in the sheep's nasal passages or sinuses from one to ten months. In some cases, the nasal discharge becomes streaked with blood. Fatal complications—such as an inflammation that spreads from the nasal passages to the brain—are possible. When the larvae are mature,

they migrate back down through the nasal passages to the nostrils and drop to the ground. They burrow into the soil and spend a pupal period of weeks before emerging as mature flies—and beginning the cycle again. The old-time method of treating sheep with nose bots was to spray or inject a cresol solution into the nose and sinuses. Oral drugs have now been developed to control the larvae of these flies. These drugs may leave residues in the animal's meat, so it is not advised to use them on lambs intended for slaughter. Giving the flock access to shady areas—where fewer flies congregate—is one help in controlling this problem. Applying some sort of repellent (herbal or chemical) around the animals' noses might also help. This is a problem I (fortunately) have never seen in our flock.

Several species of the tiny parasitic coccidia attack sheep. These parasites cause a localized infection in the intestines which may be mild and pass almost unnoticed or become severe enough to cause the animal's death. Symptoms of infestation include diarrhea, lack of appetite, weight loss, and general dullness. In serious cases, the chronic diarrhea becomes bloody and dark, soiling the wool around the hindquarters. Lambs are more often affected than older animals. Farm and feed-lot lambs (particularly those subjected to damp, cold environments) are the most susceptible. Animals are infected by oocysts shed in the feces of other animals which have been previously infected by the coccidia. These oocysts contaminate feed dropped on the ground and are thus ingested by the animal. The oocysts contain sporozoites—tiny bodies that invade the animal's intestinal lining and begin to reproduce. If the animal is in good condition, it may have a brief case of coccidiosis with erratic attacks of scouring, and then recover spontaneously. Animals that cannot resist the coccidia as well become more severely ill. Oral treatment with sulfonamides (sulfamethazine, for instance) gives good results. The sick animal should be moved to dry, sheltered quarters and given fresh water and high-quality feed. Recovery may take a week or longer.

EXTERNAL PARASITES

While liver flukes, tapeworms, and coccidia infect the interior of your sheep, flies, ticks, and lice threaten the exterior. Keeping your animals in clean, well-drained areas and feeding them well is the first step in avoiding all of these parasites. Attention to docking, tagging, and shearing will add a second step. Treating all wounds and watching for infestations of exterior parasites is the third important step. Your flock—whether a growing commercial size or just a few around the farm—will produce better fleeces, better lambs, and live longer, healthier lives if you keep them free of external parasites too.

Sheep Ticks: The first and most readily apparent parasite on your sheep's body may be the sheep tick or ked. This singularly unattractive, large insect is found walking about on the newly shorn sheep, or clinging to the skin when you part the wool of an unshorn animal. It is brownish in color and up to a quarter inch in size with six legs and an oval, flattish body. The sheep tick is not really a tick, but a wingless, blood-sucking fly. It spends its entire life on the sheep's body and should it fall off, rarely survives to infect another animal. These ticks pierce the sheep's skin to suck blood, causing an irritation that can lead the sheep to rub and bite at its fleece. Almost any sheep will have at least some of the parasites. A severe infestation may ruin the fleece and cause the animal to lose condition. Sheep ticks may be controlled by dipping or spraying or dusting the animals with rotenone or other insecticides. Sheep are usually treated just after shearing, and an insecticide that lasts three to four weeks will kill not only the present adult ticks, but those that emerge later from their pupal states. If you treat all animals regularly at shearing time, you should be able to control these pests fairly well.

Our sheep have very few sheep ticks even though we never dust or spray for them. They share the farm with a flock of free-ranging chickens who dine on the sheep's parasites in the following (often observed) ritual: a chicken will approach a sheep who is busily grazing or cleaning up windfallen apples and begin to circle it purposefully. The sheep invariably stops what it is doing, lowers its head, and stands motionless. The chicken walks around and around grabbing a tick here and there from shoulder or side. When the chicken is done, it strolls away in search of other food. The sheep resumes its grazing, or hurries off to catch up with its friends. The first few times I saw this happen, I watched in total amazement! I have since come to take it for granted—a perfect, simple, symbiotic relationship. I've also read of other farm flocks where the same relationship exists, sometimes involving turkeys as well.

Screwworms: Blowflies cause two kinds of problems with sheep: screwworms and wool maggots or fly strike. The screwworm fly is a common, bluish-green fly that needs a warm-blooded host for its larval stage. They are most common in the southern and southwestern states. These flies lay their eggs along the edges of wounds—even the tiniest wound caused by a tick bite will do. When the larvae hatch, they burrow into the wound and live there, feeding on the fluids produced by the wound. Their presence causes the animal irritation which corresponds to the number of maggots present (which relates, in turn, to the size of the wound). They cause the wound to ooze a reddish-brown fluid which attracts more flies of different species. They may also cause secondary infections. After about a week, the maggots drop off the animal to continue their life cycle in the soil. An animal that is badly infested with screwworms may be found in dark, shady areas trying to escape attacks by other flies. It should be treated at once, and other animals in the area checked carefully for these parasites.

Screwworm larvae may be killed with an application of a suitable insecticide (consult with your farm adviser or veterinarian, or check farm supply catalogues). These are smeared directly into the wound. The *Herbal Handbook* mentions smearing maggots with linseed oil. Preventive measures should also be taken. All wounds—from shearing, castrating, docking, etc.—should be treated with a protective insecticide. Use either a specific type against screwworm infestation, or a general-purpose type, such as derris (rotenone). Animals should be watched carefully during warm (fly) seasons as they may have small wounds you haven't noticed which become infested.

Wool Maggots or Fly Strike: Blowflies of other species cause a similar problem in sheep. The green-and-blue bottle fly and the black blowfly cause a condition known as "fly strike" or "fleece worms" or "wool maggots." These flies lay their eggs in wool. Usually they lay in soiled wool around the anus. An animal that has been scouring for any reason is especially susceptible to these flies. Sheep with long wool around their tails and hind legs will likewise be susceptible (as will sheep with long tails). Sometimes, though, the flies will lay in clean wool—usually on a hot day following a rain. Once the eggs are laid, maggots hatch and burrow about the sheep's body. They excrete a substance that attracts other flies, and the infestation becomes more and more severe. As the maggots burrow under the skin, the animal begins to show signs of irritation and restlessness. The wool may become discolored as serum oozes from maggot-caused breaks in the skin. The animal may be badly infested before symptoms begin to show.

Treatment for this problem involves clipping the wool away from the "struck" areas and treating the areas with an insecticide or combination insecticide/salve. During warm weather, all sheep should be kept tagged (the area around tail and legs clipped short) and watched as carefully as possible.

Mange: Sheep are susceptible to four types of mange. Sarcoptic mange affects only the sheep's head—forming thick, crusted areas after an initial period of intense itching. It may be treated with lime-sulfur or lindane solution. Sheep scab (psoroptic mange) is now considered rare in this country. It affects the heavily wooled areas of the body, causing itching and then the formation of scaly, crusted areas. It is treated with lindane. Derris would have a similar effect. Leg mange (chorioptic mange) affects the hind legs and the scrotum of rams. Nicotine, lindane, and creosote are possible treatments. The fourth type of mange is demodectic—which causes small lesions almost anywhere on the body. These lesions are filled with a waxy, grayish material. They are treated by cutting the lesions and filling them with tincture of iodine. This fourth type seems to be more common in goats than in sheep.

Lice: Like the ked or tick, the louse spends its entire life cycle on the body of the sheep. They suck on or bite the sheep, causing itching and irritation. The sheep's wool will become dry and broken and may be badly damaged by the rubbing and biting the sheep does in response to the irritation caused by the lice. If you part the wool and look carefully at the skin, you may be able to see the lice or their eggs. Affected animals may be sprayed, dipped, or dusted with rotenone or chemical insecticides. Two weeks later, they should be treated again to kill the young lice that have hatched since the first treatment. If you neglect this second treatment, your animal(s) will become reinfested. Lice are most common during the winter months when animals are confined together. The lice pass directly from animal to animal. If one of a group of animals has lice, others are probably affected too, though perhaps to a less-obvious degree. All animals in the group should be treated, and the barn or shed cleaned after each treatment.

337

BUTCHERING SHEEP, DEER AND GOATS

Slaughtering the Animal: Practicality comes in big chunks on the small farm. If it is a farm based on low income from the outside, the practicality may become crushing at moments. If you just cannot call a veterinarian, if you cannot afford to feed any of the extra critters that have accumulated from a couple of bumper crops of offspring, or if you just want to be more self-sufficient and raise your own meat, you will have to face the reality/practicality of slaughtering your animals. Surplus meat produced on the farm is of value as a food for humans as well as for the family dog or cat and the chicken flock. The following was written about my experience with my goat herd; I believe the approach and feelings and knowledge will be useful to you for many species of animals and birds.

The absolute necessity of preventing the overpopulation of a dairy goat herd is fundamental to the program of a milking herd for a small farm. One or two does that milk a gallon each will consume hardly as much as a dozen poor milkers with bad temperament who go around wasting their time butting everybody else out of the feeders. When a fat goat who gives no milk stares you in the eye some early morning and dares you to try to budge her, it is time to dust off the trusty old rifle.

I know that a goat is intelligent. I know that we live close to our small animals on our small farms. We deliver them, raise them, train them, pamper them, love them. They respond by becoming individuals almost as special to us as our own human children or friends. It is inconceivable that we would slaughter our own children; consequently it is a justice problem to decide to slaughter a milk goat. Weigh the matter carefully from the moment the kid is born. A good way to go about it is to examine the kids minutely at birth, evaluating their parents and their parents' parents, if possible. Then taking all the physical conditions into account, give each a percentage of chance to survive as a useful member of the herd, a certain given number of years, for a certain use. If it is known from the moment that the little animal gasps its first gasp under your gentle hands that it is meant for needed meat supply someday, and if this thought is steadily maintained with a sense of the fundamental sacredness of mere survival, then this knowledge and a regard for the comfort of the animal will help to make slaughtering day of infinite value as a ritual almost amounting to an act of reverence.

Very few slaughter houses have people who know the proper technique for slaughtering your milk goat. Also, the goat is too smart to fool. She or it or he will know what is going on. It is best to do this job yourself, or have a compassionate neighbor who does not mind shooting animals do the job for you. Most country people who hunt deer and other game for part of their subsistence will gladly help you. But you must supervise this out of regard for the animal. Talk with this person about the fact that the goat must be handled with great care at this culmination of her life. She must not know what is going to happen, so that she might be spared unnecessary terror and upset. She must be led out toward the back pasture away from the other does, if possible many times in the days before slaughter. It is a good habit to have all of your does accustomed to going out to the woods or back pasture or away from the loafing area or pens in order to prevent their having fear when the final day comes. Take along a pan of grain on the day chosen—a pan that can be thrown away if the bullet hits it, and preferably one made of rubber or plastic. If the doe is hungry, it is easier and kinder to do the job as she gobbles her grain out of the pan set on the ground. Her head will be low and her eyes not up where they can see your gun.

The .22 Winchester Rim-Fire Magnum is a good homestead utility rifle. The .38 Special pistol with 255 foot-pounds of energy is also good. Some guns are too heavy, with a rapidly disintegrating bullet which may be dangerous in a confined space. Some guns just won't do the job. Be very careful in choosing a sensible gun. Know how to handle it. Take a course in the use of the gun. If you have an old .22 and have an unhappy experience requiring several shots to put your old goat out of her misery, you will not want to go on being a farmer, or you may be just plain horrified. Remember that death is a part of life—natural and meant for every single one of us—and if it can be dignified and painless and come quickly and unawares, it is but one of life's true adventures. An old or sick animal needs to be put to rest painlessly and efficiently, with a minimum of emotional impact. The art of raising goats is expressed in the fine art of eliminating members of the herd when the proper moment comes to do so without upsetting the rest of the animals on the farm. Do the job downwind because they can hear and smell the activities, and blood has a strong, upsetting smell to goats.

Goats play by delivering themselves hammer-like blows to the head, so it is pretty doubtful that you can deliver a humane knock-out blow with a hammer, or by shooting the bullet into the forehead. Well meant, perhaps, but impossible. The proper place to position the muzzle opening of the rifle or handgun is behind the horn area, about an inch below in the center, as if you drew a line between the horns and lowered the gun. Fire down or forward. This way you will strike the brain without fail. The straight-down entry is probably best because it minimizes penetration of the mouth or nose. If the goat is small or has a small head, do not lower the gun a whole inch below the line drawn between the horns. Use your best judgment. Move slowly but deliberately. Speak in a normal voice and try not to exhibit anxiety or tension to the animal. Once you have shot the animal, you should set your gun aside in a safe, secure place and go about butchering. If your milker gives you ninety pounds of good meat, she has done you one great final blessing besides all the lovely kids and milk she gave you. Her life ends in a spirit of respect and thankfulness.

Beginning: I learned to butcher first from my neighbors, two brothers aged seventy-nine and eighty-five. The last time we worked together was on a buck deer one of them had shot. For the first time, I did the whole skinning and gutting process myself, without help. As I finished, the older brother, Fred, said: "Well it's good to know you can do it. Who knows, we may be bringing our meat down to you in twenty more years."

I doubt I'll ever see the day when I can do better than they can, but I felt pleased that I knew what to do and could do it. Reading the preceding section on slaughtering helped me when I killed my first animal. Butchering is something that I do with great thought and much care. I am conscious throughout the process that I have taken a life and I am thankful. But I also feel good that I am doing it. I have eaten meat most of my life without facing where it came from or my (the consumer) responsibility as part of the process. When my sister, who lives here, talked about butchering one of our lambs, I agreed as long as she would do it. I didn't think I could face blood or muscle or bone connected with a real animal I'd raised—and then eat it. I have discovered that I can, that I know what I'm

doing when I butcher and what I'm eating when I eat. Butchering is one more basic link between my skills and my survival.

It's hard to describe the butchering process without it sounding gory. Nevertheless, I'll try to give the details as completely as I can. It is a *skill* like any other, with techniques to be learned. The emotional and karmic aspects are another level of the process. This article is about butchering small animals: sheep, goats, deer. I've never butchered a pig or steer, so can't describe either of those. I'll discuss sheep and goats first, since they are dealt with the same way. There are slight variations for deer, which I'll cover later.

My neighbors, after sixty years on their land, have their butchering table and meat house all set up. It's a fine system, but much more primitive conditions will do. They have a table the right height for cutting, with slats so that blood can drain through. The table is out in the barn and there's straw on the floor beneath to soak up blood and gastric juices. From the rafter four ropes are hung (half-inch thickness), and these are used to tie the legs out of the way. There is also a block and tackle for hoisting the whole carcass—this also needs a good stout rope. A metal washtub is kept on hand for the guts and head. The meat house is a small shed, set in a shady spot with screen windows on opposite sides for ventilation. This is where the meat is hung for aging (tenderizing). If you don't plan to do much butchering, you can work without such facilities. You will need a block and tackle or at least a good stout rope hung over a tree branch or beam. Gutting can be done on a tarp on the ground. Aging can be done by covering the whole carcass carefully with cheesecloth to keep flies off, and then hanging it in the shade.

The first step after the animal has been shot (especially with a head shot) is to bleed it. This keeps blood from getting into the muscle tissue and damaging the meat. It should be done as soon as the animal falls and is dead. Cut the jugular vein in the neck with a sharp knife and keep the head down. You can catch the blood in a pan to use for fertilizer or food if you wish. (This bleeding is the hardest part of the process for me. It is the moment I actually accept the fact of death.)

Skinning: With a sheep or goat, the next step is to skin the animal before gutting. Deer are usually gutted first, as will be described later. Some people handle goats like deer, but you should find what is easiest and most sensible to you. We do the initial

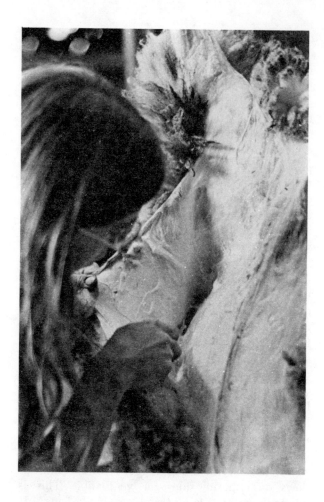

below (footward) where the joint starts to swell. This is close to the center of the widest part. Cut all the way around with your knife, as deep as you can, then bend the leg backward, and it should snap off. It's a nice feeling when this cut goes easily. Do this on both hind legs. At the genitals, cut around the vulva or testicles so they can be pulled into the intestinal cavity, cut down the center of the tail. You'll begin skinning at this point.

Be careful to cut through the skin only! You don't want to puncture the guts with your knife or to bring muscle tissue off with the skin. There are four layers here: skin, connective tissue, fat, and muscle or intestinal wall. On a good skinning job, the hide should come off clean of any fat, connective tissue, or muscle and be ready to be tanned.

On the front legs, cut around and up them just as you did on the hind legs and pull the skin back to expose the knee. Cut up each leg to the center of the breast, then cut through the skin from the breastbone to the head, exposing the windpipe. Saw through the breastbone with the crosscut saw while the legs are still tied out of the way. This will expose the chest cavity. Next cut and break the front legs at the knee joint, just as you did the rear ones.

The last step before hanging is to slit the skin from the breastbone to the tail down the center of the belly. Be very careful not to cut through the intestinal wall. The best way to cut is to hold two fingers under the skin in a V shape, raising the skin up. Then cut the raised section with the knife, sliding your fingers as you go.

Now you are ready to hang the animal for the actual skinning. Place a gambril (metal bar with hooks) or any metal bar or stout stick you have through the hamstring on each hind leg. Then attach a strong rope to the center of the bar. A block and tackle will help greatly, but you can just throw the rope over a tree branch or beam if you have to. Raise the animal quite high so you can cut off the head. First cut through the skin as close to the base of the skull as you can. Then, using the butcher knife, cut through the neck muscle all the way around, clear through to the ball-and-socket joint at the top of the neck. Twist the head while someone else holds the front legs and it will snap off. Put it in the washtub for burial later. Then lower the carcass so you can begin skinning at the hindquarters.

When skinning, you should work from the center

skinning in the barn on the worktable. You can also do it on a tarp on the ground. You can tie the legs in a stretched out position or have a friend hold them for you. A good sharp knife is needed and a whetstone should be handy to keep it sharp during the process. Your knife does not need to be large—a small pocket or paring knife will do most of the work. You will also need a larger butcher knife and a crosscut saw.

Begin by cutting around the hind legs just at the knee joint. Then slit the skin up along the center of the leg, toward the genitals. With all skinning procedures, try to cut with the blade of the knife upward, so you cut out through the skin. Cutting down toward the flesh tends to get hair on the meat. Pull the skin back on both legs where it has been cut, exposing the knee joint. The hamstring is on the upper part of the leg and you want to break the legs off at the knee below where the hamstring joins. The hamstrings are used to hang the carcass, so you don't want to sever them by cutting too high. There is an easy place to get through the joint—about one eighth to one quarter of an inch

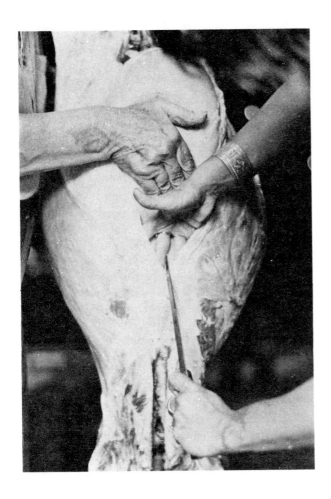

your knife to cut it off the skin. The neck is the toughest part, and you may have to use the knife quite a bit there. Once the skin is free, set it aside for salting and tanning later.

Gutting: Begin gutting by cutting through the muscle around the rectum and the genitals. Cut down the center of the belly a short way. Cut with the knife blade facing upward so that it cannot cut any of the intestines. If these spill they will ruin the flavor of the meat in that area. After you have cut six to ten inches down the intestinal wall, you should be able to reach in and pull the genitals and rectum into the body cavity. Cut more if you need

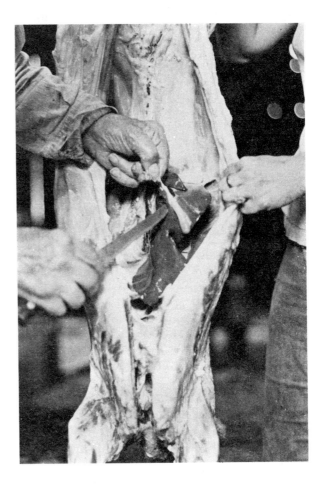

of the hide out in both directions. Wherever possible, you want to use just your hands and pull the hide off. It will come clean that way. With a knife, you are more likely to cut down into connective tissue or fat. You will need the knife at the beginning, however, and for any hard to work spots. Keep the blade flat against the skin and cut it free from the connective tissue. Start at the tail and work the hide downward, off the tail and the hindquarters. As soon as you can, grab the hide with your left hand (if right-handed) and pull it upward and out. When it is started, use your thumb and knuckles instead of the knife. Push down with your thumb, separating the skin from the membrane. Start from the backbone and work the skin free with your thumb all the way around to the belly. Then return to the backbone and work in the opposite direction. Once you're well away from the legs and onto the back, the skin should just pull free. Grasp it in both hands and pull downward and out. You will have to keep moving your hands to pull evenly all the way across. If any muscle or fat begins to come off with the skin, use

to until you can see the bladder (a pale-colored sack, full of urine). Pinch the bladder closed where it joins the intestine and cut on the intestine side. Very carefully remove it (holding it shut) and the genitals without spilling any of the urine on the meat. Put it into your washtub. Then continue slitting the intestinal wall to the breastbone. The stomach and intestines will fall out. Hold the

341

stomach as it falls, pinch it off where the esophagus enters, and cut it off. Let the stomach and intestines fall into the washtub, but make sure the stomach doesn't spill. Now the abdominal cavity should be empty up to the diaphragm—a wall of thin muscle that divides it. Cut the diaphragm out along the rib wall. This will expose the heart, liver, kidneys, and lungs. On the liver in goats and sheep is the gall bladder, a small greenish-yellow sack. It must be pinched off and cut out just like the bladder to keep it from spilling. Then cut out the liver, heart, and kidneys and put them in a bowl as they are all edible. Cut the lungs away from the ribs, pull them and the windpipe out of the body, and discard. You are now done gutting and the body cavity should be clean. If the animal was not shot through the head, cut out any bloodshot or bullet-damaged meat. Clean any hair off the meat with a clean dry rag; don't use any water for this cleaning.

Gutting and Skinning Deer: Deer are handled slightly differently from sheep and goats. They are usually gutted where they are shot, since they often must be carried long distances before they are skinned and the extra weight is a burden. The best and safest shot for a deer is into the heart or lungs. With this shot, you will not need to bleed the animal as it will bleed itself. Once the guts are removed, you can pierce the diaphragm and drain out the blood.

I'm assuming that you will be butchering a buck deer, since they are the legal game in most areas. First you should cut the musk glands out of the hind legs. These are on the inside of the leg, near the knee. Slit the skin and remove the gland as it will give the meat an off-flavor. Then cut open the testicles and pull them out. Then slit just the skin from the breastbone to the anus, cutting to one side of the penis. Very carefully cut through the abdominal wall and let the intestines slide out. Cut the bladder and stomach as described above for sheep and goats. Make sure all of the blood drains out, too. It will be thick, brownish, and clotting by now.

Now you are ready to skin. Cut around all four legs as for sheep and goats and break them off. Slit the skin around the neck at the base of the skull and down the center of the neck to meet the cuts from the front legs and the belly. Saw through the breastbone. Hang the deer by the antlers and begin skinning. Start from the neck down. The procedure is the same—use your knife when you have to, your hands when you can. It's actually a little easier to skin from the head down, but you have to have an animal with horns to do this. After the skin is completely off, cut out the diaphragm, heart, kidneys, liver, lungs, and windpipe—deer don't have a gall bladder on the liver. Then lower the deer, rehang it by the hamstrings, and cut the head off. Now the meat is ready for aging and the process is the same as for the other animals.

Aging: The next step in butchering almost any animal is aging the meat. Goat meat, unless it is from an older animal, does not need aging. Hanging it for twenty-four hours to cool it thoroughly is all that is necessary. Kid meat shouldn't be aged at all. Aging will greatly increase the tenderness of most other meats. The longer the meat is aged, the more tender it will be. How long you can hang meat depends on the weather: in cool, dry weather it can hang seven to twelve days; in wet or hot weather it should only hang three or four days. As I mentioned before, the best place for hanging meat is in a meat house—a cool, dark shed with screened windows for ventilation. You must keep flies off the meat, and keep mold from forming on it. If you don't have a meat house, wrap the meat very carefully in cheesecloth so it is completely covered. Hang the carcass in the shade, out of the reach of animals. If any mold forms, wipe the meat with vinegar and wrap it in clean cheesecloth.

Cutting Up the Meat: The first step in cutting up the meat is to quarter the hanging carcass. This is done by sawing down the center of the backbone with a crosscut saw. Then use a butcher knife to cut each half into quarters, cutting at an angle along the last rib (between the last rib and the hindquarters). The quartered meat should then be carried into the house or to a convenient worktable.

For cutting, you need a sharp butcher knife and a butcher saw. If you don't have a special saw, a hacksaw with a new, sharp blade will do. On any cut, use the knife until you are up against the bone. Saw through the bone and then use the knife again. The saw will tear up the meat and should only be used on bone. You should cut your meat into usable pieces. It's difficult to describe exactly how to cut up each quarter, since that will vary with the size of your family and your desires. There's great flexibility, for instance, in the thickness of steaks or chops, the size of roasts, and the choice of roasts or stew meat. But these are some basic guidelines. The neck is usually used for stewing. The shoulder and upper portion of the front leg are usually used

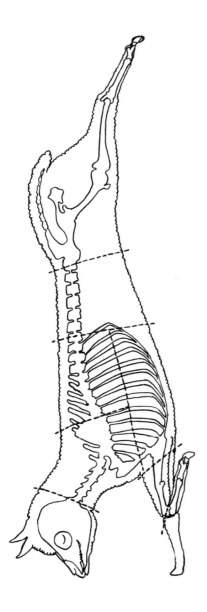

of the front leg on all three animals is used for stew or ground meat. The chops come from along the backbone. To get them, use the meat saw to cut perpendicular to the ribs where they begin to flatten out. You will be able to recognize the chops. Cut each chop with the knife, parallel to the ribs. The hindquarters of goats and sheep are used to make leg of chevon or lamb. Again, this may be one large or a small and a medium-sized roast. On deer, this section is sliced into steaks, until there is too much bone for practicality. The lower leg and the rump on all three are used for stew or ground meat. If you are going to freeze the meat, separate it into serving-size portions and wrap it tightly in freezer paper to seal out all air. If the meat is to be stored for longer than three months, it should be double-wrapped.

Home-butchered meat is wonderful—it is tender, well-flavored, and free from hormone and pesticide residues, if you so choose. Lamb has always been a favorite of mine, far too expensive for me to buy. Now I have it in my freezer—fifty pounds for the price of the grass in my meadow and my own labor. I have also made some discoveries. "Mutton" from a year-and-a-half- or two-year-old wether tastes like lamb, but butchers at two or three times the weight! Chevon (goat meat) is tender and at least as good as lamb. Goat kid is more like veal and is good breaded or cooked in a sauce. Venison, far from being gamey or strong, is a flavorful meat, "strong" only in the sense that beef is. And deer liver is the best there is.

In eating, I take the lives of other living things. Butchering a lamb is part of the same continuum as pulling up a cabbage in the garden. In a larger cycle, we all take our turns as consumer and consumed. I accept my place as consumer now with consciousness and thankfulness. I know that when I die my body will return to the soil, nourishing other lives.

for one large or two small roasts on goats and sheep. Venison lacks fat and doesn't roast well, so this section is usually used for stew or ground meat (which is excellent) on a deer. The lower portion

TANNING

I cannot pretend to be an expert at tanning. I began to learn to tan right after my first butchering, for there I was sheepskin in hand. My first hide came out stiff and dry, but since commercial tanning is a very expensive luxury, I've kept on learning. Tanning is a *lot* of work; the results in softness or usefulness seem to equal the amount of energy expended. But it is also tremendously exciting to begin to produce leather for clothing or farm use.

The first step in the tanning process is fleshing—removing all fat and tissue from the hide. This should be done as soon as possible after skinning. If you can't work on the hide immediately, salt and soak the hide (see below) and then flesh it. To flesh the hide, stretch it taut. I've found nailing it to the side of my house (out of direct sun) is the easiest way. This is a two person job, one stretches, one nails. Then use a very sharp knife (my pocket-

knife works well) to scrape off any remaining bits of flesh or fat. With a good skinning job, there should not be much. The skin itself has several layers, so be careful to only remove flesh or fat, not the top layer of skin. On my first hide, I often scraped too deeply. In some places I actually made holes, in other places the hide was weakened and wore out more easily.

Once the hide is fleshed, remove it from the wall and lay it out flat in a cool dry place. Rub the flesh side of the hide with coarse salt. A salted hide seems to tan more easily, salt also sets the hair on a hide and prevents shedding. Salt is also a preservative. If you don't want to tan immediately, the hide can be stored after salting for six to eight months. A hide soaked in a brine solution can be stored indefinitely. If you want to continue tanning or to store the hide, you should let it lie with the salt on it for about three days. Then brush off any salt that has not been absorbed. To store the hide, hang it over a rope or pole to dry and store it in a cool dry place until you are ready to continue with the next step in the tanning process—soaking.

The hide can be soaked in plain water, especially during cold weather. During warm weather, a mixture of borax, soap flakes, and water works best. My neighbors taught me that a ratio of 4 gallons water, ½ ounce borax, 2 ounces soap is about right. Whether you use plain water or the solution, the hide should be soaked only long enough to make it soft and supple (about two to four hours). If you've used soap, the hide should be thoroughly rinsed. It can be fleshed at this stage if you didn't do it first.

I have never removed the hair from the hide. My neighbors said that they soak the hide in a mixture of hard wood ashes and water (1 quart ashes to 1 gallon water) until the hair is loose. Then they stretch the hide as for fleshing and scrape the hair off. Rubbing ashes into the hair while scraping sometimes helps. I have also read that a weak lye solution can be substituted for ashes.

Once the hide is soaked, it is ready to be tanned. The tanning solution actually preserves the skin, making it into leather. There are all kinds of recipes and formulas for tanning. Solutions made from bark (especially of the tan oak tree) produce the toughest and most durable leather, as for shoes or harnesses. I've never done any bark tanning and don't have a recipe for it.

Alum is another common tanning agent, but it usually produces harsh, difficult-to-work hides. It's really only useful for damaged hides or where you want coarse, stiff leather for bindings, etc. The formula for alum tanning is to soak the hide in a mixture of:

1 gallon *soft* water (use rain water if necessary)
1 pound alum
½ pound salt

Make enough of this solution to thoroughly cover the hide, let it soak two to six days, then remove the hide and rinse.

The one thing alum is good for is tanning sheepskins. These may also be tanned in the sulfuric acid solution given below, but the alum process saves soaking the wool and may be easier. Sheepskins should be washed before tanning in warm water and soap, then rinsed thoroughly. Mix together:

1 pound alum
½ pound saltpeter
twice the total bulk of bran

This mixture should be spread over the wet skin of the hide. Then roll up the hide, wool side out. Moisten the hide each day to keep it damp and let it sit for a week. Then scrape off the mixture. This works fine for sheepskin rugs.

For really soft furs and good-quality leather, a sulfuric acid solution works best. I've tanned sheepskins, deer, and goat hides in this recipe and the fur has always come out glossy and beautiful. Sulfuric acid is dangerous, however, so be very careful not to get it on your body or clothes and do not breathe the fumes. Use a wooden barrel or stoneware crock for this solution, never metal or enamel which will be eaten by the acid. The basic formula is:

1 gallon *soft* water
1 quart salt
1 fluid ounce commercial-strength sulfuric acid
(battery acid, available at gas stations)

This amount will tan two small skins (like rabbits). I use about three gallons to do a deer or goat hide; four or five gallons for a sheepskin. It is important to have enough of the pickling solution to thoroughly tan the skin.

Start by heating the water and salt until the salt is completely dissolved. Let this cool in the barrel or crock, then add the acid, pouring it down the side of the container so it can't splash. Place the hide in the solution and weight it down with stones if necessary. The hide should be stirred and turned every few days. The length of time it needs to soak varies with the thickness of the skins and the temperature of the room. At 65° or warmer, a rabbit hide takes about ten days, horses and cattle fourteen to twenty days. Below 65°, tanning takes longer. Hides can soak in this solution for months, so leave them long enough to be sure they're tanned. Never reuse the solution for a new hide.

After removing the hide from the acid bath, it should be soaked for several hours in an alkaline solution (1½ cups washing soda to 2 gallons water). Then it should be thoroughly rinsed.

The last step in the tanning process, breaking or working the hide, is by far the most important and is my downfall. Working the hide means *working* it for hours and hours and hours. This breaks the fibers in the skin and produces soft leather. An improperly worked hide will be so stiff as to be useless. My hides fall somewhere in between—suitable for rugs but too stiff for clothing. Working takes persistence and perseverance, a few hours one day is not enough. No matter what tanning solution you used, you should begin working the hide as soon as it is rinsed and continue until it dries. If this is not enough, you should soak the hide again in water and continue. You can work and resoak the hide several days in a row or allow weeks to pass between each soaking.

There are many methods for working a hide. As every man I've asked for advice has quickly pointed out, Indian women used to chew the hides until they were soft. For form's sake, I have tried this, but neither the taste nor the texture appealed to me. Small hides can sometimes be worked by simply holding the hide in both hands and pulling, tugging, twisting, and rubbing it. Larger hides can be tied in a loop over a horizontal tree branch. A stick is then inserted in the other end of the loop and the hide is twisted tight, first one direction and then the other. Another alternative for large hides is to sharpen one edge of a $2'' \times 4''$ and nail it up horizontally at chest height. Stretch the hide over the board and have one person on each side bear down as hard as they can, working it back and forth. Keep turning the hide so every bit of it gets worked. I've used both of the last two methods on my hides, and as I've said, they're stiff. Perhaps if I resoaked them and continued working they would get softer. The sheepskins have made fine rugs after being worked for three to four hours. On the other hand, the deer hide intended for a wall hanging could be used to build the wall itself. But patience and perseverance are not my strong points, and the results may reflect more on me than on the method.

Recently a friend showed me another method of breaking hides. His hides are soft and professional looking, but I have not yet tried this way myself. He uses a wooden barrel filled with small stones which he cranks round and round, letting the stones batter the hide soft. If an old barrel is not available, one can be made by cutting the end circles out of $\frac{3}{4}''$ plywood. Then nail $1'' \times 4''$ slats all the way around, except for the last two or three which should be screwed in place. These are removable so you can get the hide in and out. Attach a crank to one end of the barrel and build a crisscross frame of $2'' \times 4''$s for it to rest on and turn in. Place the damp hide in the barrel, skin side up. Then partially fill the barrel with small stones. Keep cranking until the hide is dry. It may take several resoakings and cranking sessions to work every part of the hide or to get it thoroughly soft.

APPENDIX

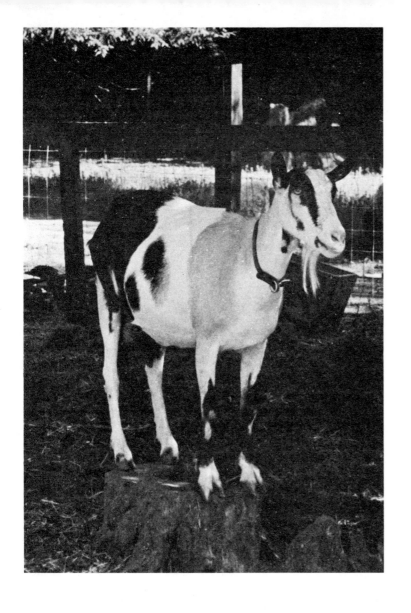

ABSCESSES

Any animal you might find yourself keeping—from your pet cat to your homestead cow—is subject to abscesses. An abscess is, most simply, an abnormal collection of fluids such as pus or serum in the body tissues. An abscess can be caused in many ways, the most common causes being injury and/or fungal or bacterial infection. It can be simple, localized, and easily treated—or part of a serious systemic infection. An abscess may appear in the obvious form of a firm swelling or it may be internal, manifesting itself in puzzling outward behavior. You can learn to recognize and treat superficial abscesses caused by injuries. With or without the help of a veterinarian's laboratory and specific diagnosis, you can deal with most abscesses caused by fungus or bacteria. The complications of internal—such as renal or cerebral—abscesses will have to be put in the hands of a skilled veterinarian.

The very simplest abscess you might have to treat is one caused by an injury. Cats commonly get abscesses on their jaws, at the base of their tails, and behind their ears. Most of these abscesses develop from injuries suffered in fights with other cats. Chickens are known to get abscesses on their feet when they jump down from too-high roosts onto hard floors. Sheep, cows, and other livestock can get abscesses from splinters of wood from feeders or fences. The barbed ends of common weeds (foxtail, for instance) and some grains (barley, rye, bearded wheat) may lodge in an animal's skin and cause an abscess. The most minor cut or tear in the flesh of any animal has the potential to become an abscess if it heals over too quickly, sealing in dirt, germs, and bacteria.

Another whole class of abscesses are those caused by bacteria and fungus in which the abscess is an external sign of an internal infection. An example of this type of abscess is that found on the jaw or shoulder of a goat suffering with coryne bacteria infection. This bacteria is highly infectious and causes all sorts of problems ranging from abscesses to swollen knees to mastitis. It is found to cause abortion in horses and cattle, jowl abscesses in swine, and so on. Actinomycosis (commonly called "lumpy jaw") is a systemic infection caused by a fungus-like material. It affects cattle, swine, horses, sheep and goats. The animal affected usually gets a characteristic swelling on or between the lower jaw bones. This swelling develops into an

abscess. Goats may also get abscesses there which are caused by staphylococcus albus bacteria, rather than by the actinomyces. Abscesses that develop without any sign of injury—prevalent in a herd of animals—or reoccur should be cause for investigation. These abscesses may not only contain highly infectious material (as in the case of the actinomycosis abscess) but may indicate a general or systemic involvement (as in coryne bacteria) which can ultimately lead to the death of the animal. You should consider taking a sample of the puslike material inside the abscess and sending it through a veterinarian to a laboratory for diagnosis. Or you should have your veterinarian take the sample. The material must be drawn under the most sanitary conditions and into a sterile tube which is immediately sealed shut. I've used a disposable syringe (without the needle) to draw material out of a just-lanced abscess. Our veterinarian was able to draw out enough fluid from another unopened abscess to use for laboratory diagnosis. The laboratory can run culture tests on the abscess material to tell you what bacteria or fungus is involved. In some cases (coryne, for one) you will then be able to vaccinate your affected goat to arrest the symptoms, and vaccinate other susceptible animals for protection. They can also run sensitivity tests that show which drugs the particular bacteria is susceptible to. This is the only way to get a positive diagnosis of the cause of an abscess and be able to treat the animal properly.

Whether the abscess is caused by a cat fight or a bacteria, it can and should be handled as follows. Once the infection has begun, tissue starts to break down under the skin and pus forms. The pus cells accumulate, stretching and disintegrating the tissue—and the abscess begins. At this point you will probably notice a small swelling. It may be painful or totally painless, soft or firm, hot to the touch or not, all depending upon the animal and organisms involved.

Most abscesses develop rather slowly. As long as the animal is not running a fever or showing other complications, the abscess may be allowed to develop until part of it begins to soften, or come to a head. This soft part would, if left alone, eventually rupture. There are two dangers in letting this happen: one, the skin can become so destroyed it begins to putrefy, making a bad infection and a wound that is difficult to heal; second, the material inside the abscess, if allowed to contaminate feeders, water troughs, and such, might infect other animals. Generally it is not a good idea to let an abscess head and rupture. One exception to this rule is in the case of the cat, who is very sensitive to the pain of lancing. If the cat begins running a fever, not eating, and so on, it should be taken to a veterinarian and the lancing done under even a local anesthesia. But if the cat seems healthy, let the abscess rupture on its own and then clean and treat it.

You can hasten the development of the abscess by applying moist, hot cloths to it. These may be soaked in a hot antiseptic solution or in plain water. Another alternative is the herbal poultice. Following the advice of a friend studying herbal medicine, I used a poultice to bring to a head an abscess developing on one of our goats. First I boiled a few teaspoons of flaxseed meal and powdered slippery elm bark in a little water until it formed a thick paste. I spooned this paste, still hot, into a little cloth bag and held this against the abscess until the bag cooled. Repeating this two or three times a day for three days caused the abscess to head. Common sense and the animal will tell you if your poultice or cloth is too hot when applied. Hold it against your own skin for a test.

Whether an abscess is hastened by fomentation or allowed to develop naturally, it is ready to lance when the fur or hair pulls away from it easily. It will usually be shiny and pink in color. Lancing prematurely may result in another abscess forming in the same place, so wait for the right stage. You will need rubbing alcohol or similar antiseptic; a quantity of absorbent paper (toilet paper works well for small abscesses; paper towels for larger ones) or cotton; a sterilized single-edge razor blade; an antiseptic or cleansing solution (boric acid solution; weak iodine solution) and a syringe top (or anything similar) with which to squirt in the solution; antibiotic ointment or tincture of iodine; a paper bag. You may want to wear rubber gloves—particularly if the abscess you are treating is one caused by actinomycosis (transmissible to people). You may also want to pack the wound with sterile gauze soaked in iodine or antibiotic solution, so have that on hand just in case.

First secure the animal well—usually a friend will be able to hold down a small animal, such as a sheep or goat. A cow or horse will have to be tied. Pull away all of the loose hair or fur and wash the area with alcohol. The head or softest part of the abscess will usually be obvious—the skin is more discolored. You should make your cut there, as that section is preparing to rupture anyway and there is less danger of damaging blood vessels or nerves. At this point you have two choices in lancing. One is to cut through the skin and right into the pus pocket. The other is to cut the skin and then use a blunter instrument to break through into the pocket. The latter method is supposed to do less damage to blood vessels. I have always made just one cut right through using the razor blade and have not had problems with excessive bleeding.

Once you have made your cut, force out as much of the puslike material as you can. Be careful to catch all of this with absorbent paper and put it directly into a paper bag (all to be burned later). Next you should flush out the cavity with antiseptic solution. Then fill the cavity with tincture of iodine, antibiotic cream or liquid, or gauze soaked in one of these (leave a small tail of the gauze sticking out so that you can remove and replace the gauze the next day). This same procedure is followed if the abscess has ruptured already. Try to clean any puslike material off the animal and its environment to avoid contaminating others.

For the next few days you should reopen the abscess, squeeze out the contents, flush, and refill with iodine or ointment. It is essential that you keep doing this until there is no more pus being formed. Any pus left in the cavity will reinfect and you will find yourself with another abscess to deal with in a few weeks.

Treating an abscess is a simple procedure. If it is treated carefully, and the conditions causing it can be understood and changed, it should not reoccur in that animal for the same reason. But you will probably find yourself with more than one opportunity to put this new skill to work.

BLOAT

If you are keeping goats, sheep, or cattle on your homestead, you should know how to recognize, treat, and prevent bloat. While a mild case of bloat might just make your animal uncomfortable, a severe case can kill it. Knowing how to prevent bloat involves learning a few basic factors and making use of them in your feeding and management. Treating bloat, should it happen, is quite simple if you catch it in the early stages.

Any sudden change of feed can cause an animal to bloat. Changing, for example, from hay feeding to green pasture (particularly new and succulent) is dangerous if it is abrupt. You should make the change slowly—give the animal its hay, then put it out to pasture for an hour or two. Next day, give the hay again, and turn the animal out slightly longer. Continue until the animal's system has had time to adjust to the new feed. Make *all* feed changes this slowly, whether you are adding new things, changing a balance of grains, or trying some new hay or silage. Never turn animals out onto wet green pasture. Wait until the dew dries off. If possible, always feed some dry hay first. Never graze your animals on a pasture that is predominantly leguminous plants (especially clover or alfalfa) without taking certain precautions. Such pasture is dangerous (bloat causing) and best avoided. You should always plant 50 per cent grass and 50 per cent legumes if you plan to use this land for grazing ruminants.

If you think your pasture is dangerous but you must use it, you can prevent bloat by feeding the animals dry hay (ten pounds per head for cattle; correspondingly less for sheep and goats) before putting them out. Or you can give each animal two to four ounces of peanut oil or tallow (this is considered the most effective measure). You can restrict pasture so the animals must eat *all* of the plants, not just the succulent parts. Some people dose with antibiotics. Procaine penicillin, for instance, controls activity of the bacterial flora and limits gas production. This seems *least* desirable to me. Your pasture may only be dangerous a few weeks (when very succulent), and it may be worthwhile to cut the pasture for hay or not use it at that time rather than risk bloat.

Bloat may occur if an animal breaks into the barn and gorges on grain (an argument in favor of tightly lidded garbage cans for grain storage and for good latches on all doors). An animal may bloat if it gets out of the pen or pasture and eats a lot of unusual feed—from your own food to stored apples to natural vegetation. Spoiled hay or grain, feed that is too finely ground, and chilled or frosted vegetation may cause bloat. Insufficient roughage in a diet (too little hay, too much grain) can also cause bloat. In young animals, bloat is commonly caused by drinking milk too rapidly or by sucking air at the end of a bottle. Bloat is also part of the development of certain diseases: young cattle, for example, often bloat in the early stages of tetanus.

In all of the above cases, bloat is considered to be "primary." "Secondary" bloat may occur if an animal gets some solid object (an apple, corncob, etc.) lodged in its esophagus. The symptoms are the same. Treatment is slightly different, as the lodged article must be removed and severe cases may have to be punctured (see below).

Bloat happens when the animal fails to expel the gas in its rumen. This leads to an excessive accumulation of gas in the first two compartments of the stomach. Failure to expel the gas is due to a complex interaction of factors. The plants or grains eaten (their pectin, protein, and roughage content), the microflora of the rumen (which affects its pH), and the amount and composition of the saliva flow are involved. In bloat, a foamy, fermenting mass forms. This prevents the working of the normal gas-expelling mechanism. The animal begins, quite literally, to swell up. Distention of the *left* side is marked. The animal will stop eating, begin to breathe more and more heavily. It is obviously in distress. Sometimes it will begin to salivate profusely. Goats will often hold their ears back strangely and grind their teeth.

Act immediately if you think your animal is bloated. A goat or sheep should be given a dose of non-toxic vegetable oil to act as a defoaming agent. For a grown sheep or goat, a couple of ounces of peanut, corn, or soybean oil is good. Give proportionally less to younger animals—an ounce or less for a kid or lamb. You can repeat these doses in half an hour or so. Mineral oil or castor oil also works. Cream is slightly less effective, but will do. *Aids to Goatkeeping* suggests you give "half a pint of raw linseed oil with a teaspoon of turpentine added." Give any of these as a drench (a liquid given forcibly to an animal, though carefully, so as not to get it down the windpipe). If the bloat is mild, give a teaspoon of baking soda dissolved in four ounces of water; then hold the head up and the mouth open to allow the gas to escape. It also helps to massage the animal's sides vigorously but gently. Another standard remedy is two ounces of turpentine in a quart of milk. A cow should be given a dose of any of the drenches suggested, but the proportions are eight ounces to a pint of oil, etc., rather than a couple of ounces. It helps to keep the sheep or goat moving. Don't let it lie down. For a cow, it is suggested that you stand the animal with its front feet up, give it the oil drench, and tie a bit of some sort (a stick of wood, a rope of twisted straw) in its mouth. The bit will make the animal keep its mouth open and chew a lot and this helps expel the gas.

If an animal is so badly bloated it has collapsed, you may have to puncture the body side to allow gas out. This is a *last resort*. You can try a stomach tube first. This is a special tube that is passed through the animal's mouth, down the throat to the stomach. It opens the passage, but in primary bloat, very little frothy gas will be able to come up. If it is secondary bloat caused by obstruction, the tube will reveal this as well as help relieve the gas. If you have to resort to puncturing the animal's side, use a thin-bladed sharp knife or a trocar (a special instrument for this purpose) to puncture the left side of the animal in the center of the triangle formed by the hip bone, last rib, and the back muscle. Give the knife a twist to let the gas escape

but don't let it escape too rapidly. This method too works best in cases of secondary bloat. The frothiness of primary bloat may not come out with this method. Defoaming agents given at once are the best way of treating primary bloat.

When we were raising our goat kids on calf replacer (powdered cow's milk) we had many cases of bloat. We treated them by giving the kid a little mineral oil, then some baking soda dissolved in water, and forcing it to walk around. We spent many an anxious night pushing a reluctant kid around the kitchen floor before it began to deflate. Now, we raise our kids on goat's milk only and have no problems with bloat.

Death from bloat may result from a number of causes—heart failure, a burst blood vessel, or literal rupture of the intestines due to internal pressure of the gas. It may also be caused by absorption of toxic gases. Bloating and death may occur really rapidly, but usually you will have ample time to recognize the symptoms and treat the animal. Don't be afraid to drench the animal even if the case looks mild—you won't hurt it and you may be saving its life. Remember to be extra-careful when drenching not to get liquid down the windpipe. This could cause inhalation pneumonia or choking. Don't hold the animal's head up too high and don't force too much liquid at once. After any case of bloat, feed the animal lightly (rolled oats or bran mash, a little good hay) for a few days. And keep a special eye on that animal—it may be unusually susceptible to future cases.

ENTEROTOXEMIA

Anyone who keeps animals has suffered through or feared the nightmare of losing them to an unknown, unpredictable, *sudden* disease. We went out to milk our goats one morning and found one of our best young Nubian does on her side in a pool of diarrhea. This doe had been fine the night before. She was in beautiful health, fat and sleek, her coat shiny. Her appetite was excellent and two weeks after kidding she was already milking ten pounds a day. When we found her she was only a few minutes from dying. She could hardly move or lift her head and her body was cold all over. She died with little struggle. This was the beginning of a really heartbreaking experience for us, and one shared here in the hopes that no one else will have to go through it.

The sudden death of this healthy, top-condition goat was a shock but not a mystery to us. There is a common disease that affects goats, sheep, cattle, pigs, and even foals. We had read of it many times in livestock magazines and veterinary books, and talked with other goat breeders who had lost animals to it. We even suspected that some newly purchased kids we had lost the summer before had died of it. The disease is called enterotoxemia. It is also known as pulpy kidney disease, overeating disease, and struck. It usually affects the very best animals—prime lambs, animals on excellent pasture, kids that are growing the fastest. The course of this disease can be as short as an hour or two, and it is not uncommon for animals to die overnight. All of this led us to believe that this was what had killed our doe. We knew that she had no access to anything poisonous, and in all our goatkeeping experience and reading there was nothing else we'd heard of that kills a healthy animal so fast. We decided to call in a veterinarian to do an autopsy, send in lab samples, and confirm our diagnosis.

The other ten does living with this one all seemed normal, rambunctious, and hungry. We cut down a little on their grain that morning and gave them a little oat hay with their alfalfa. We knew that enterotoxemia can literally be caused by overfeeding or by an overly rich diet. It seemed strange that the same feed (quantity and type) the does had been getting for years could suddenly be toxic. We were careful not to cut down too much, knowing that a sudden feed change can cause ketosis in a pregnant doe or milk fever in a just-freshened

doe. The does were all young (the oldest was three) and most had either just kidded or were due soon. There was one milker who had been fresh a year and was not rebred, and another doe who hadn't taken and was long dry. These variables were to confuse us in later attempts to fix what had triggered this disease.

The veterinarian who came to autopsy our doe had very little experience with goats or goat diseases. We told him what we thought, asked him to have the lab check specifically for the *Clostridium perfringens* bacteria that causes enterotoxemia. The autopsy itself revealed very little—besides the excellent general condition of our goat. There were a few minor ulcers on the intestinal lining, but everything else looked fairly normal. A little fluid in the lungs and some odd-colored pinkish fluid in the intestine were unusual. The veterinarian took samples of liver and lung tissue; sections from the intestine, and so on for laboratory diagnosis. We asked him to order antitoxin (for immediate use on affected animals) and toxoid (for long-term protection) if the laboratory confirmed our suspicions. In the meantime we began reading up on enterotoxemia and really thinking about our goat-feeding program. We spent a lot of anxious hours with the goats themselves, but they all seemed fine—*too* fine, we thought grimly. We looked appreciatively at the two does who had really lost weight after kidding—we'd been trying to fatten them up; now it looked as though their poor condition would protect them.

Enterotoxemia is known all over the world, and occurs under all sorts of animal-keeping conditions. It is one of the most common diseases of feed-lot lambs in this country. Most references stress the susceptibility of *young* animals to this disease. Foals, for instance, die in their first few days. Calves under ten days old are especially vulnerable. Lambs and kids are very susceptible when they begin eating grains and hay (or good pasture) in addition to their milk. Kids and lambs nursing on very heavy-milking does or ewes are likely to be affected. Older animals that are being especially well fed, grazing lush pasture, or being fattened up for slaughter are all susceptible. This disease may be seasonal, occurring mostly in late winter and spring. All the factors that cause animals to get enterotoxemia are not understood. But generally it seems that if your animals are well taken care of and fed you should take preventive measures *before* you have to deal with an outbreak.

There are a number of types of enterotoxemia. Type A has been found in lambs in California, though it is mostly confined to cattle in the United Kingdom and sheep and cattle in Australia. Types B and C affect lambs, calves, pigs, and foals, causing dysentery with a great deal of pain. The animals frequently die, though some may recover after a period of general debility. Type C causes "struck" in adult sheep and enterotoxemia in goats—both are usually fatal. Type D causes what is commonly called "pulpy kidney" disease because of the rapid degeneration of these organs after the animal dies. This type is very common in this country in feedlot lambs and probably in well-fed young goats. All of these types of enterotoxemia have similar symptoms and run a similar course in the affected animal.

The simplest description of what actually happens when an animal gets enterotoxemia is as follows. Clostridia bacteria are found everywhere in the soil and are part of the intestinal contents of normal animals. Usually when the bacteria are ingested, they are destroyed in the rumen (first compartment of the ruminant stomach) and abomasum (fourth compartment, or true stomach). A few that survive may reach the duodenum (small intestine) where they can multiply and produce toxins. Normally the movement of food ingested will keep these bacteria (and their produced toxins) down to a low level. If something interferes with this normal movement (for example, a heavy feeding of milk to kids or lambs), the bacteria have an opportunity to multiply and produce enough toxins to, in effect, poison the animal. Ruminants need a bulky, fibrous diet. A feeding of lots of concentrates and/or grain could slow down normal digestion. Once the digestive system slows down and the clostridia bacteria begin to multiply and produce toxins, the animal will show outward signs. These signs—or symptoms—vary. One common first sign is a profuse diarrhea which has, as one reference puts it, "a peculiarly evil smell . . ." This diarrhea may have mucus or blood in it. Animals may have a less acute form of enterotoxemia which lingers for days or even weeks. Diarrhea or dysentery is chronic and the animal, if young, fails to grow normally. It may or may not recover. A common secondary effect of the toxins is a stimulation then a depression of the central nervous system. Lambs have been observed to stagger about;

older sheep grind their teeth and stagger. Older animals will refuse to eat; lambs, calves, and kids refuse to nurse. The affected animal may have colic and be in obvious pain, or it may bloat. Animals may die in convulsions or may be depressed and just lie down. Some cases show respiratory symptoms—rapid, shallow breathing, fluid in the lungs, and so on. An animal's temperature may remain normal, as no fever per se is involved. Sometimes the temperature will rise slightly then fall.

Because of the variability of the signs enterotoxemia is sometimes difficult to diagnose. It may be confused with other diseases or with bad feed, etc. Usually in the case of sudden death of healthy, strong animals you should presume enterotoxemia as a probable cause and work from there. Autopsies may give definitive diagnosis, but not in all cases.

The lab results on our first doe came back negative for clostridia bacteria. We were advised not to bother vaccinating the rest of our does. A week later another doe from the same group was listless and depressed at morning milking time. She had severe diarrhea and was swaying on her feet. We called the veterinarian at once and on his advice gave her a large injection of penicillin and an oral dose of sulphamethazine. She seemed a little better a few hours later, though she still refused to eat and was obviously unhappy. We kept her on the penicillin sulphamethazine combination for about five days—she seemed to recover, began eating again, and went back up in her milking. A week later she relapsed with the same signs—depression, vertigo, and profuse diarrhea. She died within a few hours—in great pain, crying out, and struggling. A second autopsy was done. Again we said "enterotoxemia" and again our veterinarian assured us the bacteria would show up in lab tests if indeed it was enterotoxemia. Two days later a third doe died. This one was an angora who was kept in a separate pasture with our sheep in a totally different area of the farm. The only contact between the two houses was some grain that the milkers had left. We had taken it down and fed it to the sheep the night before. This doe was dead when we went out in the morning—she was bloated, her head was bent back as though she had died in convulsions. She was surrounded, as the other two had been when they died, by a pool of diarrhea. This doe was not autopsied, but a sample of her feces was sent to be analyzed. The results on her and the doe who had died before were the same: all samples showed strep bacteria and coliform bacteria, no clostridia. Both strep and coliform are very common. We asked the veterinarian if they were found in abnormal quantities. He said the lab reports didn't say this, only that they were present. By now we were dreading the ordeal of going out in the morning for chores. I remembered talking with an old-time goat breeder in Connecticut who had lost almost half her herd of fine registered Toggenburgs before enterotoxemia was diagnosed. She said she hated going out in the morning because she would invariably find one of her best young does dead or dying. Like her, we were losing our best—our biggest, healthiest, best-producing does.

We lost two more of our does before clostridia bacteria finally made a definitive showing in the lab. It had taken almost two months to establish the diagnosis we'd suspected in the beginning. If we had trusted ourselves more and vaccinated the herd in the very beginning, we might have lost only the first doe. Or if we'd had antitoxin on hand, we could have established a diagnosis with the second doe. Her response to antitoxin would have been dramatic, positive proof that what we were dealing with *was* enterotoxemia. Now we are vaccinating all does, kids, bucks, and sheep regularly. The cost is minimal compared to the loss of even one animal. The effort is nothing compared with the stress and sadness and loss of actually going through an outbreak of this disease. In retrospect, I can only wish that I'd believed the warnings in the goat books that: "Enterotoxemia is the worst goat-killer." It is indeed.

Control is the best way of dealing with enterotoxemia. If you have never had an outbreak of this disease, you can vaccinate your animals with toxoid or bacterin. These may be purchased from a farm supply, a vaccine company, or from your veterinarian. You may want to have a veterinarian give the first vaccinations, showing you how to do this, but it is not difficult to do yourself. You can use type-specific (i.e., type A, types B and C, type D) *or* multivalent vaccine (usually types B, C, and D combined;) *or* a multiple vaccine (to protect against enterotoxemia and other clostridia-caused diseases). Mature breeding animals should be vaccinated twice initially. If your ewes or goats are pregnant, vaccinate them initially six weeks before they are due to kid or lamb. A month later (or two weeks before they are due) give the second vaccination. The colostrum from these vaccinated animals will give their offspring temporary passive im-

munity. Mature sheep should then be given an annual "booster" injection. Try to plan this so that the pregnant ewe is given her booster about two weeks before lambing (to give her lambs maximum protection). Goats should also be given an annual booster. This should be given to pregnant does about two weeks before kidding. Lambs, calves, and kids that have received colostrum from vaccinated mothers will be passively protected from a few weeks to six weeks, depending upon the antibody response of their mother and the strength of the antibodies in the colostrum. They may be vaccinated with toxoid themselves when four to ten weeks old, and given their second toxoid injection two weeks to a month later. Veterinary books, animal-care books, and veterinarians themselves vary considerably on the age suggested for vaccination of very young animals. I am choosing four to ten weeks of age as the period when the animal will respond well to the injection in terms of making antibodies (extremely young animals cannot make antibodies). It is also a critical period for the young animal in terms of susceptibility to this disease, so that protection is needed. Immunity develops about ten days after the first injection; after the second, immunity is stronger and takes three to four days to develop. This young animal should be given an annual booster too. Don't forget to vaccinate your bucks and rams. They are as susceptible as your does and ewes. If you are vaccinating your animals against tetanus, you should know that there is now a combination tetanus-enterotoxemia injection available from certain supply companies.

If you haven't vaccinated and have an outbreak of this disease, you may give antitoxin or antiserum to all sick animals and all those susceptible (being kept under similar feeding program). Toxoid may be given the well animals at the same time, or shortly after (repeat the toxoid in one month) to develop active immunity. Other treatments for enterotoxemia include oral administration of sulfa drugs and injections of broad-spectrum antibiotics such as penicillin/streptomycin. The antibiotics and sulfa drugs are given to prevent the bacteria from proliferating further and producing more toxins; the antitoxin is given to counteract the toxins already produced and in the animal's system.

When you are vaccinating your animals, you should choose an area of loose skin such as that on the shoulder or neck. Some veterinarians give these injections into the loose skin along the backbone;

most goat breeders seem to prefer the neck and shoulder area. Most goats will develop a swelling at the injection site which may develop into an abscess or may simply reduce to a slight, hard area. These swellings are apparently caused by a substance called an "adjuvant," which keeps the vaccine localized and stimulates the animal's body to produce antibodies. Abscesses may be treated as discussed in this section's special article. Some goats also seem to become slightly lame for a day or two after their injection. This lameness is temporary and will pass. Whenever you give toxoid or antitoxin, watch each animal carefully for allergic reactions. An animal may go into shock. The antidote is epinephrine, which is available from veterinarians and supply companies. It is inexpensive and something you should always have on hand for such rare but possible reactions. Other cautions I've encountered in my reading about enterotoxemia were against using type D toxoid in sick or recently sheared animals, and special care in giving antiserum to Saanen goats (they are very prone to anaphylactic shock).

TETANUS

"Aren't you afraid, using all of that horse manure in your garden?" some of our new country friends asked us while visiting one day. "Of what?" we wondered incredulously. "It's full of *tetanus* and you're *handling* it!" they informed us, genuinely concerned with our obvious naïveté. This was our first—though somewhat misguided—reminder of an ancient presence on our new homestead. The spores of clostridium tetani bacteria were indeed present in the manure-dressed soil of our garden. Not only there, but all over the world, these bacteria are a persistent and potentially dangerous part of the inhabitable environment. They are common in the feces of most animals and of people, and pass with the feces into the soil. Their spores can last there for years. It takes more than handling manure, though, to contract the serious and usually fatal disease they cause. This disease is tetanus, commonly known as "lockjaw."

Tetanus spores themselves are quite harmless as an inactive part of the soil. It is when they are introduced into the animal body via an injury that they become dangerous. Tetanus bacteria are anaerobic; they cannot multiply in the presence of oxygen. A wound that tears or punctures an animal's body and introduces (along with some soil) tetanus spores provides a place where the tetanus may be able to proliferate. If the wound is deep enough or even partially sealed up, an oxygenless environment is formed. The traumatized tissue becomes the growing site for the bacteria. They may begin to proliferate at once, or they may remain inactive for months. In the latter case, another trauma to the site of the first injury may cause the oxygen in that tissue to be lowered. The bacteria then begin to multiply. In either case, the beginning proliferation of the bacteria is the beginning of the disease. The incubation period may have been a few days or a few months; the wound may be still present and obvious, or it may be completely healed and apparently gone.

When the bacteria begin to multiply, they produce toxins (poisons) that spread in one of two ways. They may be picked up by the lymph system and spread by it, or they may be absorbed by the motor nerves in the area and pass up peripheral nerve trunks to the central nervous system. The exact manner in which these toxins affect nervous tissue is not, so far as I know, fully understood.

They seem to cause a state in which the muscles of the body are continually subject to spasm. A spasm is an involuntary contraction of muscle tissue. Tonic spasms, the type caused by tetanus, are steady, uniform, and prolonged (these contrast with clonic spasms, in which the muscles contract and relax alternately and quickly). This state produces one in which normal stimuli (sound, movement, etc.) produce increasingly severe responses in the animal.

The first signs of tetanus are similar in all species of animals. First, the animal becomes slightly stiff and may have muscle tremors. The muscles of the jaw and neck are among the first to be affected. They become tense, then stiff, and finally almost rigid (hence the common name, lockjaw). The animal looks anxious—its nostrils may be dilated, its ears held in an unusually alert, stiff position. The appearance of the third eyelid (a whitish membrane that slides up over the eye) is one of the first definitive signs of tetanus. The animal may be nervous and jumpy and then absolutely still, as though it is concentrating on something. I have seen three cases of tetanus in young goat kids and in each case, the symptoms were classic, just as described. Another definitive sign is the inability to chew or swallow. The animal will try, but the spasms of jaw and throat muscles make it unable to eat or drink. It may take a mouthful and then drool or regurgitate food or water through its nostrils. One of the first signs of tetanus in a goat kid is an inability to suck. Usually it will try, eagerly butting the nipple or teat, then simply quit, as though it has lost interest. (This may be confused with a response to an overheated bottle. If the milk is too hot, though, the kid will usually try to drink then jump back, shaking its head. If it has burned its tongue or lips, it might be very reluctant to drink for the next few scheduled feedings.)

In the early stages of tetanus, the body temperature and pulse rate of the animal remain normal. Later, both may rise as muscle involvement increases. Muscle stiffness begins to affect the animal's hind legs, making it walk unsteadily. The tail is usually held out stiffly (in the case of the goat, it may be curled up against the back). As the disease progresses, the animal (particularly a horse) may assume what is called the "sawhorse" stance—hind legs stiff behind, forelegs forward. It will probably fall over and, if a large animal, have a hard (to impossible) time getting up again. Young cattle often suffer with bloat in the early stages of tetanus. Other complications, found in some animals, include an inability to urinate and/or constipation.

At this point in the progress of the disease, convulsions begin. They are stimulated at first by almost any sound or touch, or by any sudden movement. Shaking, grinding of the teeth, and profuse sweating accompany the convulsions. Often the animal will fall over on its side and kick spasmodically. These convulsions become more and more frequent and more severe as the disease progresses. The spine of the animal may begin to curve due to the pull of uneven muscle contractions. These convulsions and spasms, the prolapse of the third eyelid, and an incredible hypersensitivity to sound and movement are all classic signs of tetanus.

About 80 per cent of tetanus cases are fatal, even when treated. The animal might last five to ten days (common for horses) or three to four days (sheep) and then die. The animal dies of asphyxiation when its respiratory muscles become fixed by spasm. If the disease has had a long incubation period, it may be mild. It will take a long time in developing and recovery may take weeks or even months.

Treatment if you suspect tetanus must be prompt and diligent. Usually horses and sheep (and goats, I suspect) have a poor chance for recovery. Cattle have a much better chance. We have saved two of three goat kids that contracted tetanus. It takes a great deal of patience and dedicated care, but it can be done. The main principles in treating tetanus are:

1. To eliminate the bacteria causing the disease. This is done by intramuscular injection of large doses of veterinary penicillin (clostridia bacteria are penicillin-susceptible). We used crysticillin, an injectable form of procaine penicillin. It may be used in a variety of animals, including horses, cattle, swine, and sheep, as well as goats. The animal is treated according to its weight. Crysticillin is usually given every twelve to twenty-four hours, and treatment continued according to the animal's response—usually for forty-eight hours after the symptoms of the disease disappear. We were instructed by our veterinarian to treat our kid for at least a week. In the case of tetanus, it is difficult to say when the bacteria have been killed as the symptoms are slow to go.

2. To neutralize the toxins produced by the bacteria. This is done by giving tetanus antitoxin. The dose is according to the weight of the animal and may be given as one injection or divided into two or three doses. Horses, for example, may be given 300,000 units at twelve-hour intervals for three injections. The antitoxin may be given subcutaneously or, in cases where symptoms are already seriously developed, intravenously. Some of the antitoxin may be injected around the wound itself. Like most antitoxins, tetanus antitoxin might cause an allergic reaction. The antidote is epinephrine. Any time you give tetanus antitoxin, have epinephrine handy and watch for reactions.

If you catch the disease early enough, sometimes an injection of antitoxin will arrest the symptoms and the developing stiffness, etc. will disappear. This will only happen if the toxins are neutralized before they do significant damage to the nervous system. It might help, also, to give a series of penicillin injections. I've known of two cases of tetanus that were caught in time this way; both animals recovered completely. Being familiar with early tetanus symptoms and sensitive to your animals might enable you to avoid serious cases of tetanus.

3. To relax the tension of the muscles and maintain this relaxation as long as necessary. This must be done to prevent asphyxiation, and also to enable the animal to eat and keep up bodily functions. Various tranquilizers and muscle relaxants are used. Chlorpromazine is one of the most generally used. These drugs, of course, are given according to the weight of the animal and its condition and response to treatment.

In addition to the treatments described above, an affected animal should receive regular, consistent, special attention and feed. It should be kept in a quiet, dark stall or room away from other animals and people. An animal with tetanus reacts to even the *slightest* sound or movement. It may have a convulsion or a muscle seizure even when under tranquilizers or sedation. The less disturbed it is, the less suffering it must endure. It is good to have the same people treat and feed and stay with the animal, to avoid the tension of having to react to a stranger. Always move and speak quietly and handle the animal slowly and gently. Deep bedding and spacious quarters will help keep an animal from injuring itself during convulsions. Horses may be suspended in slings during their bad periods. Smaller animals should be turned over a few times a day and kept dry. Intravenous or stomach-tube feeding may be necessary. Animals that can eat should have light feed, including gruels and any easily digestible, tonic food they will take. Enemas and catheterization might be needed in some cases.

An animal with tetanus needs absolutely dedicated attention and loving care. I've seen animals respond to being held, stroked, and softly spoken to during some of the most severe tetanus convulsions. They know and trust you, and even though you may be the source of some noise and movement, they seem to benefit from the presence of someone who really cares about them. I've spent nights sleeping next to a goat kid who was "incurable" according to our veterinarian. I know that she was comforted by my being there, even at her sickest. After more than a week of almost twenty-four hour a day care, she was on her feet—and on her way to recovery. Her kids are part of our goat herd now. I've also shared weeks of caring for another kid who, despite all we could do and all we cared, finally died. One experience was uplifting and encouraging; the other saddening and discouraging. I would try again with any animal, knowing that there are a few who can and do live through this devastating disease. Your only real failure can be when you don't try at all.

The best way to deal with tetanus is, of course, to prevent it. Whenever you are docking, castrating, or shearing animals, use absolutely clean instruments and turn the animals out in clean pens or to clean pasture afterward. All wounds—particularly puncture wounds—should be opened and cleaned thoroughly with an antiseptic. You can also follow a regular program of immunization. Horses, sheep, and goats should all be vaccinated regularly for maximum protection.

Tetanus toxoid may be given to any animal six weeks of age or older. This first vaccination will give the animal an active immunity that takes ten to fourteen days to develop and lasts a year. You should then revaccinate. In relatively tetanus-free areas, this will give your animal basic lifelong protection (after a number of years, you may want to give a booster vaccination). If you have especially valuable animals (or value your animals especially!), or live in a bad area for tetanus (i.e., your land has had horses or mules on it, or does now), follow a "more vigorous program." Vaccinate your animal, wait six to eight weeks, and vaccinate again. Every year you should give a booster vaccination. We keep horses, sheep, and goats and give all animals the two initial vaccinations with yearly boosters. We learned the hard way that regular protection is the *least* difficult and costly way to deal with tetanus.

Any animal that is badly injured or has a deep puncture wound should be given a protective injection of *antitoxin*, even though it has been regularly vaccinated. Antitoxin gives immediate, passive, short-term (ten to fourteen days) protection. It may be given at the same time as the toxoid without interfering with the body's response to the toxoid. An animal that has been injured and never vaccinated should be given toxoid and antitoxin at once.

A young animal may receive temporary passive immunity from its mother via the colostrum (first milk) if the mother has been actively vaccinated. If she has had her annual booster during the last two to three weeks of her pregnancy, her antibodies will be high and her offspring will receive the highest possible natural protection. This protection will last from a few to six weeks, depending upon the mother's antibody response and so on. It is impossible to say exactly how long the protection lasts in any individual case, or how strong it is. If you keep horses, you should vaccinate each foal at age five to six weeks with toxoid. A younger foal cannot develop antibodies properly in response to the toxoid. If you have a foal whose mother has never been vaccinated, you can give the foal weekly shots of antitoxin to protect it until it is old enough for toxoid. A goat kid that receives passive immunity via colostrum of its vaccinated mother will also be ready for its first toxoid at about six to eight weeks of age. If you keep horses as well as goats, or have had problems with tetanus before, give your kid antitoxin when you disbud or/and castrate it. If the kid's mother has not been vaccinated; this antitoxin at the time of disbudding or castrating is absolutely necessary.

Lambs are treated similarly to goat kids. Vaccinated ewes give passive protection to their lambs. Lambs may be given antitoxin at docking time or when they are castrated. Cattle are not as susceptible to tetanus as these other animals. In the *Merck Veterinary Manual*, a table listing common immunizing agents does not even include tetanus as a disease cattle should be vaccinated for. The most susceptible of this species is the freshening cow. Tetanus may be introduced into the genital tract during calving (particularly if the calving is difficult and the cow must be assisted). If you have horses and cows both, you might consider vaccinating your breeding animals. Pigs are likewise not often affected by tetanus. Castration is one operation that might be dangerous if not done under very sanitary conditions with properly disinfected instruments.

Don't forget that you are as susceptible to tetanus as any of the animals you care for! Give yourself the same consideration and care and have a regular protective tetanus shot. If you step on a rusty nail, get a deep puncture wound, or a bad cut, go to your doctor or hospital and have an antitoxin injection. Don't be afraid of the manure in your garden—but don't take chances with this deadly disease.

ketosis

Last winter we lost a goat who was heavy with kid just a few days before her due date. She had gone noticeably off her feed about a week earlier, becoming listless and depressed. She wasn't running a fever, had no signs of scouring or other warning signals of a sick goat—she just seemed low in energy and not interested in anything. Her reluctance to get up and move about I attributed to her heaviness with kid. She was a small doe and not particularly strong to begin with. Her lack of appetite and general depression was more worrisome. The veterinarians I called had "no idea" what could be happening, and the books I had on goat-keeping similarly offered no key—her symptoms didn't seem to match any of the diseases discussed. I decided that the doe was just having a hard time carrying her kids to term. I put her in a separate stall, hand-fed her special food and brought her warm water to drink, spending many hours with her, puzzling and watching and comforting. As her condition worsened, she seemed to get weaker and weaker. She had no interest in food, could hardly hold her head up, and yet she still showed no signs of being ill in terms of body temperature, irregular bowels, and so on. To the very end, calls to local veterinarians brought no ideas about her condition, and I was afraid to subject her to the stress of a long car ride to see one of them. This doe died just days before she would have kidded, and with her died three large, well-formed, beautiful kids. I thought she had just been too weak to take the strain of her pregnancy, though it seemed strange that she had been healthy up until this point.

Less than a month later another of our pregnant does began acting oddly. About two weeks from her due kidding date, she began acting as though she were going to kid—not eating with the other does, turning to look back at her tail and calling, and so on. By that evening she was refusing to eat and seemed to be stiff in her hind legs. Like the doe we lost, this one was huge with either twins or triplets. I put her in a separate stall and spent the night there, worrying that she would abort her kids or was suffering with some peculiar disease that I couldn't identify. The next day she was still down —but willing to eat if I tempted her with her garden favorite (Swiss chard) or huckleberry branches along with her usual alfalfa. I gave her a mixture of warm water and molasses to drink, and she enjoyed

that. She also nibbled at her grain. Remembering how the first doe had gotten weaker and weaker lying down all the time, I forced this one to get up and walk around the yard with me a few times each day. I felt certain I was dealing with some sort of pregnancy-related disease now—the symptoms were too similar to be coincidental and this was a strong, healthy doe. I went back to my books, spending a good part of the next two days poring over all of my back copies of *Dairy Goat Journal* and rereading all of my goat books. By the following night I had my diagnosis, and the symptoms were classic: these does suffered with ketosis, also known as pregnancy toxemia, twin-lamb disease, or lambing paralysis in ewes.

Usually ketosis occurs within the last four weeks of pregnancy. Animals carrying multiple fetuses or carrying a particularly large single fetus are most likely to get ketosis. Susceptible animals are those overfed and underexercised, though undernourished animals are also susceptible. A brief period of fasting or starvation may spark the disease. In goats, this period may be related to voluntary starvation, as when a goat loses a friend or companion, or is moved to a new home. Ewes that are moved to a new pasture may not eat for a day or so until they become accustomed to the strange environment. Any stressful situation, such as a bad storm or being chased by dogs, may cause ketosis. Diet is very important. Carbohydrate or protein deficiency is a direct cause, particularly when combined with insufficient exercise. Feed rations must have adequate cobalt, phosphorus, and iodine. Changing feeds or a slowly declining plane of nutrition during the last six weeks of pregnancy may cause ketosis. As I read more and more about this disease, I realized that we had made a critical change of feed a few weeks before our first doe showed her initial signs. We had changed grain mixes, and though the new resembled the old generally, there were elements added and others subtracted, and the mixtures were not balanced quite the same.

The symptoms of ketosis are various enough to make it hard to recognize until you've read a lot or had some firsthand experience with this disease. One of the first signs is a declining appetite. A doe who doesn't come to eat her normal grain ration or doesn't come to eat hay or forage with the other does is suspect. General listlessness is another sign —though some animals first go through a period of nervousness. Colic-like signs (the goat turns her head back toward her stomach, cries out, etc.) may be combined with teeth-grinding or "aimless walking." Another sign is "propping": the animal stands with her head pressed against a wall or fence. Twitching ears and stiffness of the hind legs or neck may occur. Ewes may become blind, lag behind the rest of the flock, or stagger when they try to walk. The animal becomes less and less willing to move about. Its urine or milk may smell like vinegar. If you suspect that you have an animal with ketosis, there is one way to make a fairly definite diagnosis. You can buy something called Ketostix or Lastix from a veterinarian or pharmacy. These are chemically treated sticks of paper made to test the urine of the animal. If she has ketosis, the stick changes color dramatically. I bought these sticks from our local pharmacy and used them to test the second doe. As does commonly urinate when they first get to their feet after a few hours of lying down, it was easy to catch her urine and get a classic positive reaction on the paper! These sticks keep well and are a good addition to your veterinary cupboard.

Most simply, ketosis is a disease caused by blood sugar deficiency. It is similar to the condition of diabetes in humans. This deficiency may occur because of diet (insufficient carbohydrates), because of a metabolic disturbance (as when bad weather causes a fall in blood glucose), or because multiple fetuses place too great a nutritional demand on the body of the doe or ewe. If you suspect your pregnant doe or ewe to have even the mildest beginnings of ketosis, you should treat her immediately. Untreated, advanced cases of ketosis are usually fatal. I have used glycerin to treat our does, as it is readily available at the pharmacy and works well. It is given as a drench twice a day, four ounces each time, or may be mixed with molasses and warm water and given free-choice to the animal. You can substitute with a cup of molasses given each day (dissolve in warm water) or a half pound of raw sugar given each day (dissolve in water). Propylene glycol may also be given (six ounces daily). This is the simplest course, one you can do yourself, and one that is effective. In severe cases, you may want to call in a veterinarian to give cortisone injections. Concentrated A-D vitamins injected weekly are considered useful, but these should be combined with the glycerine or molasses treatment. Enforced exercise is very important. You should lead your animal around for short periods three or four times daily. Keeping her eating is also essential. Try any favorite treat you know she likes, in addition to good-quality, high-protein hay. If your animal will continue to eat, is exercised, and seems fairly alert and strong, she has a good chance of carrying her lambs or kids full term and delivering them normally. Our second doe gave birth to four healthy, good-sized kids. All of them survived, and she recovered completely. If it looks as though the doe/ewe is too weak to carry her young full term, you may consider having them removed by Caesarean section. Or you may be able to have a veterinarian induce her labor one week before her due date. These premature kids or lambs can survive if they are given special attention.

Preventing ketosis is infinitely simpler than treating it once it has developed. Be sure that pregnant does and ewes are well fed, consistently fed, and moderately exercised. Don't overfeed, but put your animals on a gradually increasing standard of nutrition in their last months of pregnancy. If they are kept in a small area, placing their feed in different places will encourage them to move around! Molasses is a good source of readily digestible carbohydrates and may be added to the diet in the last few weeks of pregnancy. It is available in blocks, mixed with grains and minerals, that the animals can nibble at as they choose. It may also be added to their drinking water. Some breeders keep a mixture of ground bone meal and molasses before their animals, allowing free-choice feeding of these. It is very important to avoid stressful situations in the latter stages of pregnancy—don't move your ewes to a new pasture just before they are due; don't transport goats to a new home during that period. Provide good shelters for your animals so that they will not be subjected to storms and bad weather. In general, treat them consistently and gently and keep a watchful eye for signs of ketosis. Those early doses of glycerin or that preventive bucket of warm water and molasses may save your favorite doe or prize ewe.

pROLApSES
1N shEEp ANô GOACS

Prolapses usually occur in the latter stages of pregnancy, and may be caused by an unusual strain on the reproductive system or be a result of a genetic weakness the animal has inherited. The common mechanical cause of a prolapsed vagina is, oddly enough, the ground itself. Sheep that are kept on hilly pastures during late pregancy are the usual victims of prolapse. If they lie to rest or sleep too often with their rearquarters downhill from their frontquarters, gravity will begin to pull at their lamb-weighted uterus, putting a pressure on the vagina that eventually weakens the muscles holding it in place. I have seen the same condition occur in a ewe kept in a flat pasture who developed the habit of standing with her front feet up on an unused old feed box. This ewe was in the last month of her pregnancy, and heavy with what turned out to be a very large single lamb. Apparently it gave her some relief to stand with her feet propped up this way; I noticed it and gave no further thought than that it must be more comfortable for her. I have since seen two goats develop a similar condition—one apparently suffering with a genetic weakness (her mother had prolapsed during previous pregnancies), and the second a small goat with no family history of prolapse but who was carrying three large kids.

Regardless of cause, all three prolapses began with a similar symptom: the protrusion of a pinkish, cylindrical mass from the vulva. The animal shows no other symptoms and doesn't seem distressed. She will eat normally and move about freely. Thus, the beginning prolapse might escape the less-watchful eye and develop to a more serious point before it is noticed. Early treatment is important if this animal is to be saved from the very critical problem that may follow the (untreated, and sometimes even the treated) vaginal prolapse: a prolapse of the uterus at lambing or kidding time. Treatment for the early-caught case is fairly simple. For part of each day, the animal must be confined in a way that places her frontquarters lower than her rearquarters. This will reverse the pressure on the womb, causing the fetal kids or

lambs to slide forward again, pulling the tissue of the vagina back into its normal place. The simplest way to accomplish this tilting action is to build a simple wooden box scaled to the size of your animal, with one end raised about a foot higher than the other end. I used an old wooden packing crate for both goats and the ewe I treated. One end was knocked out to allow the animal to be led in; once she was in place, two boards were nailed across this open end to discourage escape attempts. She was also tied by the collar to keep her confined. The box was big enough to allow her a little freedom of movement, but not enough so that she could get herself turned around. The tie rope was left slack enough so that she could stand or lie down, depending upon her mood. Once the animal got used to the box, she seemed to accept her confinement. A rack of sweet alfalfa hay made things more bearable. In the least serious cases, I simply confined the animal for a few hours in the morning and a few hours before evening chores, leaving her loose most of the day to exercise and sun. The more serious case had to be confined most of the time, with regular exercise periods—sometimes involving being led about for an enforced walk. In most cases that I have had direct experience with, or have talked with people about, the animal confined to the tilting box will improve within a few days. She may even cease prolapsing altogether, and carry her lamb or kid to full term with no further problems. Some animals will have to be treated periodically for this condition right up until they lamb/kid. Others will develop a more serious problem which may be dealt with as discussed below.

If the prolapse has developed to a serious stage (i.e., the protruding mass is two or three inches in length, and will not retract even when the animal is in the tilting-box position), I know of two ways to handle it. The first I have not had experience with. You will probably want to call in a veterinarian to help. This first alternative involves the placing of metal prolapse pins inside the vagina to hold the prolapsing tissue in place. They may be left in until the animal is ready to lamb/kid. The second alternative is one that you can do yourself, with simple materials but with some daily effort involved. You will need some warm water and mild antiseptic and/or antibiotic ointment and some lengths of clean, soft cloth or bandage. First, wash the protruding parts carefully. If you want, follow this washing with a flushing of antibiotic ointment as an extra precaution. Make a pad of cloth or bandage, gently push the prolapsed tissue back into the vulva and cover with the pad. Next comes a puzzle to challenge your inventiveness and knot-tying skills: how to keep that pad in place, with enough pressure to hold the prolapsed tissue in but not so much pressure that you make the animal terribly uncomfortable. Try running one strip of cloth or bandage across the top of the pad, up over the animal's hip bones, and tie it over the back. Run a second strip over the lower part of the pad and tie each end around the animal's flank and under the leg. You may then have to run various types of crossties and anchor pieces, depending upon the activity and energy of the animal, the elasticity of the bandage, and so on. A pad should be left in place until the animal ceases to prolapse or until she lambs or kids. It should be placed so that it does not cover the anus and interfere with defecation. Because it will become urine soaked and dirty, it should be replaced at least once daily. This animal should be placed in the tilting box each day, too. Be very attentive to signs of impending labor so that you can remove the pad and ties in time.

If the vaginal prolapse is discovered and treated in time, the doe or ewe may kid or lamb normally. Sometimes, though, the whole uterus may prolapse when the kid/lamb comes. When this happens, you may feel thoroughly intimidated by the huge, spongy, strange-looking mass hanging from the animal's vulva. Treating an animal with this problem has been one of the very hardest tasks—physically and emotionally—in all my years of animal keeping. The problem is a little like trying to push a soft basketball through a keyhole, but it is live tissue you are handling, and a sensitive creature you are reorganizing! If you are faced with this problem and have to treat it yourself, act quickly and firmly. Have a friend there to help hold the animal and give you moral support. The uterus should be flushed clean with a solution of warm water and a mild soap (the obstetrical soap used when you help deliver a lamb or kid is perfect) or a very mild antiseptic. Your hands and arms should be as clean as possible, too. When the uterus is thoroughly clean, begin with one portion nearest to the vulva and gently work it into the vagina. Work slowly and patiently. The animal I worked with was a sheep—she seemed to be uncomfortable during this process, but not in pain. It is hard to describe actually repositioning the uterus—you use a kind of kneading, pushing motion. The animal may have contractions which will send the whole uterus spilling out again. Start over and try to hold the tissue in place if her contractions begin again. Once the uterus is back inside and in place (how easy this sounds; how hard it is), break up a couple of uterine boluses (the type that disinfect and protect the uterus from infection—available from country pharmacies and veterinarians) and insert them as far up into the vagina as you can. At this point you may want to stitch the vulva partially closed to hold the uterus in place. Use sterilized thread and needle for this (or have a veterinarian do it) and, if you can get one, a local anaesthetic on the area to minimize pain. You may also make use of the pad and ties described earlier. This animal should be given a fairly heavy dose of broad-spectrum antibiotics to help prevent infection from developing inside the uterus. Continue these antibiotics for five days or so, depending upon her general condition and response. Keep a check on her temperature and watch to see that no evil-smelling discharge is coming from the vagina (a bloody or mucous and blood discharge is normal; a greenish or yellowish discharge with a bad odor indicates infection). If she develops an infection, you may want to flush out her vagina with an antiseptic solution and should most certainly keep her on antibiotics. She should be confined to the tilting box for a few days after lambing/kidding and taken out only for regular exercise or to nurse her offspring. You will probably want to keep her lamb or kids in a box near her head if you are letting them nurse—otherwise, she may fret and strain to get out of her confinement and mother them. With good care, this animal has a reasonable chance for total recovery from this traumatic problem. You will have to use your own judgment about breeding her the following year. If you think that her prolapse was mechanically caused by hilly ground, standing with her feet up, and so on, she may be safely bred again. Don't let these conditions reoccur (with this animal or any of your others). If, though, you know that this animal comes from a line with this problem a known genetic factor, you should probably cull her and cull her offspring.

milk fever

Like ketosis, milk fever is a pregnancy-related disease affecting cows, goats, and sheep (and, rarely, pigs). It may be fatal if not promptly treated. If you are keeping any of these animals—particularly heavy milking individuals—you should know that it is not an uncommon disease. Because the signs are so classic and the treatment so simple and immediately effective, this is another disease with which you should be familiar and prepared to treat.

Milk fever is known technically as parturient paresis. It is most common in high-producing cows between five and nine years of age. Jerseys have a higher incidence of this disease than other breeds. Similarly, it affects older ewes (particularly if they are high-conditioned and on pasture) and goats in their second, third, and fourth lactations. Individual animals and sometimes families of animals seem more susceptible. An animal that has had milk fever once is quite likely to come down with it again at successive parturitions.

In cows, milk fever usually occurs within forty-eight hours of calving. The danger period is up to ten days after calving, and some cases are known to happen as much as six to eight weeks later. In ewes, the susceptible period is much longer—from six weeks before lambing to ten weeks after. Goats are probably susceptible over the longer period, both before and after kidding.

In the first stages of milk fever, the animal may show some signs of muscle involvement. She may have muscle tremors and will walk unsteadily. She is often found in the second stage lying down on her sternum, often with her head drawn back against her side. A ewe may have her hind legs extended backward. The animal's eyes will usually be dull and staring, pupils dilated. Dry muzzle, cold extremities, subnormal body temperature, and constipation are other signs. The animal cannot stand. If treatment is not given at this stage, the animal passes into a coma, lying on her side, her breathing shallow. At this stage she is susceptible to bloat, regurgitation, and aspiration pneumonia. If she is not treated, she will usually die. The course of the disease from the first signs to the fatal coma is usually twelve to forty-eight hours.

Treatment for a cow, ewe, or goat with milk fever is simple and specific. She is suffering from acute hypocalcemia—which means that her serum calcium level has dropped radically from its normal level. This may happen for one of three reasons. First and most simply, she may be suffering excessive loss of calcium in her colostrum (first milk) beyond her ability to absorb calcium from her intestines or to utilize it from her skeletal structure. Second, she may be suffering impairment of her normal ability to absorb calcium from her intestine. Third, and most probably, she is suffering from a more complicated failure to use her calcium reserves. This is thought to occur when an animal suffers a parathyroid insufficiency due to decreased calcium and phosphorus metabolism during her dry period. A diet not adjusted or adequate to her needs at this time would cause this insufficiency. Oddly enough, what she needs is *not* a diet extra high in calcium, but one low in calcium and high in phosphorus. This diet would stimulate her parathyroid, conditioning it for the demands of lactation. But whatever the cause, the cure for milk fever is immediate injection with calcium salts. Farm supply houses, some feed stores, and most veterinarians have available commercial preparations of calcium gluconate with dextrose and phosphorus, or mixtures of calcium, magnesium, dextrose, and phosphorus. These are injected subcutaneously, intravenously or intraperitoneally. Subcutaneous injection is slower, with less danger of cardiac arrest (not uncommon with intravenous injections). You may give half the solution intravenously (this will have to be done by someone trained, probably a veterinarian) and half intramuscularly for the benefit of speed with a little less danger than giving all intravenously. The response to injection of these calcium salt solutions is usually dramatic. An animal may be on its feet and normal in half an hour! An animal that relapses or fails to get up after eight to twelve hours should be treated again. If you have a milk cow, goats, or sheep you would do well to invest in a bottle of calcium salts solution (less than $2.00 for enough to treat a cow) and just keep it on hand. You may not know how to give the solution intravenously but you can certainly give it subcutaneously in such an emergency and probably save your animal.

If you have a case of milk fever and no calcium solution available, you could try the old-time method of udder inflation. This method is outlined in veterinary textbooks as an alternative for animals that don't respond to the calcium injections. It involves some risk to the udder but if done gently and in absolutely as sterile a way as possible, it can save your animal's life. The method is as follows. Each quarter or half of the udder is inflated through a sterile teat tube until firm. Usually a special hand pump is used. You may have to improvise with a small bicycle pump. When each teat is full and firm, the tube is removed and the end of the teat taped or tied gently with gauze. One text I read suggested removing the tapes and massaging the udder gently every half hour, then retaping. Another source said to leave in place three to four hours, then remove and milk out the udder partially. Milk out partially again in six to eight hours. This udder inflation treatment slows down the excretion of calcium and phosphorus into the udder and allows them to build up to their normal level in the blood serum. This should alleviate the symptoms within about twelve hours.

Another alternative to the calcium salts injections in mild cases of milk fever is given in *The Herbal Handbook*. If the animal is found in the early stages of mild milk fever, she may be given a drench of "two big handfuls of powdered seaweed mixed into two pounds of molasses" (molasses mixed with water or milk to a loose, drinkable consistency). This drench, rich in calcium and iodine and other minerals, may be repeated hourly. Severe cases are to be treated with the air-inflation method described above.

General preventive measures against this disease include attention to feed• and environment. Animals should be kept on a good standard of nutrition but not a diet overly high in protein. The diet should *not* be extra high in calcium—rather, a high phosphorus, low calcium diet shortly preceding parturition seems to prevent this disease. Several texts suggest that bone meal be available free-choice (or added as 1 per cent of the concentrate ration) to pregnant and lactating goats as a preventive. Pregnant and lactating animals should not be subjected to stress, unnecessary exercise, or avoidable excitement such as that of being transported. A very important preventive is to never fully milk out a newly freshened animal. If you have taken away the calf, lamb, or kid, you should milk the mother partially a number of times each day. Following this procedure for the first few days will help prevent milk fever (and udder problems, too).

If you have an animal that suffered this disease previously or is from a family in which the disease has commonly occurred, you can take some special precautions. A phosphate-treated diet may be fed (5 per cent monosodium phosphate added to her concentrates). Administration of vitamin D is used as a preventive, but there is one major drawback. Doses must be given for five days before the cow is due to calf but may not be given for a longer period because of possible toxic affects. If administration must cease four days or sooner before the cow calves, she is likely to get parturient paresis. Because the exact calving day is difficult to know, this method seems very risky. Perhaps the best method is to watch this animal very carefully and be ready to treat her if she shows signs of this disease.

Losses of animals to milk fever can be averted by the simple precautions discussed above. If you have high-producing animals, you should be particularly sensitive to the possibility of this problem and have injectable calcium salts on hand.

MASTITIS

If you keep a milk cow or goat, beef cattle or sheep or a couple of pigs, you should be familiar with mastitis. Mastitis is, simply, an inflammation of the mammary gland. It involves a change (physical, chemical, and usually bacteriological) in the milk these animals produce. At its most minor, mastitis can be a low-level infection that makes milk production drop slightly. It can make an udder a little uneven or inflamed. At its worst, it can destroy the functional capacity of the udder and even kill the animal. Whether you are milking the animal for your own needs or a commercial dairy, or letting it raise its young, you should be familiar with this disease—its signs, treatment, and prevention.

Mastitis can be caused by an injury, by unsanitary milking conditions, by dirty pens or barns, by improper milking, and a host of other things. It may be non-infectious—as when caused by an injury, a cut, or bruise. In this case there is a period of inflammation, soreness, and possibly swelling. Disease-producing germs are not involved and the condition usually disappears in a few days. You should handle the animal with utmost gentleness and consideration during and after this period. Milking her more than twice a day will keep pressure from building up inside the udder and aggravating the condition. It may help to change her diet slightly by cutting down on concentrates (particularly corn and barley which produce heat in the udder)—not severely, but enough to relax the demands made on her udder. Perhaps you can give more of a bulkier type grain such as rolled oats. If there is a cut or scrape on the outside of the udder, keep it clean with a mild disinfectant and be extra careful when milking. Massaging the udder with antiseptic cream or with an herbal preparation (make a brew of dock leaves with a few drops of oil of eucalyptus added) will help. There seems to be some controversy over using hot or cold compresses. As far as I know, hot compresses are used when the udder feels very hard and congested. They stimulate blood flow to the area and aid in breaking up congestion. Cold packs or compresses are used when the mastitis is of a more severe type involving bacteria that produce toxic substances. The cold packs slow down absorption of toxins and help to lower the accompanying fever.

When microorganisms such as bacteria are involved, the mastitis is infectious and a good deal more serious than the non-infectious type. There are two schools of thought about bacteria and mastitis. One holds that the bacteria in themselves can cause mastitis. Once they gain entrance to the teat canal or milk cistern they may multiply rapidly and may produce toxins. The irritation of their presence or the presence of the toxins produces mastitis. The second school of thought holds that the bacteria play a "secondary invader role." They may be present in the teat canal or cistern for some time with no sign of mastitis developing until the udder resistance is lowered for some reason. Then they begin to multiply and mastitis develops. Lowered udder resistance may be due to an injury, chilling of the udder (due to drafty quarters or lack of adequate bedding), improper milking, and so on.

Detecting mastitis may be simple if there are sudden, obvious signs. More subtle signs can be picked up by following a few simple routines. When you milk you should use a strip cup—which is simply a cup with a shiny black plate set in the top. The first few streams of milk are squirted onto the plate and checked for discolouring, for flakes or clots or other abnormalities. Beginning each milking with the use of a strip cup can help you catch the very first signs of mastitis and treat it before it develops more seriously and obviously. You can buy a strip cup for two or three dollars or improvise your own.

There are ways to examine milk culturally and you can take individual samples or bulk samples from your herd to a state or municipal laboratory, some veterinarians, or technicians in dairy labs for examination. If you suspect an animal to have mastitis and want a professional check which will find and identify the organisms involved, you may choose to do this. Or you can do your own regular checking at home using a fairly simple test called the CMT (California Mastitis Test). This test shows the presence of an abnormal number of leucocytes in the milk (increased leucocyte count is one of the most important characteristics of mastitis). You can buy a CMT kit (about $10) that contains a highly concentrated reagent and a white plastic paddle. The reagent is diluted with water before use. Milk is drawn from each teat into a separate section of the paddle and mixed with an equal amount of reagent. A gel formation indicates mastitis and the amount of gel shows the severity of involvement. The test is easy to do and fairly accurate. We test all of our milking animals regularly one day a week. It seems worth the time and effort to catch any mastitis before it develops into a serious problem. Animals in their first week of lactation and in the last stages of their lactation may show a positive CMT test even when they are not diseased.

There are also brom-thymol-blue test blotters that may be used to detect mastitis. They react to the acidity of the milk. Our experience using these blotters was that they would react positively to our drinking water and very inconsistently to our milk.

One final way to detect mastitis is to palpate the udder right after milking. You should be feeling for any unusual firmness that might indicate fibrosis (increase of connective tissue) which is typical of some types of mastitis. Some animals have naturally a more "meaty" or firm udder, and this will not indicate fibrosis. You should also be able to detect localized areas of fibrosis which are hard masses ranging in relative size from a pea to a fist. Any heat or swelling should be noticeable too. After palpation, you should look at the udder for consistent size of halves or quarters. A difference in size may mean that one half or quarter is infected and atrophying.

All of these methods and tests will enable you to keep a regular check on the health of your milking animal's udder. Animals that are nursing their young should likewise be checked periodically—a visual check or palpation may be sufficient. If you notice an animal that kicks at her young and won't let them nurse, you should look at her carefully. She may well be suffering with mastitis, either beginning or advanced.

Infectious mastitis may be classified in three ways: chronic, acute, and peracute. Chronic cases are the mildest and the least dramatic. They can cause a loss of production (often very gradual) and some change in the milk produced. Off-flavor milk may be the only indication of chronic mastitis. There may be lumps in the udder or sometimes clots in the milk. This type of mastitis has no further effect on the animal's body as long as it stays at this level. It may, however, develop into acute mastitis or infect other animals. Acute mastitis is more serious. It usually comes on quickly and manifests itself in a swelling of the gland affected, loss of appetite, dramatically lowered milk production, and usually some noticeable change in the milk produced. Acute mastitis may flare up, sub-

side, and flare up again. Peracute mastitis is the most serious because it involves a severe general or systemic reaction. The animal will quite suddenly be running a high fever, be totally depressed, weak, and miserable. The udder will often be swollen, hard, hot, and painful with very abnormal milk or secretion which contains blood, clots, pus, and so on. The animal usually has a rapid heart rate and as its muscles weaken it may fall or lie down. This form of mastitis can lead to gangrene of the udder, total collapse, or death of the animal.

Treatment for chronic or acute mastitis usually involves the use of udder infusions which introduce antibiotics into the teats and milk cistern. If you can take a sample of the infected milk to be analyzed and have the implicated bacteria run through drug sensitivity tests, you can use the antibiotic or combination of antibiotics that will specifically control that bacteria. This is the best but usually the most expensive and time-consuming course. More likely, you will follow the usual course of using an infusion containing broad-spectrum antibiotics or a combination of narrow-spectrum drugs. These infusions are sold by veterinary supply companies, rural feed stores and pharmacies, Sears, Roebuck (see their farm catalogue), and so on. They come in disposable plastic tubes which you can buy singly or in boxes of twelve. Be sure to choose the type for a lactating animal if yours is milking or for a dry animal if yours is dry (this seems obvious, but we've had friends bring us the wrong type—it usually states right on the tube but you have to read the small print).

Before you use an infusion, the animal should be milked out. If you can, wait until the last milking of the day so that the material will remain in the udder overnight. For best results, an effective concentration of the drug should remain in the udder for six days. This means using multiple infusions of a water-soluble preparation or one or two infusions of a slow-release base preparation (mineral oil and/or aluminum monostearate). Which type you have will be indicated somewhere on the product you buy. Be absolutely as hygienic as possible when you use an infusion. If you are careless, you may introduce bacteria, yeasts, and fungi into the udder and do more harm than good. Have the animal stanchioned and tied securely. If she is very difficult to manage, you may need a friend to help. Clean the end of the teat with alcohol, and give the infusion tube a good shaking to mix the contents. Remove the protective cap from the tube and insert the tube gently and slowly into the teat. Some animals will try to jump or kick; others won't even seem to notice. A foot hobble or the restraining hand of a friend will keep the animal from injuring herself or you. Don't ever force the tube as you might injure the teat canal or tiny muscles. It will slide in easily once you have the right angle. Be patient and gentle. When the tube is in a good inch or so, inject the contents and remove the tube. It helps to then hold the end of the teat closed and massage the udder. You can actually feel the liquid infusing through the tissue.

During the period you are using the infusions and for ninety-six hours after the last infusion, milk from this animal should be thrown away. This is *essential* because of the antibiotic residues in the milk.

Besides using udder infusions, you should reduce feeding of concentrates, massage udder, and milk frequently to help relieve congestion and pain.

The treatment of peracute mastitis is different because of the severity of this type and the systemic involvement. The udder is usually too swollen and congested for infusions to work. Occasionally drugs may be injected directly into the affected gland through the skin of the udder. This may not be any more helpful than an infusion— again, the drug cannot be diffused because of tissue swelling and congestion. Usually peracute mastitis is treated systemically with parenterally (outside of the intestines—i.e., injectable, not oral) given drugs.

Penicillin, penicillin-streptomycin, sulfonamides, or tetracyclines may be used—usually these are given intramuscularly. Treatment should be continued for about six days even if the signs subside. In really severe cases, antihistamines and isotonic fluids may be given as supportive therapy.

Treatment for any of the forms of mastitis should be begun as soon as the condition is noticed and continue as advised. Beginning a treatment and not following through can lead to a more severe flare-up after a short period—or may lead to the development of a drug-resistant bacteria. You should also know that an animal that has once had mastitis is more suceptible to it a second time. If you have an animal who has been through one attack, con-sider using a dry udder infusion at the end of her lactation period. The infusion is given after the very last milking and left in the udder "permanently" (i.e., don't milk it out). It will not affect the colostrum when she freshens agains and it should keep her protected through her dry period.

The Herbal Handbook (see Resources) argues very strongly *against* the use of antibiotics in treating mastitis and offers some alternative treatments. These are intended, besides treating the specific problem, to have a tonic affect on the whole animal organism. If you are trying to learn to treat your animals herbally, this book will give you very detailed information.

Generally you can prevent mastitis more easily than you can effectively treat it. Keep all animals in clean pens or pastures, keep bedding in stalls and stables clean, and avoid drafty or damp housing. When you milk, be as sanitary as possible. Wash your hands, the animal's udder, and milk in a clean area. Always be gentle and thorough when you milk. Milk that is accidentally left in the teat can serve as a breeding place for bacteria. Using a teat dip after each milking is said to reduce the chance of mastitis by about 75 per cent in itself. Teat dips are available in most large feed stores or by special order. They are commonly a tamed iodine mix, and come in concentrated solution which is diluted with water for use. They are *very* inexpensive. You can use a small plastic cup, a baby food jar, or anything similar to hold the solutions for use. The end of each teat is dipped in briefly right after milking. The teat is disinfected and protected from bacteria that can cause mastitis once they enter the teat canal. Dry animals can also get mastitis, so keep an eye on their udders too. Unevenness, swelling and so on should be noticeable. The fluid in a non-lactating udder should be clear and watery. Clots, blood thickening, or flakes are all suspicious. There are special infusions for dry infected udders. If you follow all of the preventive measures and have a regular routine of mastitis detection, your animals should stay healthy and free of this disease. An occasional accident might happen, but you'll be prepared and able to deal with it.

RESOURCES

ANIMALS

GENERAL:

Feeds and Feeding (abridged)
by Frank R. Morrison
The Morrison Publishing Co., Claremont, Ontario, Can.

This is *the* basic book about animal nutrition, the composition and production of most common feeds, planning and producing your own animal rations. Includes sections on cattle, horses, sheep, swine, and poultry.

The Homesteaders Handbook to Raising Small Animals
by Jerome Belanger
Rodale Press, Inc., Emmaus, Pa. 18049; 1974.

Written by the editor of *Countryside* magazine, this book is intended as a basic guide to raising rabbits, sheep, goats, hogs, pigeons, and so on. It provides essential information in a relaxed, enjoyable text. Not a detailed reference, but a good general introduction to what is involved in keeping small stock.

Countryside Small Stock Journal (monthly magazine)
Route 1, Box 239, Waterloo, Wisc. 53594.

Countryside is one of the most consistently valuable, interesting, and delightful publications concerned with land, animals, and country living that I've seen. It is written by homesteaders for homesteaders and contains a potpourri of wisdom, experiences, resources, new and old ideas. Regular features are on gardening, poultry, small livestock, and the like. Special articles cover using draft animals, building a garden seeder, planting an orchard, and so on. *Countryside* is warm, lively, *useful*—save up your back copies and you'll have a virtual resource library for your small farm.

Dairy Production
by Clarence Bundy and Ronald Diggins
Prentice-Hall, Englewood Cliffs, N.J., 1954.

NASCO Farm and Ranch Catalog
Fort Atkinson, Wisc. 53538, or Modesto, Calif. 95352.

Provides a comprehensive catalogue for all kinds of animal and poultry equipment and health-care products. Fast service, good company.

American Supply House
P. O. Box 1114, Columbia, Mo. 65201.

Free catalogue carries supplies for "the goat and sheep owner." A good line of leather collars plus disbudding irons, milk strainers and pails, and so on.

Diamond Farm Book Publishers, Dept. S
Route 3, Brighton, Ontario, Can. KOK 1 HO.

Ask for their folder on books about sheep, cattle, hogs, and general veterinary medicine.

GOATS:

Dairy Goats: Breeding/Feeding/Management
University of Massachusetts Extension Service Leaflet #439.
Available from American Dairy Goat Assoc., Box 186, Spindale, N.C. 28160.

If you are looking for an inexpensive but fairly comprehensive introduction to goatkeeping, this booklet is a prize. It has a good feed section, diagrams for feeders, basic health care, and so on. Nicely illustrated, with particularly good photographs of udder types, hoof trimming, and breed types.

Aids to Goatkeeping
by C. A. and .C. E. Leach
Diary Goat Journal, Scottsdale, Ariz. 85252.
Available from *Dairy Goat Journal* or American Supply House.

This is a basic guide to the care and raising of goats that I recommend as one of the best. It covers all aspects of goatkeeping in simple, well-organized, and fairly thorough chapters.

Goat Owner's Scrapbook
by C. E. Leach
Dairy Goat Journal, Scottsdale, Ariz. 85252.
Available from *Dairy Goat Journal* or American Supply House.

Here is a collection of wisdom about goats—not a reference book, really, but full of valuable information.

Goat Husbandry
by David MacKenzie
Faber and Faber Ltd., London, 1970.

Goat Husbandry approaches this animal with humor, originality, and understanding. Provides a lot of food for thought and the basics for alternative ways to care for your animals. MacKenzie treats goats as individual beings—not milk-producing machines.

Dairy Goat Journal (monthly magazine)
P. O. Box 1908, Scottsdale, Ariz. 85252.

This magazine provides articles on every aspect of goat raising from·veterinary advice to neighborhood zoning problems! If you are searching for foundation stock to improve or begin your herd, the *Journal* has listings of reliable breeders all over the country. Top producers and show animals are featured with photographs and official test records. For the small homestead herd, there are plans for milking stands or goat carts, listings of animal supply companies, articles on kidding, and so on.

HORSES:

Treating Horse Ailments
by G. and W. Serth
First Aid for Horses
by Dr. Charles H. Denning, Jr.

Wilshire Book Co., 12015 Sherman Road, N. Hollywood, Calif. 91605.

These are two inexpensive paperbacks that provide basic health care for horses. They are well-illustrated, clear, and comprehensive. Wilshire Books has a long list of other books about horses—ranging from the backyard foal to dressage. Those that I've seen are of good quality and very usable.

PIGS:

Swine Production
by Clarence Bundy and Ronald Diggins
Prentice-Hall, Englewood Cliffs, N.J., 1956.

POULTRY:

Murray McMurray Hatchery
Webster City, Iowa 50595.

Theirs is the best illustrated, largest, and most interesting poultry catalogue I've seen. They will ship baby chicks anywhere with guaranteed healthy, live arrival. Special breeds have to be ordered early (sometimes a year in advance) because they sell out.

Starting Right with Poultry
Available from M. McMurray Hatchery or Countryside Bookstore.

This book is a simple, general introduction to poultry raising. It will provide you with the basics but is by no means the definitive book.

Standard of Perfection
American Poultry Assoc., 500 pages.
Available from M. McMurray Hatchery.

This book illustrates and describes all breeds of ducks, chickens, geese and turkeys.

Poultry Press (monthly magazine)
P. O. Box 947, York, Pa. 17405.

This magazine is basically for those interested in breeding and showing fine poultry. It includes up-to-date sources for birds of all types—so if you are looking for something special, this may be the best source.

Modern Waterfowl Management and Breeding Guide
by Oscar Grow
Available from American Bantam Assoc., P. O. Box 464, Chicago, Ill. 60690.

Countryside magazine calls this "the definitive book on waterfowl."

Marsh Farms
14232 Brookhurst Street, Garden Grove, Calif. 92643.

Bird and aviary equipment such as brooders, cages, thermostats, and feeders. Also sells quail and all sorts of books on game birds, waterfowl, etc. Free catalogue.

RABBITS:

Commercial Rabbit Raising
Agricultural Handbook #309, U. S. Department of Agriculture

Order from Superintendent of Documents, U. S. Government Printing Office, Washington, D.C. 20402.

This is a 68-page booklet covering all aspects of raising rabbits from choosing a breed through breeding, managing, slaughtering, and so on. The health section is very well organized for quick reference. A fairly thorough basic guide.

SHEEP:

Production Practices for California Sheep
by Spurlock, Weir, Bradford and Albaugh
California Extension Service Manual #40, 92 pages.
Order from Agricultural Publications, University of California, Berkeley, Calif. 94720.

Although this booklet is written for the California sheep raiser, it contains enough useful general information to make it valuable anywhere. The emphasis is on creating and keeping a commercial flock, though the sections on lambing, feeding, and so on are applicable to any size flock.

Sheep Production
by C. Bundy and R. Diggins
Prentice-Hall, Englewood Cliffs, N.J., 1958.

Part of the Prentice-Hall Vocational Agricultural Series, *Sheep Production* is an excellent general reference book. It covers the breeds of sheep, feeding, breeding and lambing, grading your wool, and so on. Well written and very useful.

Sheep Production Handbook
American Sheep Producers' Council, 200 Clayton Street, Denver, Colo. 80206.

This book provides basic information about sheep care and commercial flock management. Has a good section on diagnosing and treating sheep diseases.

The Shepherd (monthly magazine)
Sheffield, Mass. 01257.

The Shepherd discusses all aspects of sheep raising for the small home flock or the large-scale commercial operation. The veterinary column alone is worth the subscription rate. Articles on breeding, pregnancy-testing, feeding, shearing, and so on. Breeder listings are a source of top-quality sheep of the breed you choose.

Sheepman Supply Co.
Route I, Box 141, Barboursville, Va. 22923.

These people carry all sorts of sheep equipment and supplies. Free catalogue.

VETERINARY:

Merck Veterinary Manual
Merck and Co., Inc., Rahway, N.J.
Available from Omaha Vaccine Co.

This is an almost indispensable basic reference book if you are keeping poultry and/or animals. Over 1,600 pages covering nutritional requirements, composition of feedstuffs, immunizations, diseases and treatment, etc.

Use a dictionary to demystify the technical language. This is an exhaustive resource tool you'll use for years.

Herbal Handbook for Farm and Stable
by J. de Bairacli Levy
Faber and Faber Ltd., 3 Queen Square, London, WC1N 3AU.

The *Herbal Handbook* offers a sensible, thoughtful alternative to treating your animals and poultry with antibiotics and chemicals. It teaches you how to keep your animals healthy by duplicating as closely as possible their natural states of being and by treating illnesses with natural compounds and herbs. My copy of this book is battered and worn—a good clue to its usefulness on the small farm. It covers cows, sheep, goats, poultry, and dogs.

Diagnostic Methods in Veterinary Medicine
by G. Boddie
J. B. Lippincott Co., Philadelphia, Pa., 1969.

If you want to learn how to go about diagnosing illnesses and problems in the animals you keep, invest in this book or order it through your library. It is clear, detailed, and very usable. Well illustrated with drawings and photographs. It teaches you how to approach, observe, and understand functions and malfunctions of the different systems of the animal body. Discusses bacteriology, helminthology, etc.

Veterinary Medicine
by D. C. Blood and J. A. Henderson
Williams and Wilkins Co., Baltimore, Md., 1968.

A substantial volume covering disease, parasites, allergies, etc., in large animals. This book goes into details not covered in the *Merck Manual,* though it is not as handy as a general reference.

Veterinary Applied Pharmacology and Therapeutics
by G. C. Brander and D. M. Pugh
William and Wilkins Co., Baltimore, Md., 1971.

If you become actively interested in treating your own livestock, this book can provide a sound introduction to principles and specifics of modern veterinary therapy. There are sections on antibiotics, tetracyclines sulphonamides, etc. The text is clear, technical but not mystified, and quite useful.

Omaha Vaccine Co.
2900 "O" Street, Omaha, Neb. 68107.

Supplies for all types of livestock and poultry. Free catalogue includes veterinary supplies, instruments, and equipment, plus feed supplements and books on animal care. Prompt and reliable service on orders.

Farmade Products Handbook (free)
Kansas City Vaccine Co., Stockyards, Kansas City, Mo. 64102.

This company supplies serums, drugs, vaccines, and so on for livestock.

CARPENTRY AND SHELTER

Old Ways of Working Wood
by Alex Bealer
Barre Publishers, Barre, Mass., 1972.

If you are trying for a simpler lifestyle and want to learn about hand tools for woodworking and building, here is a wonderful source book. Shows how to saw, fell, plane, turn and shape wood. Explains the use, making and care of wood-related tools. Full information you've wanted but couldn't find elsewhere.

How to Work with Tools and Wood
Robert Campbell, ed.
Pocket Book Edition, New York, 1971.

This book will introduce you to tools, their uses and care. It is a thorough and detailed enough book to be a very valuable tool itself. The 488 pages include drawings for projects with wood. A wonderful help in demystifying carpentry!

Home and Workshop Guide to Sharpening
by Harry Walton
Popular Science Publishing Co., Harper & Row, New York, 1971.

If you want to learn how to keep your woodworking and gardening tools in top condition, make the investment in this book. It shows clearly how to sharpen everything from lathe gouges to your garden hoe. Explains how sharpening works, how and when to grind, the different tools for sharpening and their uses.

Woodcraft
313 Montvale Avenue, Woburn, Mass. 01801.

This is a catalogue of the most beautiful tools you would ever want to dream over. Reasonably priced and well-made tools, plus books on all aspects of woodworking and building.

Carpenters and Builders Library
Theodore Audel and Co., 4300 West Second Street, Indianapolis, Ind. 46206.

A four-volume set covering all aspects of carpentry from foundations through roof framing and finishing. Individual volumes may be purchased, too. These are among the best basic references in carpentry that you could find.

Wood Frame House Construction
U. S. Department of Agriculture Agricultural Handbook #73, Superintendent of Documents, U. S. Government Printing Office, Washington, D.C. 20402.

A small book covering house construction from excavation for foundation through ceiling and roof framing. This book includes chimneys and fireplaces, maintenance and repair, installing gutters, and so on. Fairly clear and would serve as a basic guide.

Dwelling (paperback)
by River
Freestone Publishing Co., P. O. Box 357, Albion, Calif.
95410.

Before you set out to build the dwelling you've dreamed of, you may discover much in the pages of this book. River has taken a thoughtful journey into the meaning of building a dwelling. Her book is rich in poetry and ideas, blended with the practical wisdom of country women and men who have built the homes they live in. Illustrations by Leona Walden and photographs by Sally Bailey.

Shelter
Shelter Publications
Available from Mountain Books, P. O. Box 4811, Santa Barbara, Calif. 93103.

Shelter takes a look at simple homes all over the world, with a clear eye for the imaginative use of basic materials and forms. A book you can browse through a hundred times and still find new images and possibilities.

Wiring Simplified
by H. P. Richter
Park Publishing Inc., P. O. Box 8527 (Lake Street Station), Minneapolis, Minn. 55408.

A good, inexpensive introduction to electrical wiring, with enough details to explain the installation of ordinary home wiring.

How to Be Your Own Home Electrician
by George Daniels
Popular Science Publishing Co., Harper & Row, New York.

FOOD

American Supply House
P. O. Box 1114, Columbia, Mo. 65201.

Source of dairy thermometers, wooden butter molds, butter cartons, wrappers, and wooden paddles for churns.

Making Cheese at Home
and
Making Butter on the Farm
C. Hansen's Laboratory, 9015 Maple Street, Milwaukee, Wisc. 53214.

These free booklets are good home guides to buttermaking and rennet cheesemaking. Hansen's also sells rennet and commercial cheese cultures.

Home Cheesemaking
Dairy Goat Journal, P. O. Box 1908, Scottsdale, Ariz. 85252.

Home Freezing of Fruits and Vegetables
Home and Garden Bulletin #10, Superintendent of Documents, U. S. Government Printing Office, Washington, D.C. 20402.

This is a good basic guide to freezing techniques, though the recommended cooking times can be shortened considerably.

Home Canning of Fruits and Vegetables
Home and Garden Bulletin #8, Superintendent of Documents, U. S. Government Printing Office, Washington, D.C. 20402.

This booklet and the two listed below contain similar information on canning techniques and cooking times. Either one would be sufficient for the home canner, but each contains slightly different recipes and for the price, it's worth getting two or all three.

Ball Blue Book
Ball Brothers Co., Muncie, Ind.

Another basic canning book by one of the main canning jar companies.

Kerr Home Canning
Kerr Glass Co.,
Sandy Springs, Okla.

The other major canning jar company's publication.

Slaughtering, Cutting and Processing Pork on the Farm
Farmer's Bulletin #2138, Superintendent of Documents; U. S. Government Printing Office, Washington, D.C. 20402.

Slaughtering, Cutting and Processing Beef on the Farm
Farmer's Bulletin #2209, Superintendent of Documents; U. S. Government Printing Office, Washington, D.C. 20402.

Morton Salt Co.
110 N. Wacket Drive, Chicago, Ill. 60606.

A good booklet on home butchering.

GARDENING

The Basic Book of Organic Gardening
edited by Robert Rodale
Organic Gardening/Ballantine Books, New York, 1971.

This really is *the* basic book. In one inexpensive edition, it provides you with all the basic information you need to plant and grow an organic garden.

How to Grow Vegetables and Fruits by the Organic Method
edited by J. J. Rodale
Rodale Press Inc., Emmaus, Pa., 1974.

This enormous volume must come close to being the definitive work on organic gardening and orcharding. It is divided into seven sections that cover the needs and propensities of all types of vegetables and fruits, the basics of soil structure and development, care of nut trees, etc. This book even includes a section of "exotic tropical fruits." I found this book worth much more than its initial price—it can be used year after year on the homestead or farm.

The Complete Book of Composting
by J. I. Rodale and staff
Rodale Books Inc., Emmaus, Pa., 1972.

This gigantic volume covers almost everything you would want to know about improving and sustaining your soil for farming and gardening. It can teach you how to apply organic methods to large-scale crop production, how to improve your pasture land, and so on. Useful for the small home garden, too, as it discusses composting principles in depth.

Organic Farming Methods and Markets—An Introduction to Ecological Agriculture (paperback)
edited by Robert Steffen, Floyd Allen, and James Foote
Rodale Press Inc., Emmaus, Pa., 1972.

This paperback covers crop rotations, organic orcharding, alternatives to chemical pesticides in larger scale farming, and so on. It includes a directory of organic farms and information for certifying yourself (and your farm or ranch) as an organic farmer.

Companion Plants
by Helen Louise Philbrick and Richard Gregg
Devin-Adair Co., Old Greenwich, Conn., 1966.

This book lists alphabetically most common vegetables, herbs, flowers, and fruits, explaining their relationships to and affect on other plants. It is a useful and important tool for anyone trying to grow plants organically. Order from your library.

The Secret Life of Plants
by Peter Tompkins and Christopher Bird
Harper & Row, New York, 1973. Paperback edition: Avon Books, New York, 1974.

This is popularized reporting of scientific research into plant sensitivities, awareness, and responses. It is important to anyone exploring their relationship to living beings on the earth and, more specifically, deepens and clarifies knowledge a gardener comes to intuitively anyway. A must for anyone who takes gardening seriously and with love.

Organic Gardening and Farming (monthly magazine)
Rodale Press, 33 East Minor Street
Emmaus, Pa. 18049.

Here is a monthly compilation of people's experiences gardening and farming all over the country—using organic, ecologically sound methods. Equally invaluable for the hundred-acre homesteader or the backyard gardener! A source, too, for up-to-date information on alternative energy systems, legislation affecting food production, etc. Articles cover everything from choosing a garden tractor to controlling cabbage moths.

Sudbury Laboratory
Sudbury, Mass. 01776.

Sells soil test kits to determine soil pH and mineral deficiencies. The kits range from moderately priced to very expensive.

Perfect Garden Co.
14 East 46th Street, New York, N.Y. 10017.

Sells litmus tape for determining soil pH. This is less accurate than the Sudbury kits but a good deal less expensive and probably adequate for the average home garden.

Sources of ladybugs and mantids for pest control:
Eastern Biological Control Co., 104 Hackensack Street, Woodridge, N.J.

Gothard Inc., Box 367, Cautillo, Tex.

California Bug Co., Route 2, Auburn, Calif.

GENERAL COUNTRY RESOURCES

Country Women (magazine)
P. O. Box 51, Albion, Calif. 95410.

This magazine was the mother of this book. Half of every issue explores some topic relevant to country feminists (for example, older women, natural cycles, living with children, women working). The other half of each issue is devoted to practical information about country skills like beekeeping, engine repair, trimming horse hooves, etc. Please send a self-addressed stamped envelope with any request for information.

Countryside Small Stock Journal (magazine)

See the Animal section for information about this excellent homestead magazine.

Countryside Book Store and Countryside General Store
Countryside Publications, Route 1, Box 239, Waterloo, Wisc. 53594.

Here is a mail order bookstore with a list of good-quality writing on all types of livestock, feeds and feeding, alternate energy sources, planning a homestead farm, and so on. The General Store sells incubators, Aladdin lamps, harness for draft animals, sheep shears, spinning wheels, etc. The *Countryside Catalog* is published quarterly for both stores.

Vermont Garden Cart
Garden Way Research, Dept. 50072, Charlotte, Vt. 05445.

This excellent cart is useful for any kind of hauling, from bales of hay to loads of rock or fence posts. It is expensive but is much more efficient than a conventional wheelbarrow and well worth the cost.

PLUMBING AND WATER SYSTEMS

The Practical Handbook of Plumbing and Heating
by Richard Day
Fawcett Publications Inc., New York.

I picked up this inexpensive little hardback in a hardware store one day and have used it many times since. Covers pipes and fittings, connections, repairs, waste disposal systems and so on. Clearly and simply illustrated.

Plumbers and Pipefitters Library
Theodore Audel and Co.
4300 West Second Street, Indianapolis, Ind. 46206.

If you want to go beyond simple plumbing and repairs, Audel's three-volume set provides a comprehensive, usable guide. A good reference to obtain from the library.

Rife Hydraulic Rams
Rife Hydraulic Engine Manufacturing Co., P. O. Box 367, Millburn, N.J. 07041.

This is the only source of hydraulic rams in the United States that we've been able to find. Their catalogue explains how rams work and what conditions are necessary for a ram to work efficiently.

Composting Privy
research and text by Sim Van der Ryn
Technical Bulletin #1, January 1974,
available from Natural Energy Design Center, University of California, Berkeley, Calif. 94720.

Provides complete instructions for building a composting privy from simple, inexpensive materials.

The United Stand Privy Booklet
United Stand, P. O Box 191, Potter Valley, Calif. 95469.
15 pages; updated 1975.

This booklet provides continually updated information on designs for safe, sanitary, ecological home-built privies from pit privies to composting ones. Best information currently available on alternatives to the conventional septic tank.

WINDMILLS AND WINDPOWER

Sources for commercially made windmills:
O'Brock Windmill Sales
North Benton, Ohio 44449.
Source for pumping windmills, towers, hand pumps, repair parts, and so on.

Baker Windmills
The Heller-Allen Co.
Napoleon, Ohio 43545.
West Coast dealer: Shaw Pump and Supply Inc.
9660 East Rush Street, South El Monte, Calif.

Plans for homemade windmills:
Windmill Design
by A. Bodek
The Director, Brace Research Institute, P. O. Box 221, MacDonald College, Quebec, Canada.

Wind and Windspinners
by Michael Hackleman and David House
Peace Press, 3828 Willat Avenue, Culver City, Calif. 90230.
Good, clear simple plans for homemade wind generators.

Sources for information on wind generators:
WINCO
P. O. Box 3263, Sioux City, Iowa 51102.

Real Gas and Electric
P. O. Box A, Guerneville, Calif. 95446.

Solar Wind Company
East Holden, Me. 04429.
Has a pamphlet that deals with the electric-power windmill.

WOOL: SHEARING, SPINNING, DYEING, WEAVING

Wool Away; The Art and Technique of Shearing
by Godfrey Bowen
Van Nostrand-Reinhold Co., New York, 1974.
Good technical book on shearing with electric clippers.

Anyone Can Build a Spinning Wheel
by W. C. West
Thresh Publications, 443 Sebastopol Avenue, Santa Rosa, Calif. 95401.
Plans for an inexpensive homemade spinning wheel. Thresh has a free catalogue of spinning and weaving books.

An Introduction to Natural Dyeing
by Robert and Christine Thresh
Home Dyeing with Natural Dyes
by Margaret S. Furry and Bess M. Viemont
Both from Thresh Publications (see above).

The first has color plates and basic recipes; the second is a reprint of an older book with more detail but no color plates.

Dye Plants and Dyeing—a Handbook
Brooklyn Botanic Garden, Brooklyn, N.Y. 11225.
Best basic guide to natural dye plants from around the world. Includes recipes. Color plates.

The Art of Weaving
by Else Regensteiner
Van Nostrand-Reinhold Co., New York, N.Y.

My favorite weaving book with clear instructions for all weaving styles from simple backstrap to eight-harness looms.

The New Key to Weaving
by Mary Black
The Macmillan Co., New York, 1961.

An excellent detailed text for weaving on two-, four-, and eight-harness looms. Has warping diagrams for pattern weaving.

Sometimes in my dreams
I still see
my Kentucky grandmother
thin, strong and hungry
holding her egg money
out to me
saying
buy land, Mary,
buy land,
buy land while it lasts
they stopped making it.